T0145360

Smart Innovation, Systems and Technologies

Volume 96

The Smart Innovation, Systems and Technologies book series encompasses the topics of knowledge, intelligence, innovation and sustainability. The aim of the series is to make available a platform for the publication of books on all aspects of single and multi-disciplinary research on these themes in order to make the latest results available in a readily-accessible form. Volumes on interdisciplinary research combining two or more of these areas is particularly sought.

The series covers systems and paradigms that employ knowledge and intelligence in a broad sense. Its scope is systems having embedded knowledge and intelligence, which may be applied to the solution of world problems in industry, the environment and the community. It also focusses on the knowledge-transfer methodologies and innovation strategies employed to make this happen effectively. The combination of intelligent systems tools and a broad range of applications introduces a need for a synergy of disciplines from science, technology, business and the humanities. The series will include conference proceedings, edited collections, monographs, handbooks, reference books, and other relevant types of book in areas of science and technology where smart systems and technologies can offer innovative solutions.

High quality content is an essential feature for all book proposals accepted for the series. It is expected that editors of all accepted volumes will ensure that contributions are subjected to an appropriate level of reviewing process and adhere to KES quality principles.

More information about this series at http://www.springer.com/series/8767

Gordan Jezic · Yun-Heh Jessica Chen-Burger
Robert J. Howlett · Lakhmi C. Jain
Ljubo Vlacic · Roman Šperka
Editors

Agents and Multi-Agent Systems: Technologies and Applications 2018

Proceedings of the 12th International Conference on Agents and Multi-Agent Systems: Technologies and Applications (KES-AMSTA-18)

Editors
Gordan Jezic
University of Zagreb, Faculty of Electrical Engineering and Computing
Zagreb, Croatia

Yun-Heh Jessica Chen-Burger
The Heriot-Watt University
Edinburgh
Scotland, UK

Robert J. Howlett
Bournemouth University
Poole, UK

and

KES International
Shoreham-by-Sea, UK

Lakhmi C. Jain
Centre for Artificial Intelligence, Faculty of Engineering and Information Technology
University of Technology Sydney
Sydney, NSW, Australia

and

Faculty of Science, Technology and Mathematics
University of Canberra
Canberra, ACT, Australia

and

KES International
Shoreham-by-Sea, UK

Ljubo Vlacic
Griffith Sciences - Centres and Institutes
Griffith University
South Brisbane, QLD, Australia

Roman Šperka
Department of Business Economics and Management and Silesian University in Opava, School of Business Administration in Karvina
Karvina, Czech Republic

ISSN 2190-3018 ISSN 2190-3026 (electronic)
Smart Innovation, Systems and Technologies
ISBN 978-3-030-06353-5 ISBN 978-3-319-92031-3 (eBook)
https://doi.org/10.1007/978-3-319-92031-3

Printed on acid-free paper

This Springer imprint is published by the registered company Springer International Publishing AG part of Springer Nature
The registered company address is: Gewerbestrasse 11, 6330 Cham, Switzerland

Preface

This volume contains the proceedings of the 12th KES Conference on Agent and Multi-Agent Systems: Technologies and Applications (KES-AMSTA 2018) which will be held in Gold Coast, Australia, between 20 and 22 June 2018. The conference was organized by KES International, its focus group on agent and multi-agent systems and University of Zagreb, Faculty of Electrical Engineering and Computing. The KES-AMSTA conference is a subseries of the KES conference series.

Following the success of previous KES conferences on Agent and Multi-Agent Systems: Technologies and Applications, held in Vilamoura, Portugal (KES-AMSTA 2017), Puerto de la Cruz, Tenerife, Spain (KES-AMSTA 2016), Sorrento, Italy (KES-AMSTA 2015), Chania, Greece (KES-AMSTA 2014), Hue, Vietnam (KES-AMSTA 2013), Dubrovnik, Croatia (KES-AMSTA 2012), Manchester, UK (KES-AMSTA 2011), Gdynia, Poland (KES-AMSTA 2010), Uppsala, Sweden (KES-AMSTA 2009), Incheon, Korea (KES-AMSTA 2008) and Wroclaw, Poland (KES-AMSTA 2007), the conference featured the usual keynote talks, oral presentations and invited sessions closely aligned to its established themes.

KES-AMSTA is an international scientific conference for discussing and publishing innovative research in the field of agent and multi-agent systems and technologies applicable in the digital and knowledge economy. The aim of the conference is to provide an internationally respected forum for both the research and industrial communities on their latest work on innovative technologies and applications that is potentially disruptive to industries. Current topics of research in the field include technologies in the area of mobile and cloud computing, big data analysis, Internet of Things (IoT), business intelligence, artificial intelligence, social systems, computer embedded systems and nature-inspired manufacturing. Special attention is paid on the feature topics: agent interaction and collaboration, modelling and simulation agents, social networks, business informatics, intelligent agents and multi-agent systems.

The conference attracted a substantial number of researchers and practitioners from all over the world who submitted their papers for main track covering the methodologies of agent and multi-agent systems applicable in the digital and knowledge economy, and three invited sessions on specific topics within the field. Submissions came from 15 countries. Each paper was peer-reviewed by at least two members of the International Programme Committee and International Reviewer Board. Thirty-four papers were selected for oral presentation and publication in the volume of the KES-AMSTA 2018 proceedings.

The Programme Committee defined the following main tracks: intelligent agent interaction and collaboration, modelling, simulation and mobile agents, and agent communication and social networks. In addition to the main tracks of the conference, there were the following invited sessions: design and implementation of intelligent agents and multi-agent systems, business informatics and business process management.

Accepted and presented papers highlight new trends and challenges in agent and multi-agent research. We hope that these results will be of value to the research community working in the fields of artificial intelligence, collective computational intelligence, health, robotics, dialogue systems and, in particular, agent and multi-agent systems, technologies, tools and applications.

The Chairs' special thanks go to the following special session organizers: Prof. Lenin G. Lemus-Zúñiga, Universitat Politècnica de València, España; Prof. Arnulfo Alanis Garza, Instituto Tecnológico de Tijuana, México; Prof. Setsuya Kurahashi, University of Tsukuba, Tokyo; Prof. Takao Terano, Tokyo Institute of Technology, Japan; and Prof. Hiroshi Takahashi, Keio University, Japan, for their excellent work.

Thanks are due to the Programme Co-chairs, all Programme and Reviewer Committee members and all the additional reviewers for their valuable efforts in the review process, which helped us to guarantee the highest quality of selected papers for the conference.

We cordially thank all authors for their valuable contributions and all of the other participants in this conference. The conference would not be possible without their support.

April 2018

Gordan Jezic
Jessica Chen-Burger
Robert J. Howlett
Lakhmi C. Jain
Ljubo Vlacic
Roman Šperka

KES-AMSTA-2018 Conference Organization

KES-AMSTA 2018 was organized by KES International—Innovation in Knowledge-Based and Intelligent Engineering Systems.

Honorary Chairs

L. Vlacic	Griffith University, Gold Coast, Australia
I. Lovrek	University of Zagreb, Croatia
L. C. Jain	University of South Australia, Adelaide

Conference Co-chairs

G. Jezic	University of Zagreb, Croatia
J. Chen-Burger	Heriot-Watt University, Scotland, UK

Executive Chair

R. J. Howlett	Bournemouth University, UK

Programme Co-chairs

M. Kusek	University of Zagreb, Croatia
R. Sperka	Silesian University in Opava, Czech Republic

Publicity Chair

P. Skocir	University of Zagreb, Croatia

International Programme Committee

Koichi Asakura	Daido University, Japan
Ahmad Taher Azar	Faculty of Computers and Information, Benha University, Egypt
Marina Bagić Babac	University of Zagreb, Croatia
Dariusz Barbucha	Gdynia Maritime University, Poland
Grażyna Brzykcy	Poznań University of Technology, Department of Control and Information Engineering, Poland
Frantisek Capkovic	Slovak Academy of Sciences, Bratislava, Slovakia
Matteo Cristani	Universita di Verona, Italy
Ireneusz Czarnowski	Gdynia Maritime University, Poland
Paulina Golinska-Dawson	Poznan University of Technology, Poland
Arnulfo Alanis Garza	Instituto Tecnológico de Tijuana. México
Mirjana Ivanovic	University of Novi Sad, Serbia
Dennis Jarvis	Central Queensland University, Australia
Piotr Jedrzejowicz	Gdynia Maritime University, Poland
Dragan Jevtic	University of Zagreb, Croatia
Vicente Julian	Universitat Politecnica de Valencia, Spain
Arkadiusz Kawa	Poznan University of Economics and Business, Poland
Adrianna Kozierkiewicz-Hetmańska	Wroclaw University of Science and Technology, Poland
Konrad Kułakowski	AGH University of Science and Technology, Krakow, Poland
Setsuya Kurahashi	University of Tsukuba, Tokyo
Mario Kusek	University of Zagreb, Croatia
Kazuhiro Kuwabara	Ritsumeikan University, Japan
Jooyoung Lee	Innopolis University, Russia
Marin Lujak	IMT Lille Douai, Douai, France
Evgeni Magid	Kazan Federal University, Russia
Manuel Mazzara	Innopolis University, Russia
Daniel Moldt	University of Hamburg, Germany
Ngoc Thanh Nguyen	Wroclaw University of Technology, Poland
Vedran Podobnik	University of Zagreb, Croatia
Radu-Emil Precup	Politehnica University of Timisoara, Romania
Nafees Qamar	Southwest University, China

Victor Rivera	Innopolis University, Russia
Ewa Ratajczak-Ropel	Gdynia Maritime University, Poland
Katka Slaninova	School of Business Administration in Karvina, Silesian University in Opava, Czech Republic
Silvia Rossi	University of Naples Federico II, Italy
Roman Šperka	Silesian University in Opava, Czech Republic
Darko Stipaničev	University of Split, Croatia
Ryszard Tadeusiewicz	AGH University of Science and Technology, Krakow, Poland
Hiroshi Takahashi	Keio University, Japan
Yasufumi Takama	Tokyo Metropolitan University, Japan
Takao Terano	Tokyo Institute of Technology, Japan
Krunoslav Tržec	Ericsson Nikola Tesla, Croatia
Taketoshi Ushiama	Kyushu University, Japan
Jordi Vallverdú	Universitat Autonoma de Barcelona, Spain
Toyohide Watanabe	Nagoya Industrial Science Research Institute, Japan
Izabela Wierzbowska	Gdynia Maritime University, Poland
Mahdi Zargayouna	University of Paris-Est, IFSTTAR, France
Lenin G. Lemus-Zúñiga	Universitat Politècnica de València. España

Invited Session Chairs

Business Process Management

Roman Šperka — Silesian University in Opava, Czech Republic

Agent-Based Modelling and Simulation

Roman Šperka — Silesian University in Opava, Czech Republic

Business Informatics

Setsuya Kurahashi	University of Tsukuba, Japan
Takao Terano	Tokyo Institute of Technology, Japan
Hiroshi Takahashi	Keio University, Japan

Anthropic-Oriented Computing

Manuel Mazzara	Innopolis University, Russia
Jooyoung Lee	Innopolis University, Russia
Victor Rivera	Innopolis University, Russia

The design and Implementation of Intelligent Agents and Multi-agent Systems

Lenin G. Lemus-Zuniga	Universitat Politecnica de Valencia, Spain
Arnulfo Alanis Garza	Instituto Tecnologico de Tijuana, Mexico

Contents

Intelligent Agent Interaction and Collaboration

Human-Agent Collaboration: A Goal-Based BDI Approach 3
Salma Noorunnisa, Dennis Jarvis, Jacqueline Jarvis, and Marcus Watson

Evolution Direction of Reward Appraisal in Reinforcement Learning Agents 13
Masaya Miyawaki, Koichi Moriyama, Atsuko Mutoh, Tohgoroh Matsui, and Nobuhiro Inuzuka

A General Framework for Formulating Adjustable Autonomy of Multi-agent Systems by Fuzzy Logic 23
Salama A. Mostafa, Rozanawati Darman, Shihab Hamad Khaleefah, Aida Mustapha, Noryusliza Abdullah, and Hanayanti Hafit

Agent-Based System for Context-Aware Human-Computer Interaction 34
Renato Soic, Pavle Skocir, and Gordan Jezic

Agent-Oriented Smart Factory (AOSF): An MAS Based Framework for SMEs Under Industry 4.0 44
Fareed Ud Din, Frans Henskens, David Paul, and Mark Wallis

Modeling, Simulation and Mobile Agents

Agent-Based Approach for Energy-Efficient IoT Services Discovery and Management 57
Petar Krivic, Pavle Skocir, and Mario Kusek

Agent-Based Modeling and Simulation for Two-Dimensional Spatial Competition 67
Masashi Miura and Hidetoshi Shiroishi

Agent Based Simulation of Network Routing: Reinforcement Learning Comparison 76
Krešimir Čunko, Marin Vuković, and Dragan Jevtić

Dispatching Strategies for Dynamic Vehicle Routing Problems 87
Besma Zeddini and Mahdi Zargayouna

Securing Mobile Agents, Stationary Agents and Places in Mobile Agents Systems 97
Donies Samet, Farah Barika Ktata, and Khaled Ghedira

How Research Achievements Can Influence Delivering of a Course - Siebog Agent Middleware 110
Milan Vidaković, Mirjana Ivanović, Dejan Stantić, and Jovana Vidaković

Agent Communication and Social Networks

Sending Messages in Social Networks 123
Matteo Cristani, Francesco Olivieri, Claudio Tomazzoli, and Guido Governatori

ER-Agent Communication Languages and Protocol for Large-Scale Emergency Responses 134
Mohd Khairul Azmi Hassan and Yun-Heh Chen-Burger

Towards a Logical Framework for Diagnostic Reasoning 144
Matteo Cristani, Francesco Olivieri, Claudio Tomazzoli, and Margherita Zorzi

Scalability of Dynamic Lighting Control Systems 156
Leszek Kotulski and Igor Wojnicki

Automatic Detection of Device Types by Consumption Curve 164
Claudio Tomazzoli, Matteo Cristani, Simone Scannapieco, and Francesco Olivieri

Business Process Management

Advantages of Application of Process Mining and Agent-Based Systems in Business Domain 177
Michal Halaška and Roman Šperka

Modelling the Validation Process of Enterprise Software Systems 187
Robert Bucki and Petr Suchánek

Design and Implementation of Intelligent Agents and Multi-Agent Systems I

Multi-agent System for Forecasting Based on Modified Algorithms of Swarm Intelligence and Immune Network Modeling 199
Galina A. Samigulina and Zhazira A. Massimkanova

Multi-Agent System Model for Diagnosis of Personality Types 209
Margarita Ramírez Ramírez, Hilda Beatriz Ramírez Moreno, Esperanza Manrique Rojas, Carlos Hurtado, and Sergio Octavio Vázquez Núñez

Towards a Multi-Agent System for an Informative Healthcare Mobile Application 215
Carlos Hurtado, Margarita Ramirez Ramirez, Arnulfo Alanis, Sergio Octavio Vazquez, Beatriz Ramirez, and Esperanza Manrique

Proposal of a Bootcamp's User Activity Dashboard Based on MAS 220
Lenin G. Lemus-Zúñiga, Valeria Alexandra Haro Valle, José-V. Benlloch-Dualde, Edgar Lorenzo-Sáez, Miguel A. Mateo Pla, and Jorge Maldonado-Mahauad

Toward to an Electric Monitoring Platform Based on Agents 231
Jorge E. Luzuriaga, Guillermo Cortina Rodríguez, Karolína Janošová, Monika Borova, Miguel Ángel Mateo Pla, and Lenin-G. Lemus-Zúñiga

Design and Implementation of Intelligent Agents and Multi-Agent Systems II

A Cooperative Agent-Based Management Tool Proposal to Quantify GHG Emissions at Local Level 243
Edgar Lorenzo-Sáez, José-Vicente Oliver-Villanueva, Jorge E. Luzuriaga, Miguel Ángel Mateo Pla, Javier F. Urchueguía, and Lenin-G. Lemus-Zúñiga

A Proposal to Improve the Usability of Applications for Users with Autism Considering Emotional Aspects 253
Ángeles Quezada, Reyes Juarez-Ramirez, Arnulfo Alanís Garza, Bogart Yail, Sergio Magdaleno, and Eugenia Bermudez

Towards a Model Based on Agents for the Detection of Behavior Patterns in Older Adults Who Start Using ICT 261
Consuelo Salgado Soto, Maricela Sevilla Caro, Ricardo Rosales Cisneros, Margarita Ramírez Ramírez, Hilda Beatriz Ramírez Moreno, and Esperanza Manrique Rojas

Intelligent Agents as Support in the Process of Disease Prevention Through Health Records 269
Hilda Beatriz Ramirez Moreno, Margarita Ramírez Ramírez, Esperanza Manrique Rojas, Nora del Carmen Osuna Millán, and Maricela Sevilla Caro

Agent-Based Model as a Provider of Medical Services in Tijuana Mexico 275
Ricardo Rosales, Nora Osuna-Millan, Consuelo Salgado-Soto, Carlos Flores-Sanchez, Juan Meza-Fregoso, and Arnulfo Alanis

Business Informatics

Understanding the Potential Value of Digitization for Business – Quantitative Research Results of European Experts 287
Christopher Reichstein, Ralf-Christian Härting, and Pascal Neumaier

A Method of Knowledge Extraction for Response to Rapid Technological Change with Link Mining 299
Masashi Shibata and Masakazu Takahashi

Agent-Based Gaming Approach for Electricity Markets 311
Setsuya Kurahashi

Finding the Better Solutions for the Smart Meter Gateway Placement in a Power Distribution System Through an Evolutionary Algorithm 321
Ryoma Aoki and Takao Terano

Japanese Health Food Market Trend Analysis 331
Yoko Ishino

Simulation of the Effect of Financial Regulation on the Stability of Financial Systems and Financial Institution Behavior 341
Takamasa Kikuchi, Masaaki Kunigami, Takashi Yamada, Hiroshi Takahashi, and Takao Terano

Author Index 355

Intelligent Agent Interaction and Collaboration

Human-Agent Collaboration: A Goal-Based BDI Approach

Salma Noorunnisa[1], Dennis Jarvis[1](✉), Jacqueline Jarvis[1], and Marcus Watson[2]

[1] Centre for Intelligent Systems, Central Queensland University, 160 Ann Street, Brisbane 4000, Australia
{s.noorunnisa, d.jarvis, j.jarvis}@cqu.edu.au

[2] School of Psychology, University of Queensland, St. Lucia 4067, Australia
m.watson2@uq.edu.au

Abstract. The Belief-Desire-Intention (BDI) model of agency has been a popular choice for the modelling of goal-based behaviour for both individual agents and more recently, teams of agents. Numerous frameworks have been developed since the model was first proposed in the early 1980s. However, while the more recent frameworks support a delegative model of agent/agent and human/agent collaboration, no frameworks support a general model of collaboration. Given the importance of collaboration in the development of practical semi-autonomous agent applications, we consider this to constitute a major limitation of traditional BDI frameworks. In this paper, we present GORITE, a novel BDI framework that by employing explicit goal representations, overcomes many of the limitations of traditional frameworks. In terms of human/agent collaboration, key requirements are identified and through the use of a representative but simple example, the ability of GORITE to address those requirements is demonstrated.

Keywords: Human-agent collaboration · BDI · Multi-agent systems

1 Introduction

At a recent workshop on Human-Autonomy Teaming (HAT), having a shared mental model was identified as being essential if HAT systems are to deliver the levels of trust and explanatory capability required for military operations [1]. Furthermore, the Situation Awareness-Based Agent Transparency (SAT) Model developed by Chen et al. [2] was identified as providing a suitable conceptual framework for future HAT research. SAT visualisation agents [2] are able to provide operators with situation awareness of an evolving mission environment. This support is at three levels:

(1) What's going on and what is the agent trying to achieve?
(2) Why does the agent do it?
(3) What should the operator expect to happen?

In addressing these questions, a SAT agent draws on its desires and intentions at Level 1 and its beliefs at Level 2. However, while the SAT model is inspired by the

G. Jezic et al. (Eds.): KES-AMSTA-18 2018, SIST 96, pp. 3–12, 2019.
https://doi.org/10.1007/978-3-319-92031-3_1

Belief-Desire-Intention (BDI) model of agency, the creation of SAT agents grounded in the BDI model of agency is a research issue yet to be explored. Rather, the focus of Chen's work has been the visualisation of information pertaining to the SAT levels and the demonstration through controlled experimentation that operator performance and trust in automation is enhanced through such visualisation. Chen has proposed that the research community should continue with that agenda and we are in agreement. However, we see a major opportunity for a complementary research program with a focus of developing a BDI software framework that explicitly supports human/agent collaboration through the use of the SAT model. As explained below, this will require a framework that provides explicit representation of beliefs, desires and intentions in order to enable the agent to reflect on and explain its actions and to enable humans to dynamically modify agent behaviour.

This represents a significant departure from traditional BDI agent frameworks and we propose to use GORITE, a novel open-source BDI framework developed by Rönnquist [3] for this purpose. GORITE itself is a mature, open source and fully functional software framework, as evidenced by the case studies presented in [3]. In particular, the manufacturing control case study is a reimplementation of an earlier commercial application developed using the JACK Teams BDI framework. However, the version of GORITE and the case studies in [3] support and focus on autonomous BDI behavior. For effective human/BDI agent collaboration, extensions to both the BDI model and to the GORITE framework are required. Our intent is to tackle this iteratively, with each iteration involving model extension, framework realization and application development. The application (waypoint traversal) presented in this paper represents the key functionality for our domain of interest – war gaming using semi-automated computer generated forces. As such, it provides an ideal example for demonstrating the effectiveness of the extensions to both the BDI model and the GORITE framework.

In the remainder of this paper, we will first discuss the BDI model, its limitations with respect to human/agent collaboration and how these can be overcome with GORITE. We then demonstrate how effective human/agent collaboration can be achieved for SAT Levels 1 and 2 using the GORITE framework. In this regard, a simple but representative example (waypoint traversal by a platoon) is employed, in order to maintain focus on the key collaboration requirements. SAT Level 3 functionality and agent/human collaboration are out of scope for the current research activity. Note that the innovation in this work lies in its extension of the BDI Model of agency to accommodate human/agent collaboration and the realization of this extended model in the GORITE BDI framework.

2 The BDI Model

The BDI model is concerned with how an agent makes rational decisions about the actions that it performs through the employment of

(1) Beliefs about their environment, other agents and themselves,
(2) Desires that they wish to satisfy and
(3) Intentions to act towards the fulfilment of selected desires.

The model has its origin in Bratman's theory of human practical reasoning [4]. Bratman's ideas were first formalised by Rao and Georgeff [5] who subsequently proposed an abstract architecture in which beliefs, desires and intentions were explicitly represented as global data structures and where agent behaviour is event driven. However, while this conceptualisation faithfully captured Bratman's theory, it did not constitute a practical system for rational reasoning. In order to ensure computational tractability, they proposed the following representational changes [6]:

- Only beliefs about the current state of the world are represented explicitly
- Information about the means of achieving certain future world states (desires) and the options available to an agent are represented as plans.
- A particular desire may be realizable by multiple plans but an agent must select one plan to pursue.
- Plans either succeed or fail; if a plan fails, then the desire which is being pursued may be reconsidered.
- Intentions are represented implicitly by the collection of currently active plans
- Desires are referred to as goals, which are represented as events. Goals have only a transient representation, acting as triggers for plan invocations.

These considerations led to the following execution model:

```
repeat
  wait for the next goal event;
  select (on the basis of current beliefs) a plan to
  achieve the current goal;
  execute the selected plan;
  update beliefs;
end repeat
```

This execution model has provided the conceptual basis for all major research and commercial BDI implementations, in particular PRS, dMARS and JACK [3].

In the traditional BDI execution model outlined above, plans consist of steps that are specified using a framework dependent plan language; these steps may involve the posting of further goal events (or the reposting of the current goal event). More than one plan may be applicable for the achievement of a particular goal – this set of plans is called the applicable set. The selection of a plan to execute from the applicable set is based on the currently held beliefs of the agent and may involve explicit (meta-level) reasoning.

Since its inception, the Belief-Desire-Intention (BDI) model of agency has underpinned many successful agent applications and has been identified as one of the preferred vehicles for the delivery of industry strength, knowledge rich, intelligent agent applications [7]. As originally conceived by Bratman, the model was intended as a means to determine how an agent should act in a situated environment. The early applications of the model reflect this focus on situated, autonomous behaviour, but within constrained technical domains, e.g. space shuttle fault diagnosis [8]. While this has continued to be a focal point for applications, as evidenced by its use in manufacturing system and UAV control, commercial success has been achieved in its application to human behaviour modelling in war gaming, where credible entity behaviour is an essential requirement for an effective military game [3]. The wargaming examples have necessitated significant extensions to the BDI execution model as originally formulated by Rao and Georgeff. In particular, the model has been extended in JACK Teams [9] to accommodate teams (such as platoons and companies) as distinct entities with their own beliefs, desires and intentions and in CoJACK [9] to provide agents with an explicit cognitive architecture to ground reasoning. However, these extensions retain a key feature of the Rao and Georgeff model, namely that the goals are not represented as explicit, persistent entities, but rather as transitory events. This makes reasoning about an agent's intentions – for example, whether to continue or discontinue with the current goal or how a goal should be resourced – problematic.

While the extensions to the BDI execution model embodied in both JACK Teams and CoJACK have significantly extended the range of problems that can be effectively addressed by BDI agents, these problems remain characterized by a requirement for autonomous execution potentially supported by delegation. The applicability of the BDI execution model becomes problematic when human/agent collaboration is required. If the collaboration involves only simple delegation, with execution being managed by either the human or the agent, then the delegation model supported by JACK Teams will suffice. However a more comprehensive collaboration model is required if goal/belief inspection/management is required of the collaboration. For example, the provision of such functionality would significantly increase the amount of behaviour that could be delegated to agents (and teams of agent) in theatre-level wargames, as the puckster would have the ability to dynamically interact with the agents.

To summarise, in terms of human/agent collaboration, the BDI execution model exhibits the following limitations:

- Interruption of plans is not supported.
- Goal representation is implicit and transient, with goals modelled as events that are not persisted. Consequently, goals are not inspectable.
- Depending on how beliefs are stored, they may be inspectable. However, no distinction is made in the BDI model between individual agent beliefs, shared agent beliefs and beliefs that are shared by agents that are collaborating on a particular goal execution.

Additional insight can be gained into the requirements of human/agent collaboration by considering the more general problem of Activity Based Computing (ABC), where human/human collaboration is mediated by a shared computational workspace. Furthermore, one can reasonably expect the key requirements for human/agent collaboration to be a subset of the requirements for ABC. Activity Based Computing was conceived by Norman, one of the pioneers of HCI. However, realisation of the concept was left to others, most notably Bardram and his colleagues [10]. ABC is of particular relevance to the current research problem because it is concerned with the support that people need when working on a shared computational activity. While our interest is in human/agent collaboration rather than human/human collaboration, the key requirements that ABC imposes on shared activities, namely suspension and resumption, context awareness and inspectability also apply to goal executions that are to be shared with humans. It is also of interest that Norman [11] has identified goals as being an appropriate conceptualization for reasoning about activity. However, this conceptualization is not present in Bardram's current work.

Based on Chen's SAT model, Bardram's work and reflection on our experience in developing multi-agent applications for military wargaming, the following key functional requirements for effective human/agent collaboration have been identified. In particular, in order to satisfy SAT Level 1 and 2 functionality, a human must be able to

(1) Delegate goal execution to an agent
(2) Suspend and resume a particular goal execution
(3) Determine why an agent has chosen to pursue a particular course of action.
(4) Inspect beliefs relevant to a particular goal execution and if appropriate, make modifications.
(5) Inspect the goals that an agent has committed to pursue and if necessary, add new goals, delete existing goals or modify the execution order

Additional functionality such as goal replay, goal re-execution with modified context and persistence of goal execution state may be beneficial in some circumstances and particularly at SAT Level 3. However, as our immediate focus is SAT Levels 1 and 2, such functionality is deemed to be out of scope. Also note that requirement 2 (suspension and resumption of goal execution) is a prerequisite for requirements 3-5 and hence constitutes the key focus of this paper.

3 Gorite

GORITE is an open source Java framework that provides class level support for the development of agent applications that involve teams of BDI agents. Agents in GORITE are modelled as Java classes that extend the structural framework classes (Performer and Team). Agent behaviour is specified in terms of goal-based process models, which are code-level constructs that employ the behavioural framework classes (Goal and its sub-classes). Below is the specification from the case study for the platoon performer's path traversal goal.

```
private Goal traversePath() {
    return new SequenceGoal(TRAVERSE_PATH, new Goal[]{
        new Goal("process percept") {
            @Override
            public Goal.States execute(Data d) {
                System.err.println("Execution started");
                Path p = (Path) d.getValue(PERCEPT);
                d.setValue(PATH, p);
                String ename = (String) d.getValue(EXECUTION);
                Execution e = etable.get(ename);
                e.state = State.RUNNING;
                String m = timeStamp()+ename+" : Execution Started";
                etable.inform( ename, m, e.state.name() );
                etable.inform( "log", m, null );
                return Goal.States.PASSED;
            }
        },
        new LoopGoal("visit waypoints", new Goal[]{
            traverseSegment(),
            trackProgress(),
            checkpoint(),
        })
    });
}
```

Note that the path traversal goal specifies both the activities required to achieve the goal and the coordination requirements for those activities. Both facets are specified explicitly and uniformly using GORITE goal class instances (e.g. `Goal`, `SequenceGoal`, `LoopGoal` in the method above). The resulting traversal goal instance can then be executed on behalf of the goal owner by a separate executor object, which traverses the goal instance graph and at each node (which is an object of type `Goal`) invoking its `execute()` method. BDI execution semantics are preserved, with the agent still able to choose between courses of action to achieve a goal or to reconsider how a goal might be achieved. Team goals are specified in terms of roles which are filled by team members; team goal execution is then managed by a single executor on behalf of all the participating team members. During the goal graph traversal process, the executor makes available to the participants in the execution a shared data context, thus providing for a clear separation between an agent's individual beliefs and those that it shares with other agents involved in the goal execution.

In traditional BDI frameworks, execution is agent focused – an individual agent (or an agent team) determines what plan to perform in order to achieve its current goal. This determination is done in the absence of any explicit representation of currently active or future intentions. In GORITE, the focus is shifted from an individual agent pursuing its current goal to an individual agent being a participant in the achievement of a larger system-level goal. We would argue that any practical agent framework should provide support for both perspectives. That is, an agent needs to be able to operate as part of a larger whole while at the same time, progressing its own goals if appropriate. In GORITE, this latter behaviour is supported through the concept of a ToDo group. Each performer can maintain a ToDo group, which is a list of the intentions that it is currently pursuing or has decided to pursue in the near future. Within a ToDo group, only one intention is progressed during a time slice – that is the intention that is at the top of the list. However, prior to the executor progressing the top intention at the beginning of a time slice, meta-level reasoning can be invoked to determine which intention is to be progressed in the next time slice. ToDo groups can also be used to model reactive behaviour, including user interaction. In this respect, GORITE provides a `Perceptor` class that can be used by a performer to add goals to its ToDo group when particular events occur. In [3], perceptors were used to model incoming manufacturing orders and requests for sensor team reformation. However, they also provide a convenient mechanism for user-initiated goal execution. The traversal goal is added to the company performer's ToDo group via the `Company.start()` method:

```
public void start(String ename,String gname,Object percept,Data d)
{
      System.err.println("Execution to be started");
      Execution e = new Execution(ename);
      e.request = Request.START;
      etable.put(ename, e);
      String m = timeStamp()+ename+" : Traversal requested";
      etable.inform( ename, m, e.state.name() );
      etable.inform( "log", m, null );
      d.setValue(EXECUTION, ename);
      Perceptor perceptor = perceptors.get(gname);
      perceptor.perceive(percept, d);
 }
```

The percept object is of type `Path` and contains the waypoints that the platoon is to visit. This object is added to the data context (`d`) for the goal execution within the `Perceptor.perceive()` method as a data element with the default name of `PERCEPT`. For convenience, the path object was made available to the goal execution as an element in the data context called `PATH` in the traversal goal definition provided earlier. The `perceive()` method also adds the goal to the company's ToDo group. In this instance, `start()` is invoked by a method chain originating in the action listener for the Start button in the application GUI. Note that multiple goals can be added to the ToDo group and that these goals can be executed either sequentially, or through the use of meta-goals, concurrently. For a more complete description of the GORITE execution model, the reader is referred to [3].

4 Human/Agent Collaboration

In the previous section, it has been demonstrated how goal execution can be initiated by a human using the GORITE preceptor. While this constitutes an essential first step in the achievement of human/agent collaboration, it constitutes only one of the requirements identified in Sect. 2. As indicated in Sect. 2, the key requirement in terms of providing effective human/agent collaboration is requirement 2 – suspension and resumption of goal execution.

In GORITE goal execution is time-sliced and within a ToDo group, meta-goals can be employed to determine which goal in the ToDo group should be progressed next. However, one then has the problem of determining how to suspend goal execution, perhaps by executing a blocking goal and more importantly, when to suspend execution. The ideal is for goal execution to be interrupted at well-defined points which we refer to as checkpoints and this is the approach that we have employed in the waypoint traversal example. As indicated in the traversal goal definition in the previous section, a checkpoint goal is performed whenever a waypoint is reached. This goal passes if there are no user requests. If there is a suspension request, then the goal blocks until a resumption request is issued by the user:

```
Goal checkpoint() {
    return new Goal( "checkpoint" ) {
        @Override
        public Goal.States execute(Data d) {
            String ename = (String) d.getValue(EXECUTION);
            Execution e = etable.get(ename);
            switch (e.request) {
                case PAUSE:
                  if (e.state == State.RUNNING) {
                    System.err.println("Execution paused");
                    e.state = State.PAUSED;
                    String m1 = timeStamp()+ename+" : paused";
                    etable.inform( ename, m1, e.state.name() );
                  }
                  return Goal.States.BLOCKED;
                case CONTINUE:
                  if (e.state == State.PAUSED) {
                    System.err.println("Execution resumed");
                    e.state = State.RUNNING;
                    String m2 = timeStamp()+ename+" : resumed";
                    etable.inform( ename, m2, e.state.name() );
                  }
                  return Goal.States.PASSED;
            }
        }
    };
```

While goal execution is blocked, the user is able to inspect and modify the data context, which contains the data elements relevant to the goal execution. In our example, the relevant data are the waypoints, and a simple GUI is provided to enable the user to modify future waypoints:

The data context is also an appropriate structure in which to hold explanations as to how the agent has arrived at the current execution point. In this regard, providing a goal trace would constitute a good starting point. If the ToDo group contains multiple goals, then in a similar way, the ordering of the goals can be modified, existing goals removed and new goals added. An example of ToDo group manipulation is provided in [3].

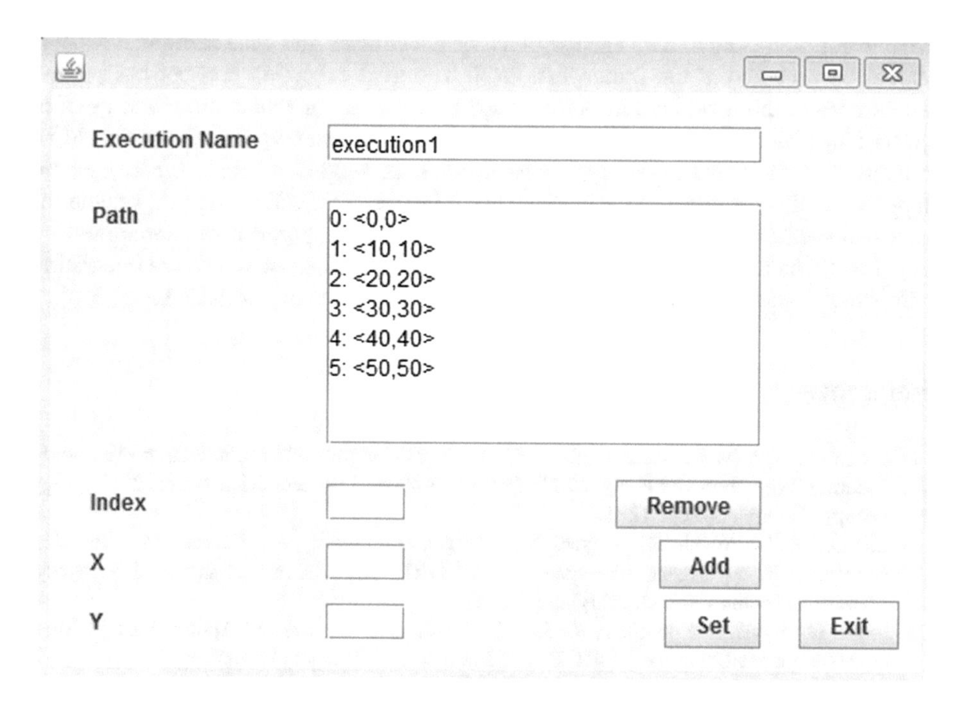

Fig. 1. The waypoint modification GUI

5 Conclusion

The motivation of this work has been to demonstrate that the GORITE BDI framework, through its explicit goal representation and corresponding execution model, supports the key requirements for human/agent collaboration and can be used to develop SAT agents. Through the use of a simple but representative example, the ability for humans to initiate, suspend and resume GORITE agent activity has been demonstrated. The ability to inspect and modify the data associated with the goal execution has been demonstrated. If a GORITE agent is intending to pursue multiple goals (either concurrently or sequentially), the goals in its ToDo group can be manipulated by a collaborating human. This particular aspect of human/agent collaboration was not included in the simplified example used in this paper. From a modelling perspective, the concept of goal execution underpins the requirements identified in Sect. 2 for effective human/agent collaboration. This concept is captured in the application through the `Execution` class and the checkpoint goal, which now form the basis of a generic goal execution capability for the GORITE framework.

Using GORITE to develop effective SAT agents will be an ongoing activity. Chen et al. have demonstrated that the transparency provided by SAT-enabled agents is beneficial in terms of human operator effectiveness. The tasks involved in these studies were relatively straightforward; a key challenge, we believe, will be in the scaling up of human/agent collaboration to address more complex problems. In particular, while GORITE may provide a basic set of building blocks for creating transparent agents, what is not clear is how these agents should be constructed and what additional support should be provided at the framework level. The goal execution concept has proven useful both in this work and in other related applications (manufacturing and medical prescribing) which suggests that such an abstraction is generally useful and should be supported at the framework level. Visualisation is another example where generic support could be provided – for instance using interactive Gantt charts as a vehicle for goal management rather than a conventional GUI-based approach as exemplified by Fig. 1 could be beneficial in terms of the user experience. Also we would see integration with simulation as a key element in the delivery of functionality at SAT Level 3.

References

1. Hill, S.G., Ling, M.F.: Human-Robot Bi-Directional Communication and Human-Autonomy Teaming Workshop Summary. DST-Group-GD-0965, Defence Science and Technology Group, Fishermans Bend (2017)
2. Chen, J.Y.C., Procci, K., Boyce, M., Wright, J., Garcia, A., Barnes, M.: Situation Awareness-Based Agent Transparency. ARL-TR-6905, Army Research Laboratory, Aberdeen Proving Ground, Maryland (2014)
3. Jarvis, D., Jarvis, J., Rönnquist, R., Jain, L.: Multiagent Systems and Applications. Volume 2: Development Using the GORITE BDI Framework. Springer, Heidelberg (2012)
4. Bratman, M.E.: Intention, Plans, and Practical Reason. Harvard University Press, Cambridge (1987)
5. Rao, A.S., Georgeff, M.P.: Modeling rational agents within a BDI architecture. In: Proceedings of the Second International Conference on Principles of Knowledge Representation and Reasoning. Morgan Kaufman (1991)
6. Rao, A.S., Georgeff, M.P.: BDI agents: from theory to practice. In: Proceedings of the 1st International Conference on Multi-Agent Systems (ICMAS 1995), San Francisco (1995)
7. Jones, R., Wray, R.: Comparative analysis of frameworks for knowledge-intensive intelligent agents. AI Mag. **27**, 57–70 (2006)
8. Georgeff, M.P., Lansky, A.I.: Procedural knowledge. Proc. IEEE **74**, 1383–1398 (1986)
9. AOS Group: AOS Group. http://www.agent-software.com.au
10. Bardram, J.E., Jeuris, S., Houben, S.: Activity-based computing: computational management of activities reflecting human intention. AI Mag. **36**(2), 63–72 (2015)
11. Norman, D.: The Design of Everyday Things: Revised and Expanded Edition. Basic Books, New York (2013)

Evolution Direction of Reward Appraisal in Reinforcement Learning Agents

Masaya Miyawaki[1(✉)], Koichi Moriyama[1], Atsuko Mutoh[1], Tohgoroh Matsui[2], and Nobuhiro Inuzuka[1]

[1] Department of Computer Science, Nagoya Institute of Technology, Nagoya, Japan
m.miyawaki.474@nitech.jp, {moriyama.koichi,atsuko,inuzuka}@nitech.ac.jp
[2] Department of Clinical Engineering, Chubu University, Kasugai, Japan
TohgorohMatsui@tohgoroh.jp

Abstract. Humans appraise the environment in daily life. We are implementing appraisal mechanisms into reinforcement learning agents. One of such mechanisms we proposed is the utility-based Q-learning, which learns behaviors from subjective utilities derived from payoffs the agent gains and a utility-derivation function the agent has. In the previous work, we know that payoff-based evolution brings utility-derivation functions that facilitate mutual cooperation in iterated prisoner's dilemma games. However, the evolution process itself has not yet been known well. In this work, we investigate the process in terms of what determines the evolution direction. We introduce two metrics showing preference of actions based on the evolved subjective utilities, which divide the evolution space into four regions. In each region, the metrics will explain the evolution directions.

Keywords: Multi-agent reinforcement learning · Reward appraisal
Prisoner's dilemma · Genetic algorithm · Evolutionary process

1 Introduction

We humans do not accept the surrounding environment as it is, but appraise it in daily life. Such appraisal mechanisms are changing our behaviors. For example, humans are able to cooperate with each other because we create a kind of rewards for cooperation in our brains [1]. On the other hand, let us consider human-like autonomous computer programs called agents learning their behaviors with reinforcement learning. It is difficult for them to learn cooperative behaviors from given rewards [2] because they are reward-maximizers.

Moriyama [3] proposed the utility-based reinforcement learning concept where an agent learns behaviors from not rewards but *utilities* derived by a utility-derivation function it has. The function is a kind of appraisal mechanism. That work showed a condition of the function giving mutual cooperation in iterated prisoner's dilemma (IPD) games. Later, Moriyama et al. [4] showed

G. Jezic et al. (Eds.): KES-AMSTA-18 2018, SIST 96, pp. 13–22, 2019.
https://doi.org/10.1007/978-3-319-92031-3_2

that such mutually cooperative utility-derivation functions were given by evolutionary computation whose fitness was the sum of rewards. In other words, the cooperative appraisal mechanisms were evolved in an environment where the agents should be reward-maximizers. In addition, interestingly, the evolved functions had a specific structure in the space of utility-derivation functions. However, unfortunately, that work did not yet investigate in detail the evolution process itself showing why and how such a structure evolved.

Therefore, this work investigates the evolution process itself. In order to make it easier, we first introduce two metrics showing preference of actions under an assumption that the opponent takes actions evenly. After that, we investigate the process using the metrics.

Human appraisal mechanisms are fast, intuitive methods for decision making. Hence, for example, similar mechanisms will be needed in robot control in an open environment that requires immediate decisions one after another. This work helps us understand the mechanisms.

This paper consists of six sections. Section 2 is a preliminary section introducing IPD games, Q-learning, and the utility-based Q-learning. In Sect. 3, we propose the metrics. In Sect. 4, we show the experiment where the utility-derivation functions evolved and analyze the result in detail with the metrics. Section 5 refers to some related works. Finally, this paper is concluded in Sect. 6.

2 Preliminaries

2.1 Iterated Prisoner's Dilemma Games

A prisoner's dilemma game [5] is a game where two players simultaneously choose their actions, either cooperation (C) or defection (D), and receive payoffs $r \in \{T, R, P, S\}$, respectively. The payoffs satisfy the following conditions: $T > R > P > S$. The relation between their actions and their payoffs is shown in Table 1. This table shows that if each player pursues individual rationality, mutual defection occurs and both of them receive a payoff P smaller than R, the payoff of mutual cooperation.

An iterated prisoner's dilemma (IPD) game is that the players play a prisoner's dilemma game iteratively. The payoffs satisfy $2R > T + S$. IPD is fascinating researchers for decades and gives us a good example where appraisal should be different from payoffs.

Table 1. Payoff table of a prisoner's dilemma game. The players are given payoffs determined by the combination of their actions, i.e., C and D. The row player receives the left payoffs, while the column receives the right ones.

Row\Column	C	D
C	R, R	S, T
D	T, S	P, P

2.2 Q-Learning

Q-learning [6] is one of the most famous reinforcement learning algorithms. In Q-learning, an agent chooses its action a_t based on an action value $Q(s_t, a_t)$ from available actions in the current state s_t at each time step t. After that, the agent receives a reward r_{t+1} from the environment and the state changes to the next s_{t+1}, and then the action value $Q(s_t, a_t)$ is updated as follows:

$$\begin{aligned} Q(s_t, a_t) &\leftarrow Q(s_t, a_t) + \alpha\delta_t, \\ \delta_t &\equiv r_{t+1} + \gamma \max_a Q(s_{t+1}, a) - Q(s_t, a_t), \end{aligned}$$

where α is a learning rate and γ is a discount rate. The agent will learn the optimal behavior without any explicit instructions, i.e., labels.

2.3 Utility-Based Q-Learning

Utility-based Q-learning [3] is an extension of Q-learning where a *subjective utility* u derived from a reward r from a utility-derivation function $u(r)$ is used as follows:

$$\begin{aligned} Q(s_t, a_t) &\leftarrow Q(s_t, a_t) + \alpha\delta_t, \\ \delta_t &\equiv u(r_{t+1}) + \gamma \max_a Q(s_{t+1}, a) - Q(s_t, a_t). \end{aligned} \tag{1}$$

In IPD games, Moriyama et al. [4] evolved coefficients of a cubic utility-derivation function and obtained subjective utilities leading to mutual cooperation. We will examine it in Sect. 4.

3 Action Tendency

Moriyama et al. [4] showed that the evolved subjective utility-derivation functions had a specific structure. However, unfortunately, they did not yet investigate in detail the evolution process itself showing why and how such a structure evolved. Therefore, this work investigates the evolution process itself.

Since the evolution changes the subjective utilities of each agent and the utilities change the agent's behaviors, we focus on the relation between the subjective utilities and behaviors in each agent. We first define two metrics, collectively called "Action Tendency", which shows preference of actions under an assumption that the opponent takes actions evenly. The metrics are called "cooperativeness" and "conformity". We next define "Dominant Action Tendency", which determines which metric gives more influence on an action choice.

3.1 Cooperativeness

Let the agent's action be $X \in \{C, D\}$. We define a function u_{cp} for an action X as follows:

$$u_{cp}(X) \equiv \begin{cases} \dfrac{u(R) + u(S)}{2} & \text{if } X = C, \\ \dfrac{u(T) + u(P)}{2} & \text{otherwise, i.e., } X = D. \end{cases}$$

It shows a subjective utility of each action under an assumption that the opponent chooses one of the actions evenly. We say that the agent is cooperative if $u_{cp}(C) \geq u_{cp}(D)$ and defective otherwise.

Next, we define the cooperativeness metric m_{cp}, which shows the relation between $u_{cp}(C)$ and $u_{cp}(D)$, as follows.

$$m_{cp} \equiv \begin{cases} \dfrac{u_{cp}(C)}{u_{cp}(D)} & \text{if } u_{cp}(D) \neq 0, \\ u_{cp}(C) & \text{otherwise.} \end{cases}$$

We say that the agent is cooperative or defective using the metric as follows.

$$\text{The agent is} \begin{cases} \text{cooperative} & \text{if} \begin{cases} m_{cp} \geq 1 \text{ if } u_{cp}(D) > 0, \\ m_{cp} \geq 0 \text{ if } u_{cp}(D) = 0, \\ m_{cp} \leq 1 \text{ otherwise,} \end{cases} \\ \text{defective} & \text{otherwise.} \end{cases} \quad (2)$$

Note that m_{cp} shows the strength of preference. For example, if $m_{cp} \gg 1$ when $u_{cp}(D) > 0$, the agent is strongly cooperative; if $m_{cp} \ll 1$ when $u_{cp}(D) > 0$, it is strongly defective.

3.2 Conformity

Let us consider a statement I that means the agent's action and the opponent's are identical. We define a function u_{cf} for the statement I and $\neg I$ as follows:

$$u_{cf}(I) \equiv \frac{u(R) + u(P)}{2},$$
$$u_{cf}(\neg I) \equiv \frac{u(T) + u(S)}{2}.$$

It shows a preference for taking a same action with its opponent under an assumption that the action pairs happen evenly. We say that the agent is conforming if $u_{cf}(I) \geq u_{cf}(\neg I)$ and anticonforming otherwise.

Next, we define the conformity metric m_{cf}, which shows the relation between $u_{cf}(I)$ and $u_{cf}(\neg I)$, as follows.

$$m_{cf} \equiv \begin{cases} \dfrac{u_{cf}(I)}{u_{cf}(\neg I)} & \text{if } u_{cf}(\neg I) \neq 0, \\ u_{cf}(I) & \text{otherwise.} \end{cases}$$

We say that the agent is conforming or anticonforming using the metric as follows.

$$\text{The agent is} \begin{cases} \text{conforming} & \text{if} \begin{cases} m_{cf} \geq 1 \text{ if } u_{cf}(\neg I) > 0, \\ m_{cf} \geq 0 \text{ if } u_{cf}(\neg I) = 0, \\ m_{cf} \leq 1 \text{ otherwise,} \end{cases} \\ \text{anticonforming} & \text{otherwise.} \end{cases} \quad (3)$$

Note that m_{cf} shows the strength of preference. For example, if $m_{cf} \gg 1$ when $u_{cf}(\neg I) > 0$, the agent has strong conformity; if $m_{cf} \ll 1$ when $u_{cf}(\neg I) > 0$, it has strong anticonformity.

3.3 Dominant Action Tendency

There are two metrics m_{cp} and m_{cf} showing the strength of preference for cooperativeness and conformity, respectively. Thus, we define "Dominant Action Tendency (DAT)" as a stronger metric of them. It is cooperativeness if the following condition is satisfied; otherwise it is conformity.

$$\begin{cases} |m_{cp}| > |m_{cf}| & \text{if } u_{cp}(D) = 0 \text{ and } u_{cf}(\neg I) = 0, \\ |m_{cp}| > |m_{cf} - 1| & \text{if } u_{cp}(D) = 0 \text{ and } u_{cf}(\neg I) \neq 0, \\ |m_{cp} - 1| > |m_{cf}| & \text{if } u_{cp}(D) \neq 0 \text{ and } u_{cf}(\neg I) = 0, \\ |m_{cp} - 1| > |m_{cf} - 1| & \text{otherwise.} \end{cases} \tag{4}$$

4 Evolution of Subjective Utilities

In this section first we review the result of the original experiment of Moriyama et al. [4] we conducted again. After that, we conducted a new cell-based experiment where the initial set of chromosomes is in a small space called a *cell* in order to investigate why and how the evolution of utility-derivation functions progressed to the result in detail.

4.1 Original Experiment

The experiment started from generating 100 agents each of which had its own utility-derivation function. The utility-derivation function was formed as $u(r) \equiv ar^3 + br^2 + cr + d$, where r was a received payoff and the coefficients a, b, c, d were real-values in the domain $[-10, 10]$. The coefficients were randomly initialized and evolved by the simple genetic algorithm [7], where its chromosomes had four genes each of which was one of the coefficient. The fitness of each chromosome was the result of an IPD game played with other chromosomes, i.e., agents. The payoffs of the IPD game were $T = 5, R = 3, P = 1, S = 0$. There were three nested loops: gameplays, pairs, and generations.

In each step of a gameplay, each agent learned its Q-values with the utility-based Q-learning (Eq. 1) using utilities derived from received payoffs and its own utility-derivation function. In this experiment, the number of states of Q-learning was 1; that is, states were not considered. This inner loop was repeated until a gameplay was finished after 1000 steps.

After the gameplay, every agent stored the sum of gained payoffs, formed a new pair with a different peer, and started a new gameplay with the zeroed Q-values. This middle-level loop was repeated until every agent formed a pair with all of the other agents; that is, there were $\binom{100}{2} = 4950$ pairs.

When all agents finished gameplays with all of the others, the chromosomes were evolved by the simple genetic algorithm, where the fitness of each chromosome was the total payoffs the agent obtained in gameplays. The evolution generated offspring for the next generation. This outer loop was repeated in 10000 generations.

We conducted this experiment 100 times. In 79 out of the 100 trials, the average payoff per step in the IPD game was more than 2.5 ($= (T+S)/2$). Hence, we concluded that at least one mutual cooperation occurred in most of the trials. In addition, as shown in Fig. 1, the best chromosomes of the final generation projected onto the a-b plane formed a line. Note that a and b are higher-order coefficients and give more significant influence on the utility-derivation function than c and d.

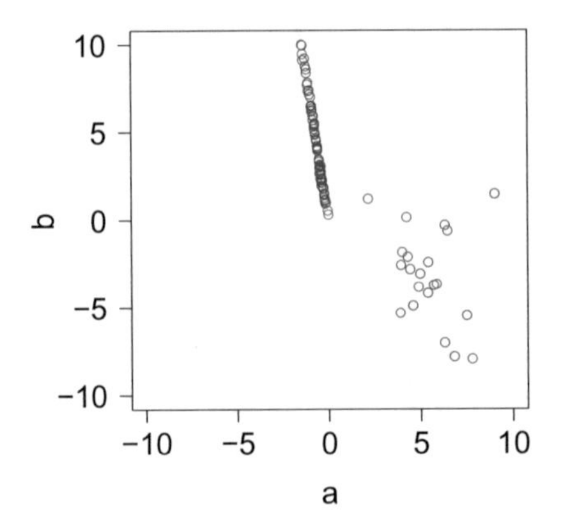

Fig. 1. The best chromosomes at the 10000th generation in all trials. Horizontal and vertical axes show a and b, respectively. Trials whose average payoff was more than 2.5 are plotted in blue, and otherwise red.

4.2 Cell-Based Experiment

Hereafter we investigate the relation between the higher-order coefficients a and b and the direction of the evolution. To see the directions from various values of the coefficients, the space initial chromosomes existed was divided into $21 \times 21 = 441$ cells, i.e., each of the axes of the a-b plane is divided into the intervals $[-10, -9.5)$, $[n - 0.5, n + 0.5)$ where $n = -9, -8, \ldots, 8, 9$, and $[9.5, 10]$. The domains of the remaining coefficients c and d were original, i.e., $[-10, 10]$. At the first generation, 100 utility-based Q-learning agents were randomly distributed in a cell. Although they were in one cell, their actions might be different because their learned Q-functions were different in accordance with their places in the cell, i.e., their utility-derivation functions. We conducted this cell-based experiment 100 times in every cell, where the number of generations was reduced to 100 because we wanted to see the evolution from a cell to neighbor cells.

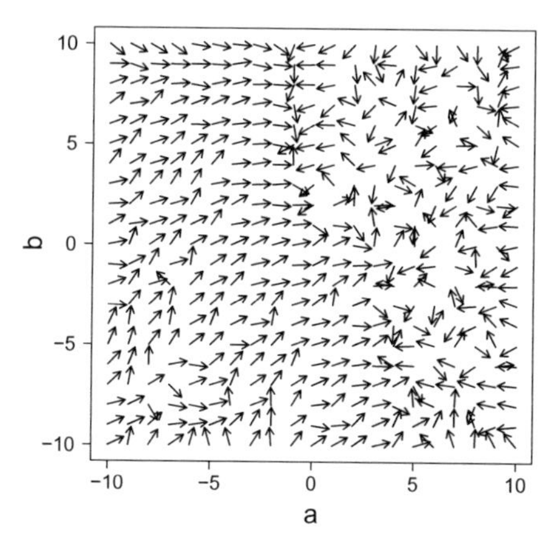

Fig. 2. The evolution direction from each cell obtained from cell-based experiments.

The result is shown in Fig. 2. An arrow from each cell shows the direction of evolution from the cell. The result suggests that b increased to positive and a approached 0 as the utility-derivation function evolved.

4.3 Action Tendency of Each Cell

In this subsection, we derive the Action Tendency of each cell by using the average of the coefficients $\bar{a}, \bar{b}, \bar{c}, \bar{d}$ of the utility-derivation functions in the cell.

First, since $\bar{c}$ and $\bar{d}$ become 0 because their domains were $[-10, 10]$, the utility-derivation function becomes $u(r) = \bar{a}r^3 + \bar{b}r^2$. It gives us that $u(T) = u(5) = 125\bar{a} + 25\bar{b}$, $u(R) = u(3) = 27\bar{a} + 9\bar{b}$, $u(P) = u(1) = \bar{a} + \bar{b}$, and $u(S) = u(0) = 0$. Hence, the functions u_{cp} and u_{cf} become,

$$u_{cp}(C) \equiv \frac{u(R) + u(S)}{2} = \frac{27\bar{a} + 9\bar{b}}{2}, \quad u_{cp}(D) \equiv \frac{u(T) + u(P)}{2} = 63\bar{a} + 13\bar{b},$$

$$u_{cf}(I) \equiv \frac{u(R) + u(P)}{2} = 14\bar{a} + 5\bar{b}, \quad u_{cf}(\neg I) \equiv \frac{u(T) + u(S)}{2} = \frac{125\bar{a} + 25\bar{b}}{2}.$$

The average coefficients $\bar{a}$ and $\bar{b}$ are ones of the following: $-9, -8, \ldots, 8, 9$.[1] Therefore, we know that the signs of $\bar{a}$, $u_{cp}(D)$, and $u_{cf}(\neg I)$ are identical in most cells.

Next, the Action Tendency metrics m_{cp} and m_{cf} become (when the denominators are not zero),

$$m_{cp} \equiv \frac{u_{cp}(C)}{u_{cp}(D)} = \frac{27\bar{a} + 9\bar{b}}{126\bar{a} + 26\bar{b}}, \quad m_{cf} \equiv \frac{u_{cf}(I)}{u_{cf}(\neg I)} = \frac{28\bar{a} + 10\bar{b}}{125\bar{a} + 25\bar{b}}.$$

As a result, we know that $m_{cp} < 1$ and $m_{cf} < 1$ in most cells.

[1] Here we ignore the border areas.

Applying $u_{cp}(D)$ and m_{cp} of each cell to Formula 2, and applying $u_{cf}(\neg I)$ and m_{cf} of each cell to Formula 3, Action Tendency of each cell is derived as shown in Fig. 3. Note that, since m_{cp} and m_{cf} are not determined at $(a, b) = (0, 0)$, the following discussion does not include the cell at the point. It can be seen that almost all agents having $a < 0$ are cooperative and conforming, and that most of agents having $a > 0$ are defective and anticonforming.

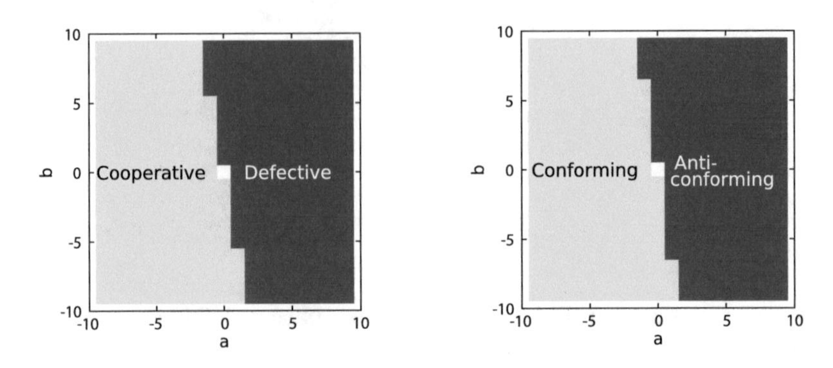

Fig. 3. Action Tendency of each cell. Left: Cooperativeness. Right: Conformity.

Due to Formula 4 with the result of the above formulae, DAT of each cell is derived as shown in the left of Fig. 4. Preference of the DAT will give more significant influence on action choice. In addition, we can categorize all cells into 4 areas by using the Action Tendency with the DAT, as shown in the right of Fig. 4. Let us call the areas by the preferred actions, the dominant one is first: Cooperative with conforming (Cp-Cf), Conforming with cooperative (Cf-Cp), Defective with anticonforming (D-A), and Anticonforming with defective (A-D).

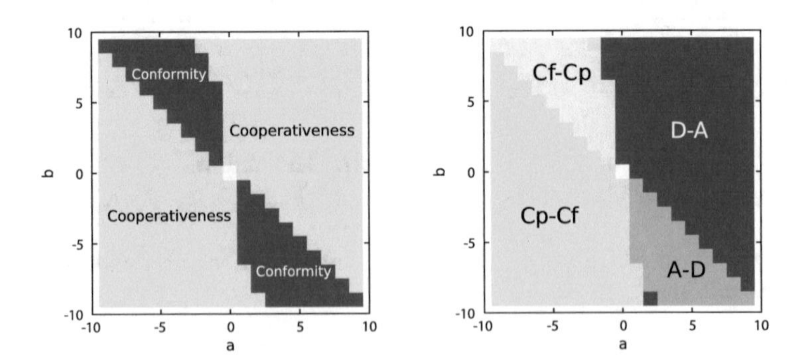

Fig. 4. Left: Dominant Action Tendency. Right: Areas divided by preferred actions.

4.4 Explanation of Evolution Directions

In this subsection, we try to explain the evolution directions by using Action Tendency of each area. Figure 5 makes it easy to confirm the gradient of Action Tendency metrics.

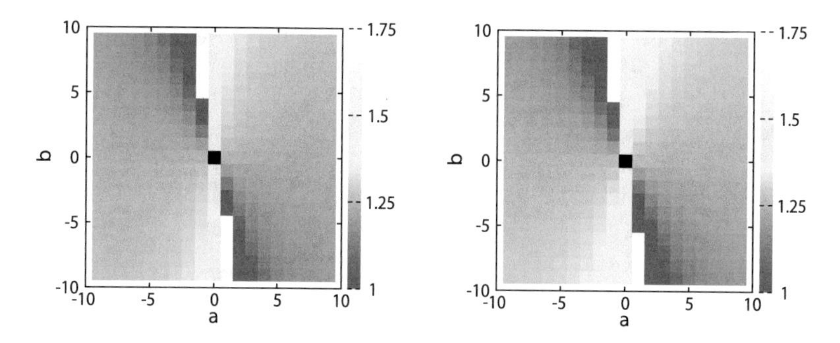

Fig. 5. m_{cp} and m_{cf} in each cell. Left: $\exp(m_{cp})$. Right: $\exp(m_{cf})$.

Let us see what happens in the areas defined in the previous subsection. Due to limited space, here we see only the D-A area.

In the D-A area, agents prefer defective and anticonforming actions. Since the average of first coefficient $\bar{a} > 0$, $u_{cp}(D) > 0$ and $u_{cf}(\neg I) > 0$ in most cells. Hence, according to Eqs. 2 and 3, the agents become more cooperative and conforming as the corresponding metrics m_{cp} and m_{cf} become larger. Then, let us see Fig. 5. It shows that the metrics become larger as the chromosomes move from lower-right to upper-right, and to upper-center. Since agents choosing cooperative conforming actions, i.e., mutual cooperation, can gain high total payoffs, new chromosomes having larger metrics (possibly resulted from a chance) will survive and climb up the slope of the metrics. Finally, they arrive at the upper-center of the space as shown in Figs. 1 and 2.

5 Related Works

After his famous competitions [5], Axelrod [8] investigated what strategies in an IPD game were evolved by genetic algorithm, where a strategy is a map connecting previous choices to the next choice. Thereafter, many works discussed the evolution of strategies in IPD games, e.g., [9]. This work is different from them because it discussed the evolution of reward appraisal instead of strategies.

The utility-based reinforcement learning concept is similar to intrinsic motivated reinforcement learning (IMRL) [10]. Although most IMRL works are for single-agent, there are a few works for multiagent scenarios. Vassiliades and Christodoulou [11] investigated evolved payoffs for Q-learning in an IPD game. They used as a fitness function the number of mutual cooperation, while this work evolved the utility-derivation function from the payoff every agent gained.

6 Conclusion

This work investigated the evolution process of the utility-derivation function in the utility-based Q-learning [3,4]. We proposed the Action Tendency metrics that show preference of actions based on the utilities each agent was given from the utility-derivation function. The metrics divided the space of the function into four areas having different properties. The evolution direction of the utility-derivation functions was explained by using the metrics.

In the future, we will investigate the Action Tendency itself. The cooperativeness and the conformity in this paper depended on the assumption that the opponent took the actions evenly. That is, we did not consider the opponent's preference at all, but the outcome will depend on it. Hence, we need to investigate the effect of the opponent's preference on the outcome and to revise the metrics to include it if necessary.

Acknowledgments. This work was partly supported by JSPS KAKENHI Grant Number JP16K00302, Kayamori Foundation of Informational Science Advancement, and the Hori Sciences & Arts Foundation.

References

1. Rilling, J.K., Gutman, D.A., Zeh, T.R., Pagnoni, G., Berns, G.S., Kilts, C.D.: A neural basis for social cooperation. Neuron **35**, 395–405 (2002)
2. Sandholm, T.W., Crites, R.H.: Multiagent reinforcement learning in the Iterated Prisoner's Dilemma. BioSystems **37**, 147–166 (1996)
3. Moriyama, K.: Utility based Q-learning to facilitate cooperation in Prisoner's Dilemma games. Web Intell. Agent Syst. **7**(3), 233–242 (2009)
4. Moriyama, K., Kurihara, S., Numao, M.: Evolving subjective utilities: Prisoner's Dilemma game examples. In: Proceedings of the 10th International Conference on Autonomous Agents and Multi-Agent Systems (AAMAS), pp. 233–240 (2011)
5. Axelrod, R.: The Evolution of Cooperation. Basic Books, New York (1984)
6. Watkins, C.J.C.H., Dayan, P.: Technical note: Q-learning. Mach. Learn. **8**, 279–292 (1992)
7. Goldberg, D.E.: Genetic Algorithms in Search, Optimization and Machine Learning. Addison-Wesley, Reading (1989)
8. Axelrod, R.: The evolution of strategies in the Iterated Prisoner's Dilemma. In: Davis, L. (ed.) Genetic Algorithms and Simulated Annealing, pp. 32–41. Pitman/Morgan Kaufmann, London/Los Altos (1987)
9. Nowak, M.A.: Five rules for the evolution of cooperation. Science **314**, 1560–1563 (2006)
10. Singh, S., Lewis, R.L., Barto, A.G., Sorg, J.: Intrinsically motivated reinforcement learning: an evolutionary perspective. IEEE Trans. Auton. Ment. Dev. **2**, 70–82 (2010)
11. Vassiliades, V., Christodoulou, C.: Multiagent reinforcement learning in the Iterated Prisoner's Dilemma: fast cooperation through evolved payoffs. In: Proceedings of the 2010 International Joint Conference on Neural Networks (IJCNN) (2010)

A General Framework for Formulating Adjustable Autonomy of Multi-agent Systems by Fuzzy Logic

Salama A. Mostafa[1(✉)], Rozanawati Darman[1], Shihab Hamad Khaleefah[2], Aida Mustapha[1], Noryusliza Abdullah[1], and Hanayanti Hafit[1]

[1] Faculty of Computer Science and Information Technology, Universiti Tun Hussein Onn Malaysia, 86400 Parit Raja, Johor, Malaysia
{salama, zana, aidam, yusliza, hana}@uthm.edu.my
[2] Faculty of Computer Science, Al Maarif University College, Anbar, Iraq
shi90hab@gmail.com

Abstract. Autonomous agents and multi-agent systems significantly facilitate solutions to many complex and distributed problem-solving environments due to the agents' various capabilities. These environments comprise uncertain, high dynamism or irregular workload that might prone the agents to make decisions that lead to undesirable outcomes. This paper proposes a Fuzzy Logic-based Adjustable Autonomy (FLAA) as a general framework for managing the autonomy of multi-agent systems that operate in complex environments. The framework includes a fuzzy logic technique to quantitatively measure and distribute the autonomy among operators (autonomous agents and humans) based on adjustable autonomy attributes. The FLAA framework dynamically changes the autonomy of the operators to directions that meets their ability to perform and produce desirable outcomes. The technical application of the framework is demonstrated by an example scenario. The scenario illustrates the competence performance of the collaborative operators. The advantages of using the framework are to capture relationships between discrete autonomy attributes, quantify adjustable autonomy and manage adjustable autonomous multi-agent systems.

Keywords: Autonomous agent · Multi-agent system · Adjustable autonomy Fuzzy logic

1 Introduction

The behavior of autonomous agents in discrete and deterministic environments manifest substantial autonomous actions due to their prior knowledge and sufficient perception to the surroundings. However, in complex environments that entail uncertain, high dynamism or irregular workload the agents are prone to make decisions that lead to undesirable outcomes [1]. Additionally, the tasks that are delegated to the autonomous agents might have different characteristics, such as primitive, deducible or critical which entails different autonomy constraints [2, 3]. These issues necessitate that the

G. Jezic et al. (Eds.): KES-AMSTA-18 2018, SIST 96, pp. 23–33, 2019.
https://doi.org/10.1007/978-3-319-92031-3_3

agents operate at different levels of autonomy in order for them to consistently make desirable decisions [4–6].

Adjustable autonomy is a technique that is proposed to enable autonomous agents to perform at different levels of autonomy and adhere to humans global control. It is complementary useful in systems where humans and autonomous agents collaboratively control an autonomous system [7]. Examples of such autonomous systems are air traffic control systems, E-Commerce systems, Internet of Things systems, multi-robot systems, network systems, remote-care systems, supply chain systems, and unmanned systems [1]. Adjustable autonomy gives agents the options of changing their operational behaviors. Moreover, it provides mechanisms to influence, oversight or intervene the agents or their activities based on a situation of exigency and improve their performance. Adjustable autonomy provides a number of benefits to multi-agent systems including flexibility of humans-agents teamwork in which they share control of a system, reliability of humans global control over the agents and efficiency of directing agents towards performance competency [4, 6].

There are many opinions and diverse understanding of what constitutes adjustable autonomy and how it is applied to systems. There are a number of visible attempts to formulate adjustable autonomy for agents in the literature and some of them are reviewed in [1]. One approach of formulating adjustable autonomy of a multi-agent system is built upon grading the autonomy of the agents to a particular band or extent that could be changed over time [5]. This approach entails measuring the boundaries of the autonomy parameters and estimating the extent to which autonomy is distributed and adjusted to or by the agents which is adopted in this paper.

This paper presents a general framework for an adjustable autonomy of multi-agent systems. The FLAA framework intends to manage the adjustable autonomy of systems that operate in complex environments including uncertain, high dynamism or irregular workload. The framework builds relationships of autonomy levels between operational tasks and actions of a system and autonomy degrees between the system's teamwork of humans-agents operators. It deploys fuzzy logic to dynamically perform linear measurement for the autonomy of the operators based on dimensions of adjustable autonomy attributes and maps the autonomy between the operations and operators of the system.

The rest of this paper is organized as follows: Sect. 2 reviews the literature on adjustable autonomy, and several attempts to apply fuzzy logic in managing the autonomy of agents. Section 3 presents the illustration of the FLAA framework for multi-agent systems that includes the processes of autonomy representation, measurement, distribution, and adjustment. Section 4 presents a possible application of the adjustable FLAA framework in an example scenario. Section 5 presents the research conclusions and outcomes.

2 Related Work

Our review of the literature indicates that there are many studies that propose frameworks for multi-agent systems, e.g., [8, 9]. Moreover, there have been many studies on the autonomy of agents with different qualitative and quantitative approaches [2, 6].

These approaches explicitly represent the autonomy to facilitate management and support competency of multi-agent systems [7].

However, there is a limited existing work that uses fuzzy logic to quantify and control adjustable autonomy of a multi-agent system [6]. Fuzzy logic is widely used to construct different types of inference frameworks for estimation and control problems [3]. It is mainly adopted in autonomous agents' research to help in handling incomplete data, improving their perception to the environment [10], narrowing down their decision options [11], including explicit control of their autonomy [12], and carrying out some of their decision-making aspects [13].

Jaafar and McKenzie [3] propose a fuzzy logic-based framework to control the autonomous behavior of agents. The framework includes arranging the behavior of an agent in weighted levels. The fuzzy logic-based framework manipulates the weights to defuzzify and determine suitable behaviors for the agents. The defuzzification limits the decisions choice to a specific number of possible actions. This approach is found to be useful in directing the behavior of agents to perform particular actions in unknown and complex environments. Couceiro et al. [12] propose a framework to manage and control the autonomy of agents based on fuzzy logic. The framework includes context-based evaluation metrics that describe the performance of the agents then adjusts the autonomy constraints based on their performance. The framework is applied in a swarm of a multi-robot system to coordinate the performance of the robots in complex environments. The robots are able to achieve their goals with limited coordination while successfully overcoming obstacles.

Gharbi and Samir [10] claim that in distributed and dynamic problem-solving environments; agents might lack clear perception, make undesirable decisions, and perform faulty actions. Gharbi and Samir propose a framework that directs the agents to collectively identifying problems. The framework supplements approximate reasoning capabilities to agents decision-making and guides their collaborative effort. The agents are applied to control a robot arm and the results show that the framework enables the agents to take the initiative and improves their performance. Shamshirb et al. [11] use a fuzzy logic-based framework to improve the performance of a multi-agent wireless sensor network. The agents are embedded within the network nodes to perform interactions and collaborative decisions. An agent of a node processes a number of sensory data as input parameters. The fuzzy logic is deployed to reduce the uncertainty of the sensory data, map the decision parameters and narrow down the decision options of the agents.

Lau et al. [13] address the need for a flexible and dynamic framework for interaction and collaboration among supply chain parties. They propose a framework of an adjustable autonomous multi-agent system for real-time supply chain control. The adjustable autonomy is defined as the capability and capacity of the decision-making agents. It is represented by levels and the autonomy adjustment entails changing the autonomy levels of the agents based on the changes of certain parameters of the environment. The framework deploys fuzzy logic to defuzzify the environment parameters into particular autonomy levels. The autonomy adjustment motivates the agents to increase or decrease the coalitions of their collaborative effort according to the environment dynamism. However, the adjustable autonomy of the agents is represented and controlled based on environmental parameters and the autonomy adjustment only affects the coalitions behavior of the multi-agent system.

3 The Fuzzy Logic-Based Adjustable Autonomy Framework

This section presents a general FLAA framework for a multi-agent system. This framework includes adjustable autonomy representation, measurement, distribution and adjustment mechanisms.

3.1 Autonomy Representation

Autonomy representation is an important issue in quantitative or qualitative adjustable autonomy formulation. In order to put the adjustable autonomy in its measurement context, a number of measurable attributes have to be defined [6]. There are a number of adjustable autonomy attributes that directly involved in formulating adjustable autonomy in the literature including *authority*, *complexity*, *confidence*, *consistency*, *dependency*, *goal-achievement*, *influence*, *knowledge*, *motivation*, *performance*, *risk*, *task-orientation*, *time* and *trust* as it is illustrated in [1]. These adjustable autonomy attributes give a clear insight into what and how to measure and draw the boundaries of the adjustable autonomy dimensions. Some attributes represent local autonomy desires of an operator (human or agent) and others represent global autonomy desires of a system, hence, satisfying the concept of adjustable autonomy [5]. The attributes determine the properties of the operational autonomy levels, the configuration of the operational autonomy levels and the qualification conditions to obtain particular autonomy levels. The value of the attributes can be inferred from the behavior of an operator or the characteristic of the operation that the operator involved in.

Let X be a set of independent variables of adjustable autonomy attributes and $X = \{x_1, x_2, \ldots, x_n)$ in which an x is an individual adjustable autonomy attribute and n represents the number of adjustable autonomy attributes in a system. Let each operator of a system has its own adjustable autonomy attributes, X^i, in which i is the index of an operator and $x_j^i \in X^i$. Let Y be a set of dependent variables of autonomy degrees and $Y = \{y_1, y_2, \ldots, y_m\}$ in which y_i is an individual operator autonomy degree and m represents the number of operators of the system, $i = \{1, 2, 3, \ldots, m\}$. The set of all autonomy degrees Y has the interval of [0, 1] in which $0 \leq y \leq 1$. Then Fig. 1 shows the relationships between the X and Y of the system.

The FLAA framework represents the adjustable autonomy by a set of autonomy levels that hierarchically manages the operational behavior of operators in a system. An operator might strictly correspond to an autonomy level, or switch between different autonomy levels. The roles and actions of the operators in an autonomy level are constrained by the properties of the autonomy level. The operators that attain high autonomy levels, the properties of the levels enable them to practice lesser constraint behaviors. They are capable of carrying out a wide range of tasks and performing different types of actions and vice versa.

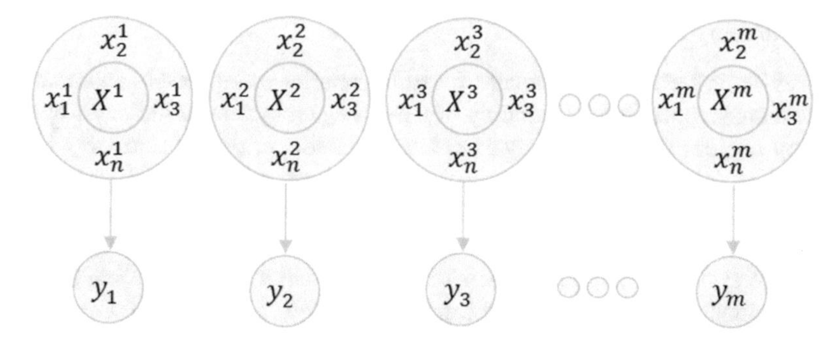

Fig. 1. The relationships between adjustable autonomy attributes and degrees.

Let the adjustable autonomy has a set of autonomy levels $L = \{l_1, l_2, \ldots, l_r\}$; A be the set of all actions sorts, $A = \{a_1, a_2, \ldots\}$ that operators can perform to complete a system's tasks. T be the possible tasks in which $\mathrm{T} = \{t_1, t_2, \ldots\}$ and a task $t \subseteq \mathrm{A}$. Then Fig. 2 shows an abstract representation of the proposed adjustable autonomy.

In the FLAA framework, an autonomy level, l, represents a band of autonomy degrees that coincides with specific set or sets of autonomy properties, P in which $\mathrm{p} \in P$ is a single autonomy property. The P include a range of fully-autonomous to non-autonomous abilities that determine the possibilities of adding, allowing, blocking, intervening, influencing, suspending, and/or terminating operational behaviors or goals of operators (i.e., the operation of adjustable autonomy). The P of a particular l are configured based on association relationship of operations (tasks and corresponding actions) and operators (humans and agents). Each l comprises sets or subsets of tasks and actions, e.g., $l_1 = \{t_1 = \{a_1, a_2, \ldots\}, t_2 = \{a_1, a_2, \ldots\}, \{\}\}$ and associations of autonomy properties as shown in Fig. 2. The l might have overlapping and distinct P, t, and/or a. We assume the setting and distribution of the X, Y, L, P, T, and A adhere to the FLAA application domain.

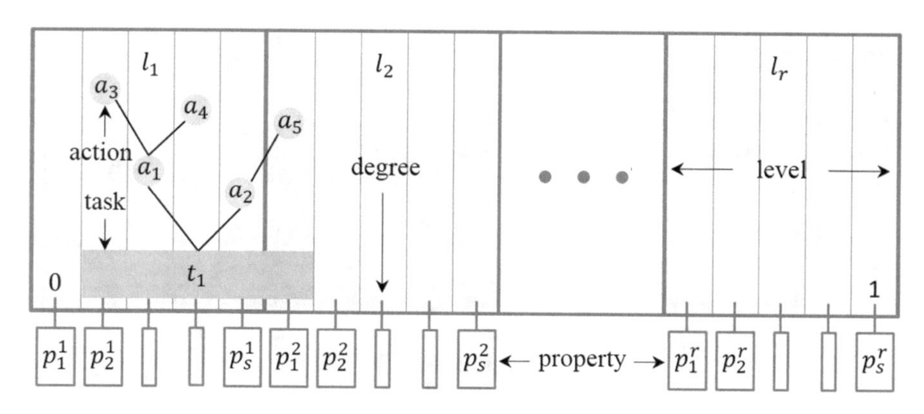

Fig. 2. The adjustable autonomy representation of the FLAA framework [6].

3.2 Autonomy Measurement

The fuzzy logic technique formulates fuzzy boundaries of adjustable autonomy attributes to manage operational autonomy of a multi-agent system. It measures operators' autonomy degrees and accordingly associates operators to proper autonomy levels.

Let Ꞓ represents a set of active operators of a system in which an agent, $\rho_i \in$ Ꞓ and a human, $\mathfrak{h}_i \in$ Ꞓ. Let the system has an operational structure of three autonomy levels that have fuzzy boundaries, $L = \{l_1, l_2, l_3\}$. Let the adjustable autonomy attributes of the system be three fuzzy sets of linguistic input variables, $X^i = \{x_1^i, x_2^i, x_3^i\}$, that have qualitative values with the labels of *High*, *Medium* and *Low* in which $x_j^i = \left\{x_j^i L, x_j^i M, x_j^i H\right\}$. Consequently, the membership for each of which is represented by trapezoidal right, middle and left functions as follows:

$$\mu_f\left(x_j^i\mu; b_1, b_2, b_3, b_4\right) = \begin{cases} 0, & x_j^i\mu \le b_1 \\ \frac{x_j^i\mu - b_1}{b_2 - b_1}, & b_1 \le x_j^i\mu \le b_2 \\ 1, & b_2 \le x_j^i\mu \le b_3, x \in \mathrm{R} \\ \frac{b_4 - x_j^i\mu}{\mathrm{d} - b_3}, & b_3 \le x_j^i\mu \le b_4 \\ 0, & b_4 \le x_j^i\mu \end{cases} \quad (1)$$

where μ_f is the membership function of the X^i sets such that $\mu_f\left(x_j^i\mu\right)$: $X\mu \rightarrow [0, 1]$ in which $X\mu$ is the adjustable autonomy dimension (universe of discourse) of a multi-agent system.

Based on Fig. 2 and Eq. (1), Fig. 3 shows a possible representation of the adjustable autonomy of the FLAA framework. It further shows the assignment of range and degree of memberships for input X^i variables and output autonomy levels respectively. Each autonomy degree corresponds to a specific set of autonomy properties as explained earlier.

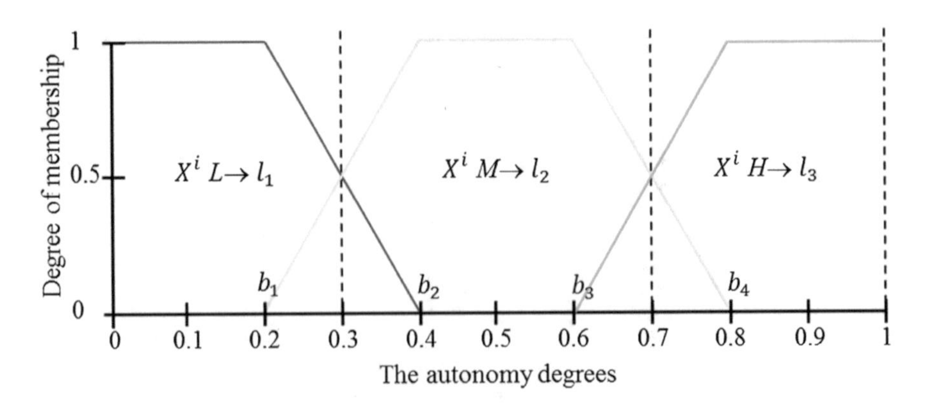

Fig. 3. The memberships of the FLAA.

Then the autonomy degree of each operator is aggregated from the adjustable autonomy attributes by using the integral of the centroid function.

$$y_i = \frac{\int_{x_1}^{x_n} \mu_f\left(x_j^i\right) x_j^i dx}{\int_{x_1}^{x_n} \mu_f\left(x_j^i\right) dx}, x_j^i \in l \tag{2}$$

where y_i is the centroid of the autonomous region that is bounded by the interval $[x_1, x_n]$ in which the y_i represents the autonomy degree of an operator. y_i value determines the l, and the P that the operator can attain in a phase of its run cycle as it is explained in the following section.

3.3 Autonomy Distribution

The FLAA framework is concerned with autonomy distribution and adjustment among operator. The FLAA consists of four main components which are Autonomy Regulator, Autonomy Fuzzifier, Autonomy Rule Base, and Inference Engine. Figure 4 shows the FLAA framework of a multi-agent system.

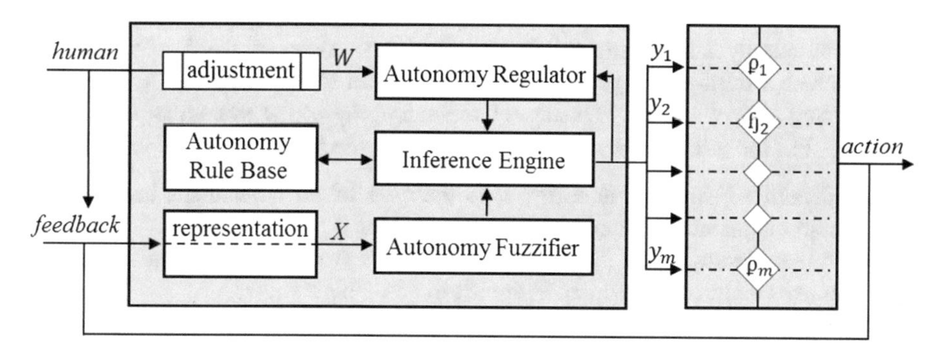

Fig. 4. The FLAA framework.

The Autonomy Regulator allows a human supervisor to manually adjust the autonomy parameters using W as explained in Sect. 3.3. The Autonomy Fuzzifier receives X^i inputs as a fuzzy set and measures the overlapping of autonomy levels (as shown in Fig. 3). The Autonomy Rule Base contains rules in which a typical rule is.

$$IF\ x_1^i L\ OR\ x_1^i M\ OR\ x_1^i H\ AND\ x_2^i L\ OR\ x_2^i M\ OR\ x_2^i H\ AND \ldots THEN\ levelX$$

The Inference Engine based on the Autonomy Fuzzifier measurements and the Autonomy Rule Base outcomes defuzzifies the autonomy of the operators. The Inference Engine then dynamically distributes the autonomy in the operational behaviors of the operators. The operators decide on the possible autonomy levels and choose a suitable level.

3.4 Autonomy Adjustment

The autonomy adjustment is a process of redistributing the autonomy among operators to redirect their performance [1]. It gives a system the flexibility to change the behaviors of the operations. This change aims to improve the performance, overcome complex and uncertain situations and/or prevent the occurrence of errors or failures [14]. The Autonomy Regulator receives the values of the autonomy adjustment parameters as a crisp set W for each of the operators, $W = \{W^1, W^2, \ldots, W^m\}$ in which $W^1 = \{w_1^1, w_2^1, \ldots, w_n^1\}$ as shown in (3).

$$W = \begin{matrix} & y_1 & y_2 & y_m \\ \begin{matrix} x_1 \\ x_2 \\ x_n \end{matrix} & \multicolumn{3}{c}{\begin{bmatrix} w_1^1 & w_1^2 & w_1^m \\ w_2^1 & w_2^2 & w_2^m \\ w_n^1 & w_n^2 & w_n^m \end{bmatrix}} \end{matrix} \quad (3)$$

The W of the system work as a bias to the autonomy degrees of the operators as follows:

$$y_i \Leftarrow w_1^i * x_1^i, w_2^i * x_2^i, \ldots, w_n^i * x_n^i \quad (4)$$

The autonomy adjustment entails changing the values of the W by a human supervisor or a third-party system. Giving low values to W decreases the autonomy while giving high values to W increases the autonomy of the system or a particular operator, W^i. The adjustment might result in one of the following autonomy modes:

- The default adjustable autonomy: It is the case of no adjustment in which the autonomy adjustment parameters are set such that $W = 1$.
- The full-autonomy: It is the case of a system is working beyond the adjustable autonomy conditions in which W are set to some high values.
- The no-autonomy: It is the case of disabling the autonomous behaviors of a system in which W are set to some low values.
- The imbalance adjustable autonomy: It is the case of disturbing the autonomous behaviors of a system in which some of the W are increased and the others are decreased.

4 Example Scenario of the FLAA Framework

The following example explains the possible application of the FLAA framework in a multi-agent system. Assume that a FLAA framework manages the adjustable autonomy of a multi-agent system that consists of three agents. These agents work independently when they have fully-autonomous properties, expose to influence of others (agents or humans) when they have semi-autonomous properties, and follow the commands of others when they have non-autonomous properties.

Table 1. Sample settings of the FLAA framework.

Level	Boundary	Property	Task	Action
l_1	$0 \leq l_1 \leq 0.3$	$l_1 : p_1^1 \rightarrow \dot{t} \vee \dot{a}$	$t_1 \in l_1 \wedge l_2 \wedge l_3$	$a_1 \in l_1$
l_2	$0.3 < l_2 \leq 0.7$	$l_2 : p_1^2 \rightarrow \dot{t} \vee \dot{a}$ $l_2 : p_2^2 \rightarrow \tilde{t} \vee \tilde{a}$	$t_2 \in l_2 \wedge l_3$	$a_2 \in l_2$
l_3	$0.7 < l_3 \leq 1$	$l_3 : p_1^3 \rightarrow \dot{t} \vee \dot{a}$ $l_3 : p_2^3 \rightarrow \tilde{t} \vee \tilde{a}$ $l_3 : p_3^3 \rightarrow \bar{t} \vee \bar{a}$	$t_3 \in l_3$	$a_3 \in l_3$

Let ρ_i be an agent that acts through implementing three operational behaviors, *tasks-selection*, ρ_i^1; *actions-selection*, ρ_i^2; and *actions-execution* ρ_i^3. Let the adjustable autonomy of a system has a set of three autonomy levels $L = \{l_1, l_2, l_3\}$. Then Table 1 shows the proposed settings of the FLAA framework for this system.

In this scenario, the three agents demonstrate fully and semi-autonomous properties in three phases as shown in Fig. 5[1]. In the first phase of their operational behavior, two agents, $\dot{\rho}_{1,2}^1$, autonomously choose $\dot{t}_1$, while the other agent, $\dot{\rho}_3^1$, autonomously chooses $\dot{t}_2$. The measured autonomy degrees in this phase are $y_1^1 = 0.66$, $y_2^1 = 0.66$, and $y_3^1 = 0.3$. The differences among these autonomy degrees can be ascribed to the differences between the states of their adjustable autonomy attributes. Subsequently, the two agents, $\dot{\rho}_{1,2}^1$, are able to influence the third agent, $\widetilde{\rho_3^1}$, to choose $\tilde{t}_1$ because $\widetilde{\rho_3^1}$ has semi-autonomous properties. In the second phase, two agents, $\dot{\rho}_{1,3}^2$, autonomously choose $\dot{a}_1 \in l_1$, while the other agent, $\dot{\rho}_2^2$, autonomously chooses $\dot{a}_2 \in l_2$. The measured autonomy degrees in this phase are $y_1^2 = 0.57$, $y_2^2 = 0.60$, and $y_3^2 = 0.62$. Although $\dot{\rho}_2^2$ has a sufficient degree of autonomy to perform in $l_2 (0.3 < y_2^2 \leq 0.7)$, $\dot{\rho}_{1,3}^2$ can push $\widetilde{\rho_2^2}$ into choosing $\tilde{a}_1$ because ρ_2^2 has semi-autonomous properties. In the third phase, the agents autonomously execute $\dot{a}_1$ because they all have a sufficient degree of autonomy to perform this action on l_1.

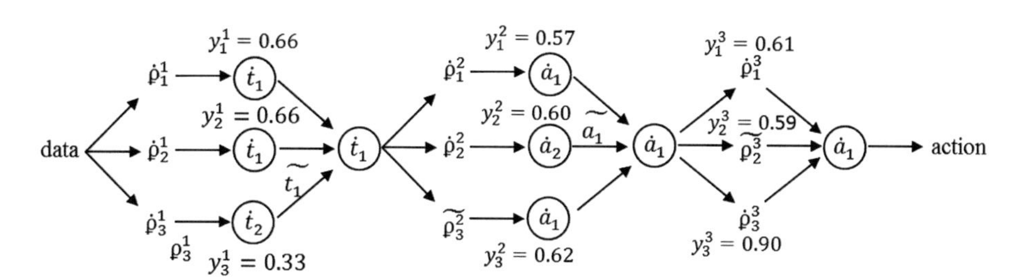

Fig. 5. The example scenario

[1] ẋ, x̃, and x̄ for each ρ, t, and a denote fully, semi-, and non-autonomous properties, respectively.

5 Conclusion

This research aims to establish an adjustable autonomy framework for autonomous systems that operate in complex and distributed problem-solving environments. Subsequently, this paper proposes a Fuzzy Logic-based Adjustable Autonomy (FLAA) framework to dynamically manage the adjustable autonomy of multi-agent systems. The FLAA employs a fuzzy logic technique to quantify and measure the autonomy degrees of operators (humans or agents) based on adjustable autonomy criteria. The FLAA segregates the adjustable autonomy to a number of autonomy levels. An autonomy level represents a band of autonomy degrees that coincides with specific set or sets of autonomy properties. The autonomy degrees associate the operators with particular autonomy properties of an autonomy level. The properties constrain the operational behavior of the operators. The paper presents an example scenario that shows a possible application to the FLAA framework. The FLAA is meant to provide efficient and flexible decision-making mechanism and reliable actions to systems. Additionally, the fuzzy logic-based control facilitates adopting, testing and comparing different adjustable autonomy attributes such as *authority*, *confidence*, and *consistency*.

Acknowledgements. This project is sponsored by the postdoctoral grant of Universiti Tun Hussein Onn Malaysia (UTHM) under Vot D004 and partially supported by the Tier 1 research grant scheme of UTHM under Vot U893.

References

1. Mostafa, S.A., Ahmad, M.S., Mustapha, A.: Adjustable autonomy: a systematic literature review. Artif. Intell. Rev. 1–38 (2017)
2. Hexmoor, H., Mclaughlan, B.: Computationally adjustable autonomy. Scalable Comput. Pract. Exp. **8**(1), 41–48 (2001)
3. Jaafar, J., McKenzie, E.: Decision-making method using fuzzy logic for autonomous agent navigation. Electron. J. Comput. Sci. Inf. Technol. **3**(1), 8–18 (2011)
4. Mostafa, S., Gunasekaran, S.S., Ahmad, M.S., Ahmad, A., Annamalai, M., Mustapha, A.: Defining tasks and actions complexity-levels via their deliberation intensity measures in the layered adjustable autonomy model. In: International Conference on Intelligent Environments (IE), pp. 52–55. IEEE (2014)
5. Bradshaw, J.M., Feltovich, P.J., Jung, H., Kulkarni, S., Taysom, W., Uszok, A.: Dimensions of adjustable autonomy and mixed-initiative interaction. In: International Workshop on Computational Autonomy, pp. 17–39. Springer, Heidelberg (2003)
6. Mostafa, S.A., Mustapha, A., Mohammed, M.A., Ahmad, M.S., Mahmoud, M.A.: A fuzzy logic control in adjustable autonomy of a multi-agent system for an automated elderly movement monitoring application. Int. J. Med. Inform. **112**, 173–184 (2018)
7. Mostafa, S.A., Ahmad, M.S., Ahmad, A., Annamalai, M., Mustapha, A.: A dynamic measurement of agent autonomy in the layered adjustable autonomy model. In: Studies in Computational Intelligence, vol. 513, pp. 25–35. Springer, Cham (2014)
8. Mahmoud, M.A., Ahmad, M.S., Yusoff, M.Z.M., Idrus, A.: Automated multi-agent negotiation framework for the construction domain. In: Distributed Computing and Artificial Intelligence, pp. 203–210. Springer, Cham (2015)

9. Jassim, O.A., Mahmoud, M.A., Ahmad, M.S.: A multi-agent framework for research supervision management. In: Distributed Computing and Artificial Intelligence, pp. 129–136. Springer, Cham (2015)
10. Gharbi, A., Samir, B.A.: Fuzzy logic multi-agent system. Int. J. Comput. Sci. Inf. Technol. **6**(4), 273 (2014)
11. Shamshirb, S., Kalantari, S., Bakhsh, Z.: Designing a smart multi-agent system based on fuzzy logic to improve the gas consumption pattern. Sci. Res. Essays **5**(6), 592–605 (2010)
12. Couceiro, M.S., Tenreiro Machado, J.A., Rocha, R.P., Ferreira, N.M.: A fuzzified systematic adjustment of the robotic Darwinian PSO. Robot. Auton. Syst. **60**, 1625–1639 (2012)
13. Lau, H.C., Agussurja, L., Thangarajoo, R.: Real-time supply chain control via multi-agent adjustable autonomy. Comput. Oper. Res. **35**(11), 3452–3464 (2008)
14. Mostafa, S.A., Ahmad, M.S., Mustapha, A., Mohammed, M.A.: Formulating layered adjustable autonomy for unmanned aerial vehicles. Int. J. Intell. Comput. Cybern. **10**(4), 430–450 (2017)

Agent-Based System for Context-Aware Human-Computer Interaction

Renato Soic(✉), Pavle Skocir, and Gordan Jezic

Faculty of Electrical Engineering and Computing,
University of Zagreb, Unska 3, 10000 Zagreb, Croatia
{renato.soic,pavle.skocir,gordan.jezic}@fer.hr

Abstract. Interaction between a human and a computer is a rapidly evolving field in computer science, with the goal to achieve efficient communication using natural language. In the recent couple of years, there have been significant accomplishments in the field, resulting with many commercially available systems capable of performing various tasks, while using natural language to communicate with human users. In this paper, we discuss the possibilities of extending the capabilities of such systems by adopting an agent-based approach. We present a model for a context-aware, intelligent, adaptive multi-agent system able to independently communicate with human users using natural language, i.e., speech. The motivation is to design an extensible system which could independently decide in which way to present information to the user, with the decision based on the user's context, retrieved from any number of devices or systems. The focus of the paper will be on describing the interaction process itself, and the significance of context regarding the human-computer interaction.

Keywords: Software agent · IoT · Context-awareness · Context resolution
Human-computer interaction · Natural language processing · BDI
Intelligent agent

1 Introduction

Human-computer interaction is a challenging research field which explores the possible ways in which human users can interact with computers. It is an interdisciplinary field combining computer science, psychology, behavioral sciences, industrial design, and several other disciplines. In scope of this paper, we will focus primarily on interaction between a human user and a computer using natural language, i.e., speech.

The idea of complex spoken communication between a human user and a computer has been an elusive goal for a long time, mostly due to complexity of speech processing and natural language understanding. However, in the last couple of years, significant advancements have been achieved, primarily due to rapid developments in computer science and artificial intelligence, and computing power available. That resulted with accelerated development of commercial services, opening a variety of possibilities where human-computer interaction using natural language could be employed in scenarios intended for the general population [1]. Therefore, systems which enable interaction between human and computer using natural language are employed in many

G. Jezic et al. (Eds.): KES-AMSTA-18 2018, SIST 96, pp. 34–43, 2019.
https://doi.org/10.1007/978-3-319-92031-3_4

different areas today. Most smart household devices and appliances respond to voice commands, as do many personal smart devices (e.g., wearables). In public transport, a synthesized voice is used to inform the passengers of their current and next station, or other important information [2, 3].

In this paper, we focus on the user-centered approach. We explore the concept of an intelligent multi-agent system relying on rich user context to provide an efficient way of communication with the user using natural language. The proposed system offers an extensible set of functionalities related to providing information on demand, communication assistance, activity management, and specific task execution. The paper is organized as follows: in Sect. 2, an overview of related work is presented; Sect. 3 provides a model of a context-aware multi-agent system; Sect. 4 elaborates types of agents involved, and their purpose; in Sect. 5, a use-case scenario within a prototype system is described; Sect. 6 provides the conclusion.

2 Related Work

The functionalities mentioned in the introduction are already available in scope of commercially available services known as Intelligent Personal Assistants. These services are present on almost every platform today: smartphones, personal computers, smart home systems, cars, etc. Such high availability is possible because these systems are developed by the most prominent IT companies, global leaders in research and development – Google[1], Microsoft[2], Apple[3], and Amazon[4]. They are relying on their own cloud services to provide all required functionalities – speech recognition and speech synthesis, natural language understanding, context resolution, decision making, etc. Furthermore, they are designed to be quite easily integrated with other devices and systems. Practically unlimited resources available (infrastructure, computing power, people, data, etc.) enabled the mentioned systems to improve and grow constantly, making the user experience better. Therefore, it is not surprising that commercial Intelligent Personal Assistants have become a very successful business opportunity, with an estimated number of unique users per month reaching over 70 million [4, 5]. While the high availability represents a strong advantage for mentioned services, that dependency on the external resource is their weak point. Availability and quality of service depend directly on the network performance, which makes the service less reliable. Even though a fast broadband or mobile connection to the Internet has become a common, widely available resource, our opinion is that a computer system should not depend exclusively on an external service to perform its core functionalities. In addition, the privacy issue is not yet regulated completely: companies may store the collected data for a certain amount of time, deleting it afterwards. However, it is disputable

[1] Google Assistant - https://assistant.google.com/.

[2] Microsoft Cortana - https://www.microsoft.com/en-us/windows/cortana/.

[3] Apple Siri - https://www.apple.com/ios/siri/.

[4] Amazon Alexa - https://www.amazon.com/Amazon-Echo-And-Alexa-Devices/.

for which purposes the data was used while it was in their possession, and what happens with new information inferred based on that data.

2.1 The Importance of Context

In the recent years, the IoT revolution sparked the creation of different smart devices and sensors able to provide various information related to user's physical and social environment, personal preferences, behavior, etc. This resulted with a possibility to create rich user context and utilize it to improve available services. Regarding human-computer interaction, the availability of rich context enabled the computer system to independently decide when, how, and why to initiate interaction with the user [6]. This introduced a significant improvement, because the interaction between computer and a human user did not have to operate in a reactive manner anymore, but rather in a proactive, or even social manner.

Intelligent personal assistants have been used before the breakthrough related to natural language processing. In the earlier stages of evolution, they were often limited to a specific domain, providing monitoring, automation and control of systems or devices, and enabling execution of simple tasks on user's behalf. Rapid advancements in artificial intelligence, paired with natural language understanding, resulted with significant improvement in capabilities of intelligent personal assistants, making them suitable for a wide spectrum of operations. The ability to understand context has enabled systems to deduce new information based on existing facts retrieved from the environment, making them more independent and able to autonomously decide which actions should be performed in response to the events in the environment. Context awareness has also enabled systems to learn the causality between seemingly unrelated events, making it possible to predict events in the future.

2.2 Intelligent Personal Assistants and Smart Environments

Integrating Intelligent Personal Assistants with smart environments is not a novel idea. Since the IoT revolution, there have been numerous attempts of integrating intelligent assistants with many smart devices or systems, achieving different goals. The results were systems providing a higher level of autonomy, sensing, and decision making, but still mostly specialized, domain-specific [7, 8].

Regarding human-computer interaction using natural language, domain-specific approaches are still the most represented today. This is not surprising, considering the complexity of natural language processing. With that perspective in mind, it is more effective to build a system which operates in well-defined borders, with little to no need of extending the domain. Such systems are widely used today, and they are present all around us. For instance, many smart home systems can comprehend various commands spoken by the user and are also able to deliver information to the user using a synthesized voice [9]. On the other hand, general purpose systems need to display "real" intelligence and be able to operate in a much broader informational domain, making them significantly more complex. Such systems need to rely heavily on the context, utilizing the available information to independently adapt their behavior to the environment. Enabling such systems to interact with human users using natural language is also a very challenging goal, especially when it comes to interpretation of human languages and their translation to the machine language [10, 11].

3 Agent-Based Model for Context-Aware Human-Computer Interaction

In scope of agent-based systems, an intelligent personal assistant could be defined as a software agent designed to assist people with certain tasks. Based on examples from the previous section, it could be implied that a single agent would need to perform a very broad set of operations, making the agent particularly complex and therefore complicated regarding the implementation. The proposed model tries to avoid this by carefully organizing the separation of concerns in the system. The result is a multi-agent system consisting of several types of intelligent agents operating on different levels of complexity. As displayed in Fig. 1, the focus is on agents related to interaction with the user, and context handling. In scope of the proposed model, context will include any information that can be used to evaluate the state of a human user in respect with its physical, social, communication and computing environment. Context awareness will be achieved by adopting a Belief-Desire-Intention architecture, where beliefs correspond to the information gathered from the environment, desires to the defined policies and preferences, and intentions to the actions which could be performed. In the proposed model, several essential agents would need to follow the BDI principles, specifically the monitoring agent, interaction agent, and context manager agent. Each of these agents would have its own set of beliefs, based on which they could carry out their reasoning, leading to selection of a goal from the set of desires, and the intention to pursue it. Finally, the intention would manifest itself by executing a plan: a sequence of actions, performed by the agent.

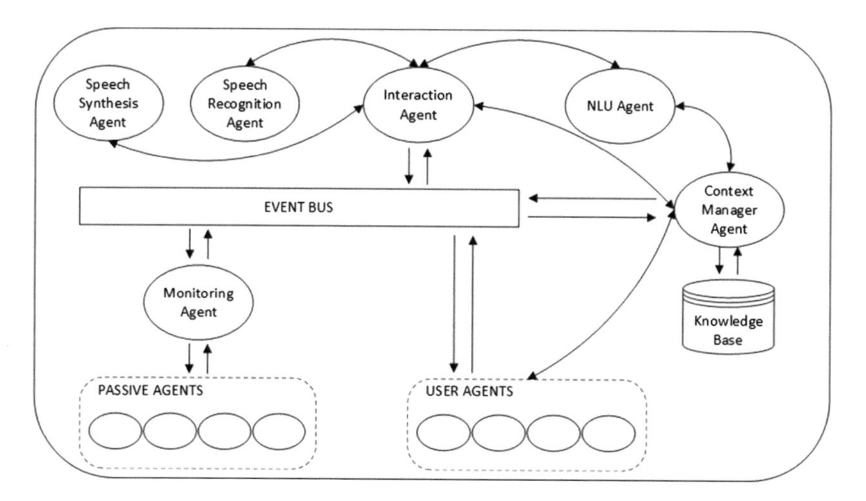

Fig. 1. The model of an agent-based context-aware system for human-computer interaction

Regarding communication in the system, an event-driven approach has been employed, enabling the system to react to changes in the environment in real-time, with possible resulting actions executed by multiple agents, if necessary. Agents can communicate between themselves directly and using a shared event bus. When communicating directly, agents will invoke each other's exposed methods, specified in each agent's external interfaces. The event bus provides communication using the publish-subscribe messaging pattern, organized into topics, based on type of information transferred. Regarding interaction between the system and a human user, we identify several important event topics: environment, interaction, context and command topic.

Environment events are the most common event types in the system. These events are important because they are feeding the context with information related to user's physical and social environment, location, activity, etc. This information enables the system to learn new facts and dynamically adapt to new conditions. Interaction events are all events related to interaction between the system and the user. When originating from the user, interaction events will result with some action in the system. When created by the system, interaction events will result with information being presented to the user. Context events are published only by the context manager agent, and their purpose is to notify all context-aware agents that user's context has changed. Command events represent actions which agents in the system need to perform. All events in the system are prioritized. This is extremely important for the context, as it helps the context manager agent to determine the impact a single event could have on the state of the entire system.

The communication flow between different agents is shown in Fig. 2. Firstly, an environment event is created and published on the environment message topic (1). The context manager agent receives the event, creating a new, or updating an existing fact in the knowledge base (2). The context event is then published on the context message topic (3). This event is received by the interaction agent, which then evaluates its own belief set, invoking the context manager agent to resolve some specific facts (4). Finally, the interaction agent decides which communication channels will be used to interact with the user (5). Messages intended for the human user will be created by NLU agent (6), and speech synthesis agent, if necessary (7). Finally, interaction agent

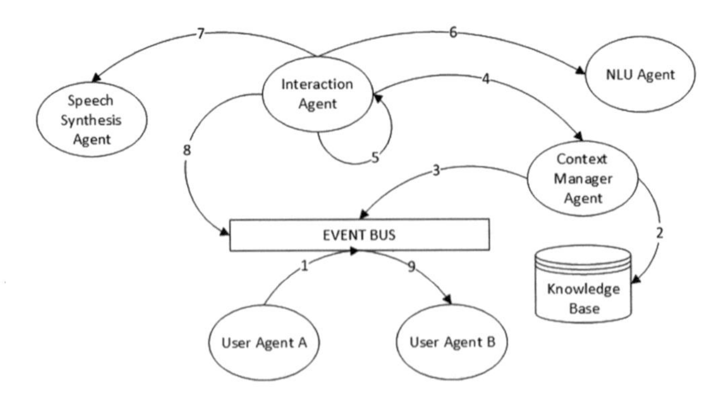

Fig. 2. An example of a communication flow between agents

will publish one or more interaction events on the interaction message topic (8). This event will be received by the appropriate user agent and presented to the user (9).

3.1 Communication Channels

Interaction between a human user and the system can be performed using different communication channels. In smart environments, many devices can be used to transfer information, in a different format. Typical communication channels in a smart home system would be devices common in the household - speakers, television, computers, tablets, smartphones, etc. Speech interaction is not always the most convenient method; it depends on multiple parameters, such as user's location and nearby devices, user's activity at the given moment, and type of information which should be conveyed. User's context has the most important role establishing interaction between the system and the user. It is used by the interaction agent to determine the most convenient moment and method of interaction with the human user.

4 Agent Types

In this section, the suggested agent types will be described. Based on their domain of operation and type of contribution in the system, agents in the system are divided into two major categories: environment and human-machine interface agents.

4.1 Environment Agents

Environment agents are all agents collecting information about the environment and performing various actions related to the environment. Several subtypes are identified here: passive agents, user agents, monitoring agent and context manager agent.

Passive agents are simple agents performing tasks strictly related to sensing, collecting information from various devices (e.g., sensors, smart devices), or other systems. They represent interfaces to the external sources of information, communicating with the environment using various protocols. Passive agents do not conform to the BDI model – they do not depend on context to perform their functions, but they are essential for context creation, as they provide information related to the environment.

User agents perform tasks for the user, varying in complexity. In comparison with the previously described commercial systems, the expected set of supported tasks would include event scheduling, traffic monitoring, personal financial management, contacting emergency services, etc. User agents may be implemented according to the BDI guidelines, but it is not mandatory, as it depends on specific user agent's purpose. User agents make a significant contribution to the user's context, providing information about the environment on different levels. For instance, a simple user agent could collect information about the weather from an arbitrary external service, providing useful information regarding the user's context.

Monitoring agent manages all passive agents in the system, collecting information from sensors, devices, and other connected systems, and publishing them on the environment event topic. The monitoring agent also performs filtering of collected information, with purpose of reducing the event load in the system. For instance, if the newly obtained sensor reading does not bring new information for the user context, it will not be published on the environment event topic. In addition, the monitoring agent can raise alarms when equipment malfunction occurs, or in cases when a parameter falls out of specified value range. For instance, an alarm will be raised if the smoke detector becomes active, or if the room temperature exceeds the defined limit.

Context manager agent is an intelligent agent, representing a BDI agent capable of providing context-related information to other agents in the system. It receives all environment events in the system and uses them to construct a set of beliefs related to the environment, which are then used by context manager's inference engine to produce a new set of learned beliefs. Based on all the available information, context manager agent can provide an even greater insight regarding the environment. This knowledge needs to be available to other agents in the system, and for that purpose the context manager agent provides a semantic query interface which can be used by all agents in the system which perform actions based on context.

4.2 Human-Machine Interface Agents

Human-machine interface (HMI) agents are all agents involved in the interaction process between a human user and the system. This includes agents performing operations related strictly to natural language processing, and interaction agent which orchestrates the communication between the user and the system, based on context.

The natural language processing agents are speech recognition agent, speech synthesis agent, and natural language understanding agent. They are critical components regarding interaction with the user, available strictly to the interaction agent, performing given tasks when requested. They are not subscribed to any message topics. The speech synthesis agent performs transforming of text to speech. This is the voice of the system itself, communicating with the user through speakers on various devices. The speech recognition agent translates speech to text. Retrieved spoken user input must firstly be transformed to text before it is sent for further processing. The natural language understanding (NLU) agent interprets the transcriptions of spoken user input, transforming the natural language to a set of commands which other agents can understand. In this process, NLU agent relies on the context manager agent to resolve possible inconsistencies.

The interaction agent handles all communication between the user and the system. To perform this assignment, it depends on the user's context; the information inferred from the state of the user's environment is used to determine the most suitable way to establish interaction with the user. Regarding communication between agents in the system, the interaction agent relies heavily on agents related to natural language processing, and the context manager agent. Interaction agent will receive, and process environment events produced by recording devices, detecting possible user intent for interaction with the system. Even though interaction agent is handling the spoken user input, it is not required to understand it. When an interaction event originating from the user occurs, it will be forwarded to the speech recognition agent, where it will be transformed into text. The text will then be passed to the natural language understanding agent, where it will be interpreted and transformed into command events, which will then be processed by user agents. In the opposite case, when an interaction event originates from the system, interaction agent will use the context manager agent to decide which communication channels should be used. Based on that decision, it may use the speech synthesis agent to produce a spoken version of the message, which would be published as a new interaction event. All available communication devices are controlled by specific user agents, subscribed on interaction event topic. Therefore, when interaction agent publishes interaction events, they are processed by appropriate user agents. In case of spoken interaction, user agents capable of sound reproduction would play the synthesized audio message.

5 Use-Case Scenario

This section will explain a use-case scenario related to the proposed system, with emphasis on explaining the process of system's interaction with the user. Firstly, the environment needs to be defined. A prototype of the described system has been implemented using Jadex[5], a software framework for development of goal-oriented agents following the BDI model (Fig. 3).

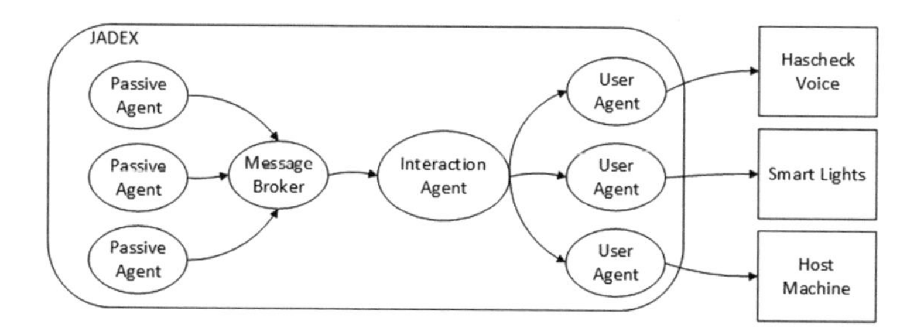

Fig. 3. A prototype system with a BDI interaction agent

The prototype has been focused exclusively on interaction agent, with purpose of validating the proposed concept. For this purpose, a simulated environment has been established, with several passive agents generating specific events. A few user agents have also been implemented, controlling the physical equipment – speakers on the host machine, Philips Hue smart lights, and an Arduino-based notification screen. For purposes of transforming text to speech, Hascheck Voice[6] has been used, an external service providing speech synthesis in Croatian language. Speech recognition and natural language understanding functionalities were not supported in scope of this prototype. Instead, predefined text messages were used for specific scenarios. The context manager agent was not implemented, as the context scope was very limited, which resulted with the interaction agent performing simple decision making itself.

[5] Jadex framework - https://www.activecomponents.org/.

[6] Hascheck Voice - http://hascheck.tel.fer.hr/voice/.

The interaction agent was implemented as a BDI agent, with its belief set related to user's location inside the apartment and noise level in the given location. These beliefs were generated and periodically updated with randomly selected values from the location and noise level value sets, simulating user's movement and activities. All environment events occurring in the system were generated by multiple passive agents simulating various household devices, such as temperature sensors, smoke detectors, movement sensors, etc. The interaction agent would receive the generated events, updating its belief set if necessary. Then it would evaluate the custom conditions related to the available environment events, possibly resulting with the execution of its interaction plan. Based on interaction agent's beliefs, appropriate communication channels would be selected, resulting with specific goals, as displayed in Table 1. These examples display only a few interesting possible combinations of beliefs, events and outcomes. For instance, an event with alarm priority happened, resulting with Philips Hue blinking in red, and with a message being spoken and displayed on the notification screen. In the other scenario, the noise level was high, so the message was not reproduced on the speakers. However, it was displayed on the notification screen. In addition, the Philips Hue was blinking in yellow, due to event priority lower than in the previous example.

Table 1. Examples of interaction outcome

Beliefs	Event	Outcome
User location: Living room Noise level: Low	Source: Smoke detector Value: On Location: Kitchen Priority: Alarm	Blinking red lights, alarm message spoken on speakers and displayed on notification screen
User location: Bedroom Noise level: High	Source: Doorbell Value: On Location: Lobby Priority: Normal	Blinking yellow lights, message displayed on notification screen

6 Conclusion and Future Work

In this paper, we presented a context-aware model of an agent-based system employing multiple interaction methods in communication with a human user. An overview of involved agent types and their roles in the system has been provided, with the explanation of communication in the system. We believe that human-computer interaction integrated with context awareness can add value to smart environments such as smart homes, smart hospitals, etc. However, this should be achieved with respect to user's privacy, but also ensuring high reliability and availability. Therefore, we attempted to avoid dependency on cloud services for core system functionalities.

The described system is interaction-oriented, providing multiple possibilities for communication with users. The focus is on interaction using natural language, which makes a serious challenge regarding the implementation, but also demanding significant processing power. Regarding future work, a detailed specification of agents in the system should be provided. This would include the specific methods and resources used for context resolution, machine learning, and natural language processing. In addition, the communication protocol between agents should be formally specified, with event types and their structure defined precisely. Finally, additional development frameworks should be evaluated, with emphasis on following the BDI model, with support for dynamic belief sets and efficient reasoning process.

Acknowledgments. This work has been supported in part by Croatian Regulatory Authority for Network Industries in scope of project "Looking to the Future".

References

1. Rise of the Machines: How AI-Driven Personal Assistant Apps are Shaping Digital Consumer Habits. http://www.vertoanalytics.com. Accessed 08 Jan 2017
2. Soda, S., Nakamura, M., Matsumoto, S., Izumi, S., Kawaguchi H., Yoshimoto, M.: Implementing virtual agent as an interface for smart home voice control. In: 19th Asia-Pacific Software Engineering Conference, Hong Kong, pp. 342–345 (2012)
3. Raux, A., Langner, B.; Bohus, D., Black, A.W., Eskenazi, M.: Let's go public! Taking a spoken dialog system to the real world. In: Proceedings of Interspeech, pp. 885–888 (2005)
4. The Rise of Intelligent Voice Assistants. http://www.wavestone.com. Accessed 05 Jan 2017
5. López, G., Quesada, L., Guerrero, L.A.: Alexa vs. Siri vs. Cortana vs. Google assistant: a comparison of speech-based natural user interfaces. In: Nunes, I. (ed.) Advances in Human Factors and Systems Interaction. AHFE 2017. Advances in Intelligent Systems and Computing, vol. 592, pp. 241–250. Springer, Cham (2018)
6. Lovrek, I.: Context awareness in mobile software agent network. RAD Croat. Acad. Sci. Arts. Tech. Sci. **513**, 7–28 (2012)
7. Santos, J., Joel, J.J.P.C.R., Silva, B.M.C., Casal, J., Saleem, K., Denisov, V.: An IoT-based mobile gateway for intelligent personal assistants on mobile health environments. J. Netw. Comput. Appl. **71**, 194–204 (2016)
8. Maracic, H., Miskovic, T., Kusek, M., Lovrek, I.: Context-aware multi-agent system in machine-to-machine communication. Procedia Comput. Sci. **35**, 241–250 (2014)
9. Noguera-Arnaldos, J.Á., Paredes-Valverde, M.A., Salas-Zárate, M.P., Rodríguez-García, M.Á., Valencia-García, R., Ochoa, J.L.: im4Things: an ontology-based natural language interface for controlling devices in the Internet of Things. In: Alor-Hernández, G., Valencia-García, R. (eds.) Current Trends on Knowledge-Based Systems. Intelligent Systems Reference Library, vol. 120, pp. 3–22. Springer, Cham (2017)
10. Gunasekera, K., Zaslavsky, A., Krishnaswamy, S., Loke, S.W.: Service oriented context-aware software agents for greater efficiency. In: Jędrzejowicz, P., Nguyen, N.T., Howlet, R.J., Jain, L.C. (eds.) Agent and Multi-Agent Systems: Technologies and Applications. KES-AMSTA 2010. LNCS, vol. 6070, pp. 62–71. Springer, Heidelberg (2010)
11. Brézillon, P.: A context-centered architecture for intelligent assistant systems. In: Faucher, C., Jain, L. (eds.) Innovations in Intelligent Machines-4. Studies in Computational Intelligence, vol. 514. Springer, Cham (2014)

Agent-Oriented Smart Factory (AOSF): An MAS Based Framework for SMEs Under Industry 4.0

Fareed Ud Din[1(✉)], Frans Henskens[1], David Paul[2], and Mark Wallis[1]

[1] University of Newcastle, Callaghan, NSW, Australia
fareed.uddin@uon.edu.au
[2] University of New England, Armidale, NSW, Australia

Abstract. For the concept of Industry 4.0 to come true, a mature amalgamation of allied technologies is obligatory, i.e. Internet of Things (IoT), Big Data analytics, Mobile Computing, Multi-Agent Systems (MAS) and Cloud Computing. With the emergence of the fourth industrial revolution, proliferation in the field of Cyber-Physical Systems (CPS) and Smart Factory gave a boost to recent research in this dimension. Despite many autonomous frameworks contributed in this area, there are very few widely acceptable implementation frameworks, particularly for Small to Medium Size Enterprises (SMEs) under the umbrella of Industry 4.0. This paper presents an Agent-Oriented Smart Factory (AOSF) framework, integrating the whole supply chain (SC), from supplier-end to customer-end. The AOSF framework presents an elegant mediating mechanism between multiple agents to increase robustness in decision making at the base level. Classification of agents, negotiation mechanism and few results from a test case are presented.

Keywords: Smart factory · Multi-Agent Systems (MAS)
Cyber-Physical Systems (CPS)
Small to Medium Size Enterprises (SMEs)

1 Introduction

A gradual evolution of the industrial revolution, which started in 18th century, is still in progression. The first version started with the incorporation of water-steam mechanical systems [1]. A second revolution arose from incorporating mass production, the division of labour and auto-mechanical implantation in the mid 20th century [2], which yielded a third version incorporating Programmable Logic Controllers (PLC) by 1970's [3]. This rapid continuation led to the fourth industrial revolution, termed as Industry 4.0 by The German Federal Ministry of Education and Research in 2011 [4]. The proliferation of research in the domain of Industry 4.0 has yielded many worthwhile answers to the questions on topics like Theories/Perspectives [5], CPS architecture [6], Interoperability/Integration [7],

G. Jezic et al. (Eds.): KES-AMSTA-18 2018, SIST 96, pp. 44–54, 2019.
https://doi.org/10.1007/978-3-319-92031-3_5

Enabling Technologies for implementing Industry 4.0 such as IoT [8], Big Data [9], Multi-Agent Systems (MAS) and their applications [10]. Although there is a huge compilation of published literature in this domain, there is still much research to be done from the perspective of integration with Industry 4.0 i.e. supply chain management (SCM) [11], service oriented architectures (SOA) [12], multi-agent systems (MAS) [10], and enterprise resource planning (ERP) [5].

MAS provides better fault tolerance by providing decision making at the local level components [10]. Many solutions are contributed to the manufacturing industry, including but not limited to enterprise integration, enterprise collaboration, process planning, scheduling and controlling shop floor [13]. Prior research has advanced efforts to provide complete autonomous systems, but none of the works focused in depth on the implementation of an agent-oriented smart factory for Small to Medium Size Enterprises (SMEs) [14]. The domain of Manufacturing Process Planning, which is a rich area, where agent technologies are implemented to provide a solution for scheduling resources, is itself an NP-hard problem [15] because of time and probability escalation. To solve the specified problems of resource allocation and scheduling, different techniques are employed in existing literature such as Petri Nets, Genetic Algorithms and Neural Networks [15,16].

Managing a warehouse is one of the many problems in industry [17]. Extensive research has been carried out in warehouse optimisation in multiple dimensions such as the works presented in [18–20]. One of the many problem in managing a warehouse is the design and structure of warehouse [18]. In a warehouse, Receiving Area (RA) is a place where products, coming to be stocked, are placed first for identification and inspection purposes only and Expedition area is place for temporary placement of products. Keeping RA and EA overloaded, causes the concerns of mismanagement in warehouse [17]. Automated solutions for warehouse management also exist in literature, where the automation is implemented through AS/RS (Automated Storage and Retrieval System via Robo-machines) to pick and ship the products using predefined trajectory and conveyor belts [21]. But in case of Small to Medium Size Enterprises (SMEs) affording such a high-tech system may be difficult [22].

This paper presents a framework of an Agent-Oriented Smart Factory (AOSF), which provides an overall supply chain layout and an agent communication mechanism for SC entities to interact together; to bring robustness in operations. Design of formal semantics and axioms for AOSF-agents follows the concept of eight orthogonal ontological constructs proposed by Kishore in [23], i.e. agent, role, goal, interaction, task, resource, information, and knowledge. The AOSF framework not only provides a generic framework for overall supply chain but also its implementation in Agent-Oriented Storage and Retrieval (AOSR) brings effective results in managing a warehouse for SMEs.

Section 2 addresses the general framework of AOSF with the classification of different constituent agents and their communication strategy. Section 3 includes the results from AOSR warehouse management system (WMS). Some future development in this project is also mentioned in Sect. 3.

2 AOSF Framework

The architecture of the Agent-Oriented Smart Factory (AOSF: shown in Fig. 1), is based on end-to-end Supply Chain (SC) model [24] and Cyber-Physical System (CPS) general framework [6], which is deeply rooted in the concept of Industry 4.0. AOSF framework is extended from an inbound supply chain, towards outbound supply chain, including an in-plant supply chain. This conceptualisation is based on our previous work [24], which explains the structural implementation of an enterprise-wide information management system where now agents are embedded from both ends: the purchase and sales sides. An Enterprise Resource Planning System (ERP) integrates all the other functional areas of a business into one central database. On the rear side it incorporates Logistical Information System (LIS) to maintain in/outbound supply chain and on the front side, it facilitates customer relationship management (CRM). This is how its structural framework provides an advancement in mechanism for a seamless flow of information. Implementing an ERP system only is not enough; pre-implementation and post-implementation factors are also necessary to seek the promised features of an ERP [24]. The next sections provide an insight into the work being presented.

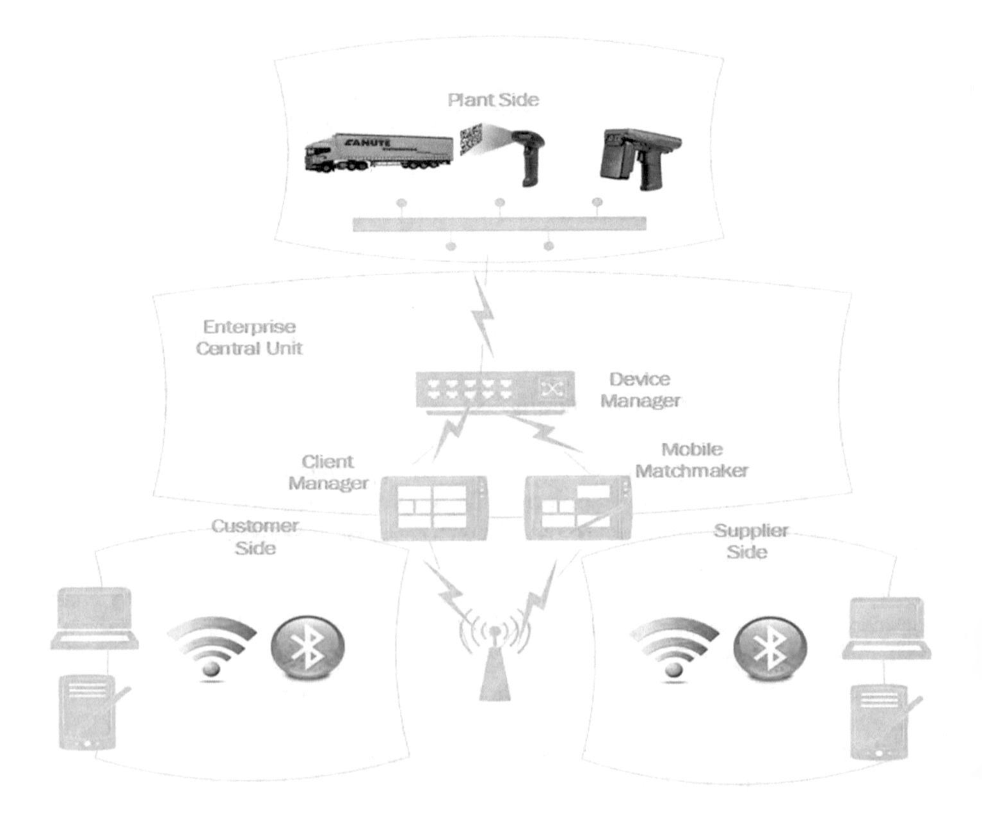

Fig. 1. Three layers of AOSF framework

The AOSF framework presents the foundation of a MAS on the basis of already contributed work in literature, to provide a unified structure for making a factory smart enough to perform the operations seamlessly. It does not only present a general framework for the whole supply chain but also provides a classification of agents with their coordination and negotiation mechanisms for resource optimisation. The objective of AOSF framework is two fold: first to provide an architecture for SMEs including end-to-end SC integration in compliance with Industry 4.0 standards; and second to provide intelligence and decision making at the base level which is achieved through AOSR-WMS planner. The AOSF framework is comprised of three main functional components: Enterprise Central Unit, Customer/Supplier Side and a Plant Side.

Enterprise Central Unit: This is a core sub-system that embeds a client manager, a device manager and a mobile matchmaker. These three elements, following the principles presented by Ruta [25], are responsible for providing domotic ambience within the system, as they provide seamless connectivity of resources to the requests coming from different clients. Enterprise Central System performs the main operations in the whole supply chain as it is linked with both ends: the Customer/Supplier Side and the Plant Side. Enterprise Central Unit of AOSF framework can be considered as a CPS, coordinating between the user interfaces and device level. Formalisation of Enterprise Central Unit as a pure CPS is intentionally left for the next phase of AOSF framework development because it comes with the concerns of security and privacy. Device Manager, which is responsible for managing the sensor devices for their current status, keeps track of overall connectivity. It maintains a fact table which includes the details of devices, including IP addresses and device properties. Client Manager keeps track of the clients of the systems, it receives the requests, ranks them and gets them fulfilled in coordination with Mediator Agent (MA). Mobile Matchmaker resolves the matching conflicts with requests coming from the user side and plant side. It is responsible for finding the appropriate correspondence for mobile clients towards a particular device function. These components are attached to the IP backbone of the enterprise including all the other devices in the network, e.g., RFID tag scanners, bar-code scanners, and GPS locators.

Customer/Supplier Side: Conventionally customers are at the front side of an enterprise and suppliers are on the rear side. For SMEs, to follow this convention, AOSF provides a general solution; not only for the integration of in-plant components but also for a Logistical Information System (LIS) at the rear side and Customer Management System (CRM) at the front side. The extension at both ends makes AOSF framework flexible enough to integrate E-Commerce based applications with CRM and procurement analytics-based applications connected with LIS on the rear side. The sub-component of Customer/Supplier Side is collectively termed as 'user level', which sends and receives multicast IP frames through Wi-Fi and Bluetooth technologies in compliance with the standardised rules. Devices connected to this sub-system may be mobile devices such as notebooks, smart-phones or Personal Digital Assistants (PDAs), able to send and receive semantic annotations. We can compare this strategy with the work presented by Loseto [26] in the domain of Home and Building Automation where the communication between clients and a static system is based on IEEE 802.11 and Bluetooth protocols.

Plant Side: The plant side is a complex domain to model, which may include different machines and devices such as production machinery, temperature/pressure sensors and image analysers. The components of this architecture belong to the same IP backbone to ensure all divisions are integrated. All the sub-systems including Plant Side, Enterprise Side and User Side (Supplier/Customer sides) are designed on the pattern of intelligent agents. Particularly, the running agent on Enterprise Central Unit is responsible for: (i) discovering and coordinating with suitable Plant Side device functionalities compatible with Customer or Supplier Side context requirements via semantic and domotic inferences; (ii) ranking and prioritisation of received requests against the best services to get activated in compliance with the requirements; (iii) finding and resolving inconsistencies in Plant Side's current status, functionalities and resources; (iv) information storage against the output of matchmaking, negotiation and coordination processes. Enterprise Central Unit is also responsible for the configuration through its Device Manager, Mobile Matchmaker and Client Manager, for example maintaining standardisations, complying with protocols and establishing new bidirectional tunnelling channels to ensure the system is ready to accept semantic requests. Categorization and communication between different agents in this architecture are defined in the next section.

2.1 Architecture of Agents

Agents in the AOSF framework are rational rather than omniscient, in nature. This means an agent in this architecture senses percepts in reaction to some sequence of actions or observations, but it is possible only when the environment change is visible. If some changes occur that are not able to be sensed by the agent, the agent is not responsible for the failure. For example, if a plant side's temperature sensor is supposed to open a valve of a machine without having prior knowledge of more production instructions from the engineering side, then an omniscient agent will complete its task to open the valve as per atmospheric conditions without catering to the instructions from the engineering side. 'Seeking instruction from engineering side', was only possible if it had been previously set in the knowledge base of the agent. Even though the agent includes four basic elements: percepts, built-in knowledge, actions and goal as per [27], it is still incomplete. For maximising performance, the AOSF agents are designed on the basis of, not only the aforementioned four basic constructs but with four additional orthogonal constructs: agent's role, interaction, resource and information. Kishore [23] mentioned these eight orthogonal constructs as a baseline definition for designing agents. An agent is not complete in its nature with an architecture only; it also needs a program to run on a defined set of instructions. This means an agent is a combination of architecture and program.

Agent type is an important part to be addressed for the design of an agent program. There are three basic types of model agents in the AOSF architecture: Smart Device Agents (SDAs), Mediator Agents (MAs) and User Agents (UAs). Smart Device Agents are generally categorised as simple reflex agents whereas Mediator Agents are knowledge-based and goal-based agents. User Agents are utility-based agents, which try to accomplish the goals of a user as per their needs through a defined coordination mechanism. SDAs are modelled on reflexes which are built into their architecture, where the possible percept-action combinations are previously set in a knowledge base for each agent. This knowledge base can be summarised on the basis of trends in the sequences because reflexes for an agent may have the same response for percepts with the same meanings, e.g. if there is a bar-code tag in the range of a scanner, then it will perform the same READ action. So in such a sequence, a Condition-Action-Rule may apply that can be written as:

if Barcode tag is in range then read the tag
if Barcode is read then compare the tag

An SDA in the AOSF framework is modelled below:

function Smart-Device-Agent(percept) {
initialise actions, condition, condition-action-rules-set
current-state = *INTERPRET-INPUT(percept)*
rule = *RULE-MATCH(current-state, rules)*
action = *ACTION-RULE(rule)*
return }

Similarly, the User Agent, which is a software side agent, can be categorised as a utility-based agent (as shown in Fig. 2), as they generate a high-quality behaviour to maximise the utility for the user. Where utility is a function that best describes the satisfaction level of the user when the user may have two or more goals to achieve, e.g. speed, accuracy or safety.

In this network, the connected resources (i.e. sensors or devices) and agents may vary as per the requirements of the environment and this increment or decrement in the network is unpredictable. A new user or device can be connected or disconnected regardless of time with no need to redefine the protocol for communication and negotiation. The User Agent, which is a specifically designed software agent, runs on a mobile device like laptop or PDA, and can make requests to an MA against a resource or functionality as per need. SDAs are responsible for providing one or more services (i.e., functional profiles, scanning or searching). This multi-agent based architecture allows SDAs to generate requests to MAs in order to support an autonomous configuration and adaptation to changes in the environment. SDAs are usually embedded within advanced smart devices (i.e. appliances with some in-built computational services and local memory storage capabilities) with employed agent planning and coordination strategies. The AOSF framework is based on Hierarchical Task Network (HTN) Planning [28]. MAS planning strategies are planned to be implemented within this framework in future e.g. PDDL [29] or IXTET-EXEC [30]. These strategies allow the provision of execution control, generation of plans, plan repair, and replanning strategies. SDAs stores device‘s current status and properties in a

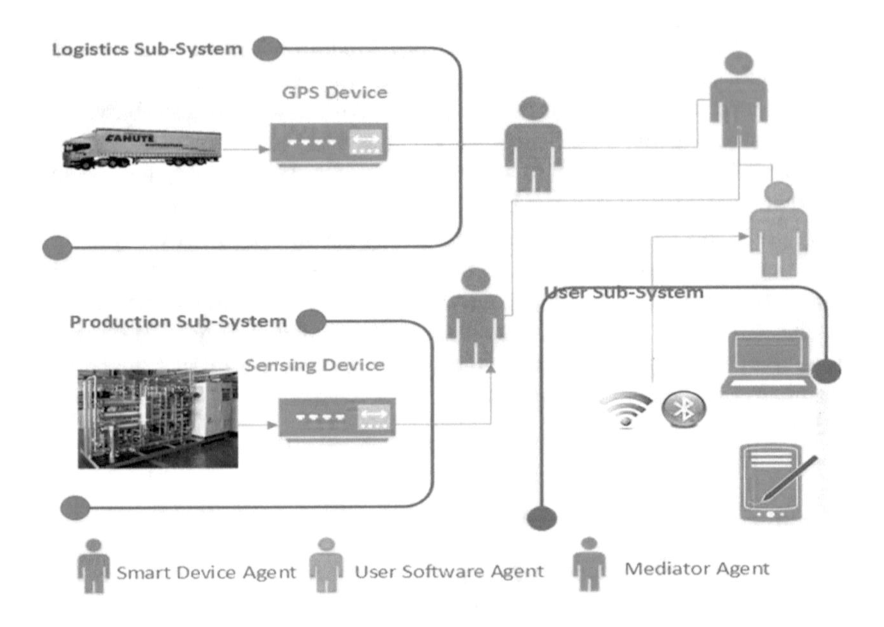

Fig. 2. AOSF agents' architecture

semantic way that is offered against the request triggered by other agents (i.e. other SDAs, MAs and UAs). SDAs semantically support elementary appliances connected to the system, in the case any agent generates a request, the request will be replied to after mediation. Conversely, if MA refers to standard SD properties, the request will be forwarded to the device seamlessly. The focus of this communication scheme is to configure a better possible situation in order to fulfil the request in the best semantic way.

3 Results and Future Work

A case of a company's distribution warehouse with constraints and limitations is applied to AOSR planner component of general AOSF framework. In contrast to a standard WMS, which provides a centralised management of tasks such as tracking location and level of products in the racks using a single logic, AOSR-Planner Agent (PA) uses a hybrid logic as per the products' characteristics to generate the placement plan. This plan is modified during runtime based on new parameters. AOSR does not rely on only one strategy; it provides a combination of different slotting and re-slotting strategies like zone logic [31], First In First Out (FIFO) [32], Put/Pick from the fewest [33], which make it hybrid in nature. After selecting the zone logic, the PA selects other suitable logic to store/sort products into the defined zone in accordance with the product specification and categorisation. AOSR planner passes through different states of the system, which are categorised as per the parameters sensed from the environment. For example preliminary states are normal initialisation states where the stocking is initiated assuming the available capacity for each product. The hybrid nature of logic selection in AOSR minimises the conflicts, hence in the initialisation states, no conflict arises, and the products are placed as per their defined racks, which are suggested by the PA. The proactive nature of AOSR, which makes it different than a standard WMS, helps to sense the upcoming conflict-states of the system. Conflict-states of the system are the states when the same parameters are sensed for a particular product, e.g. the advance shipment and delivery (AS/DN). In such a case, PA decides which products need to be re-slotted, as it can predict that more products are coming, so it re-slots the previous smaller quantity products having prior knowledge of shipment. This is how it reduces the issues of wandering items and overcrowded receiving/expedition areas (RA/EA).

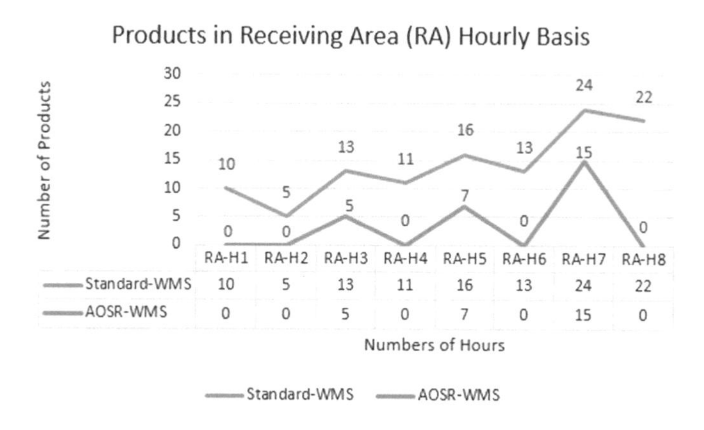

Fig. 3. Experimentational results of receiving area using AOSR

Figure 3 represents the execution results between a standard WMS and AOSR-WMS. The graph shows the average number of products in RA on a particular day on an hourly basis. PA-algorithm generates a dynamic placement plan for the products to be placed into the exact racks based on the hybrid logic. keeping RA overloaded with products, increases the concerns of lost/wandering items and may leads towards stock imbalance [17]. The AOSR algorithm is designed to utilise its auto-inspection mechanism through features of RFID scanning and weight sensing to avoid such problems. Thus the products, in their certain packing units (i.e. case/pallets, box/cases), stay in RA just for identification and are then placed in the suitable racks as mentioned in the placement plan generated by the planner algorithm. Figure 3 shows a clear difference between the results of a standard WMS and AOSR-WMS. A closer look can explain that, in case of AOSR, the time span for first two hours, RA is entirely free and products are shifted to their exact places after consecutive intervals but in case of a standard WMS the RA becomes more and more congested with upcoming products as the time passes. During the time interval of H-8, the gap is quite apparent, representing a clear performance difference.

Through properly defined zone logic there are multiple EA's defined in a warehouse to accurately place the product and to identify the exact location even when they are in EA. Figure 4 demonstrates the results of AOSR to be better than a standard WMS, by reducing the total quantity of products in EA by nearly half, on an hourly basis. The AOSR algorithm is programmed to move the products in EA only when it cannot find a suitable space in the rack for a product in both cases: minimum possible and maximum possible available space. Only those products whose shipment date is near, are placed into EA, and so, very soon they are moved from EA to the shipping area, leaving the EA free for future possibilities. Thus the objective of maintaining a minimum number of products in EA is also achieved by AOSR-WMS. This is how the AOSR algorithm keeps EA less loaded so that the demarcation lines remain evident for the unobstructed movement of forklift trucks and floor staff within the shop floor.

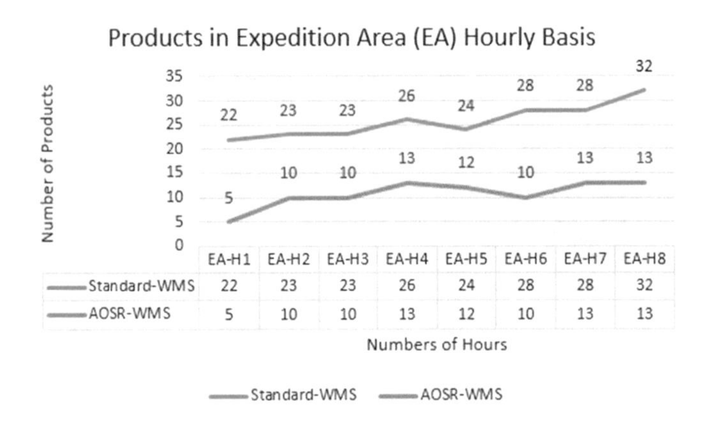

	EA-H1	EA-H2	EA-H3	EA-H4	EA-H5	EA-H6	EA-H7	EA-H8
Standard-WMS	22	23	23	26	24	28	28	32
AOSR-WMS	5	10	10	13	12	10	13	13

Fig. 4. Experimentational results of expedition area using AOSR

In future the AOSF framework and its associated AOSR-WMS algorithm are planned to be implemented with JaCaMo [34] with its associating environment and organization programming. It is expected that the AOSF framework will use an existing planner such as DOMAP [35] and IXTET-EXEC [30], to enhance the functionality and robustness of operations in SMEs. Handling tasks with the same priority can also be elegant future work in order to provide more flexibility in decision making for the user side. The implementation of Plant Side and multiple dimensions of User Side is also intentionally left for upcoming development in this particular project. The AOSF framework is planned to incorporate cloud architecture in the next phase of development in order to completely fulfil the idea of CPS.

References

1. Deane, P.M.: The First Industrial Revolution. Cambridge University Press, New York (1979)
2. Mokyr, J.: The second industrial revolution, 1870–1914. In: Storia dell'economia Mondiale, pp. 219–245 (1998)
3. Freeman, C., Louçã, F.: As Time Goes By: From the Industrial Revolutions to the Information Revolution. Oxford University Press, Oxford (2001)
4. Brettel, M., Friederichsen, N., Keller, M., Rosenberg, M.: How virtualization, decentralization and network building change the manufacturing landscape: an industry 4.0 perspective. Int. J. Mech. Ind. Sci. Eng. **8**(1), 37–44 (2014)
5. Wang, S., Wan, J., Li, D., Zhang, C.: Implementing smart factory of industrie 4.0: an outlook. Int. J. Distrib. Sens. Netw. **12**(1) (2016). https://doi.org/10.1155/2016/3159805
6. Lee, J., Bagheri, B., Kao, H.-A.: A cyber-physical systems architecture for industry 4.0-based manufacturing systems. Manuf. Lett. **3**(Suppl. C), 18–23 (2015)
7. He, W., Da Xu, L.: Integration of distributed enterprise applications: a survey. IEEE Trans. Ind. Inform. **10**(1), 35–42 (2014)
8. Majeed, A.A., Rupasinghe, T.D.: Internet of Things (IoT) embedded future supply chains for industry 4.0: an assessment from an ERP-based fashion apparel and footwear industry. Int. J. Supply Chain Manag. **6**(1), 25–40 (2017)
9. Manogaran, G., Thota, C., Lopez, D., Sundarasekar, R.: Big data security intelligence for healthcare industry 4.0. In: Cybersecurity for Industry 4.0, pp. 103–126. Springer (2017)
10. Adeyeri, M.K., Mpofu, K., Olukorede, T.A.: Integration of agent technology into manufacturing enterprise: a review and platform for industry 4.0. In: International Conference on Industrial Engineering and Operations Management (IEOM), pp. 1–10. IEEE (2015)
11. Ivanov, D., Dolgui, A., Sokolov, B., Werner, F., Ivanova, M.: A dynamic model and an algorithm for short-term supply chain scheduling in the smart factory industry 4.0. Int. J. Prod. Res. **54**(2), 386–402 (2016)
12. Voss, S., Sebastian, H.-J., Pahl, J.: Introduction to intelligent decision support and big data for logistics and supply chain management minitrack (2017)
13. Shen, W., Hao, Q., Yoon, H.J., Norrie, D.H.: Applications of agent-based systems in intelligent manufacturing: an updated review. Adv. Eng. INFORM. **20**(4), 415–431 (2006)
14. Sadeh, N.M., Hildum, D.W., Kjenstad, D.: Agent-based E-supply chain decision support. J. Organ. Comput. Electron. Commer. **13**(3–4), 225–241 (2003)
15. Shen, W.: Distributed manufacturing scheduling using intelligent agents. IEEE Intell. Syst. **17**(1), 88–94 (2002)
16. Shen, W.: Genetic algorithms in agent-based manufacturing scheduling systems. Integr. Comput. Aided Eng. **9**(3), 207–217 (2002)
17. Richards, G.: Warehouse Management: A Complete Guide to Improving Efficiency and Minimizing Costs in the Modern Warehouse. Kogan Page Publishers, London (2017)
18. De Koster, R.B., Johnson, A.L., Roy, D.: Warehouse design and management (2017)
19. Centobelli, P., Converso, G., Murino, T., Santillo, L.: Flow shop scheduling algorithm to optimize warehouse activities. Int. J. Ind. Eng. Comput. **7**(1), 49–66 (2016)

20. Ma, H., Su, S., Simon, D., Fei, M.: Ensemble multi-objective biogeography-based optimization with application to automated warehouse scheduling. Eng. Appl. Artif. Intell. **44**, 79–90 (2015)
21. Manzini, R., Accorsi, R., Baruffaldi, G., Cennerazzo, T., Gamberi, M.: Travel time models for deep-lane unit-load autonomous vehicle storage and retrieval system (AVS/RS). Int. J. Prod. Res. **54**(14), 4286–4304 (2016)
22. Llonch, M., Bernardo, M., Presas, P.: A case study of a simultaneous integration in an SME: implementation process and cost analysis. Int. J. Qual. Reliab. Manag. **35**, 319–334 (2018)
23. Kishore, R., Zhang, H., Ramesh, R.: Enterprise integration using the agent paradigm: foundations of multi-agent-based integrative business information systems. Dec. Support Syst. **42**(1), 48–78 (2006)
24. Ud Din, F., Anwer, S.: ERP success and logistical performance indicators a critical view. Int. J. Comput. Sci. Issues, 223–229 (2013). http://www.ijcsi.org/papers/IJCSI-10-6-1-223-229.pdf
25. Ruta, M., Scioscia, F., Di Noia, T., Di Sciascio, E.: Reasoning in pervasive environments: an implementation of concept abduction with mobile OODBMS. In: IEEE/WIC/ACM International Joint Conferences on Web Intelligence and Intelligent Agent Technologies, WI-IAT 2009, vol. 1, pp. 145–148. IEEE (2009)
26. Loseto, G., Scioscia, F., Ruta, M., Di Sciascio, E.: Semantic-based smart homes: a multi-agent approach. In: WOA (2012)
27. Russell, S., Norvig, P.: Artificial Intelligence: A Modern Approach. Artificial Intelligence, p. 27. Prentice-Hall, Englewood Cliffs (1995)
28. Minglei, L., Hongwei, W., Chao, Q.: A novel HTN planning approach for handling disruption during plan execution. Appl. Intell. **46**(4), 800–809 (2017)
29. Strobel, V., Kirsch, A.: Planning in the wild: modeling tools for PDDL. In: Joint German/Austrian Conference on Artificial Intelligence (Künstliche Intelligenz), pp. 273–284. Springer (2014)
30. Lemai-Chenevier, S.: IXTET-EXEC: planning, plan repair and execution control with time and resource management, Ph.D. thesis (2004)
31. Piasecki, D.: "Warehouse management systems (WMS)," Inventory Operations Consulting LLC (2005). http://www.inventoryops.com/warehouse_management_systems.htm
32. Jones, M.M., Juneja, M.O., Gnanamurthy, K., Kandikuppa, K., Sheu, J.Y.W., William, E.R.V., Hadagali, G.R., Rawat, S.S., Berry, V., Agrawal, D., et al.: Consigned inventory management system, 31 March 2016. US Patent App. 14/499,372 (2016)
33. Preuveneers, D., Berbers, Y.: Modeling human actors in an intelligent automated warehouse. In: International Conference on Digital Human Modeling, pp. 285–294. Springer (2009)
34. JaCaMo, C.: Framework for Jason and Moise, "Jacamo framework for jason, cartago and moise." (2017). http://jacamo.sourceforge.net
35. Cardoso, R.C., Bordini, R.H.: A distributed online multi-agent planning system. In: Distributed and Multi-Agent Planning (DMAP-2016), p. 15 (2016)

Modeling, Simulation and Mobile Agents

Agent-Based Approach for Energy-Efficient IoT Services Discovery and Management

Petar Krivic, Pavle Skocir(✉), and Mario Kusek

Faculty of Electrical Engineering and Computing, Internet of Things Laboratory, University of Zagreb, Unska 3, 10000 Zagreb, Croatia
{petar.krivic,pavle.skocir,mario.kusek}@fer.hr

Abstract. Internet of Things (IoT) systems are becoming omnipresent, with miscellaneous solutions in diverse domains. To release the full potential of the IoT, it is needed to have dynamic user-centric systems where services can be executed according to user demands. A growing number of devices that is being deployed need to be capable to be discovered easily by the IoT platforms, and offer their services automatically, without human intervention. Due to the complexity of IoT systems, automatic discovery and management of services offered by IoT devices is rarely aligned with user demands. In this paper we propose an agent-based approach which enables the discovery and management of IoT services. Furthermore, it enables the control of devices according to user-defined rules or direct commands.

Keywords: IoT · M2M · Agents · Service management

1 Introduction

Internet of Things (IoT) is a concept that has found its purpose in a wide range of industries. As of its first appearances in research literature and respective first prototypes, it has grown in significance and now includes billions of connected devices [12], and has an important role in improving both business and consumer processes. These two segments are being presented as two distinct directions of IoT, where consumer IoT (cIoT) aims at improving the quality of people's life by saving time and money, and industrial IoT (IIoT) focuses on the integration between Operational Technology and Information Technology and on how smart machines, networked sensors and data analytics can improve business-to-business services across a wide variety of market sectors [10]. However, in both of these sectors the distributed environment of connected devices has a goal of achieving end consumer benefits, using more of machine learning and data analytics, while minimizing human involvement.

To reach the full potential of IoT, dynamically adaptive environment of collaborating connected devices must be established first. Device discovery and

G. Jezic et al. (Eds.): KES-AMSTA-18 2018, SIST 96, pp. 57–66, 2019.
https://doi.org/10.1007/978-3-319-92031-3_6

interoperability among different communication protocols are research issues that need to be addressed in the process of achieving this goal. Device discovery integrates new incoming devices in existing smart environments without human interference, which enables a setup of self-organizing collaborating groups with support of mobile devices. Even more important for the entire IoT ecosystem and end-users is the service discovery that includes discovering the capabilities offered by IoT devices. End-users directly benefit from the services (e.g. temperature monitoring, luminosity level monitoring), while the device discovery can be regarded as prerequisite for service discovery. IoT devices are often battery-powered, and the execution of services offered by these devices (e.g. sensing, data analysis) needs to be energy efficient. Therefore, only those services for which a user's interest is shown should be executed.

In this paper we decided to tackle the aforementioned challenges concerning device and service discovery by using the agent-based approach. In the area of IoT, these challenges can be referred to as Wireless Sensor Networks (WSN) related ones, which additionally include architecture, energy efficiency, security, protocols and quality of service (QoS) [5]. Due to their appropriateness for modeling autonomous self-aware sensors in a flexible way, agent and multi-agent systems have been identified as one of the most suitable technologies to contribute to the IoT and sensor networks domain [13]. The main goal of our approach is to eliminate human interference from the IoT environment of collaborating devices. Agents are included in this environment with the functionality of enabling discovery of IoT services and their management according to user-defined needs.

Section 2 gives a short overview of related projects in the area of service discovery. Afterwards, in Sect. 3 the developed agent-based model for service discovery and management in M2M/IoT environment is described, followed by Sect. 4 where we present use case of the multi-agent system in a smart home setting. Section 5 concludes the paper.

2 Related Work

Service discovery is defined as the ability to discover the capabilities (e.g. measured attributes, location, accuracy, etc.) of the heterogeneous sensors and actuators that support a wide range of applications [6]. It maximizes the utilization of these sensors since it enables IoT platforms to automatically detect the capabilities of connected IoT devices upon their registration. It is one of the fundamental requirements of any IoT platform which minimizes, or ideally eliminates the need for external human intervention for configuration and maintenance of deployed objects [3].

Service discovery mechanisms already exist, and the majority has been conceived for local area networks (LANs) and then extended for constrained IPv6 over low-power wireless personal area networks (6LoWPAN) [1,2]. Such solution is Universal Plug and Play (UPnP) that uses TCP as transport protocol and XML as the message exchange format, which presents an overhead and makes it unsuitable for application on constrained devices in IoT networks. Similar solutions are

Service Location Protocol (SLP), Jini, and Salutation. Special solutions are suggested to be used in IoT networks which support constrained devices. Authors in [1] propose the usage of Constrained Application Protocol (CoAP) together with Domain Name System Resource Discovery protocol (DNS-SD). However, CoAP has several specification issues [3]. It does not specify how a thing should join the CoAP server first time and announce itself, there is no specification on how a remote client can look up into the resource directory and query for the resource of interest, and a centralized approach using resource discovery and CoAP suffers from scalability issues and denial of service (DoS) attacks.

Architecture for resource discovery can have the following layers [3]: proxy layer, discovery layer, service enablement layer and application layer. Proxy layer enables discovery of physical things regardless of communication technologies and protocols used by the things. Discovery layer stores the data about the things in the database, while service enablement layer provides access control policies to discovery layer, forwards results, and if applicable manages subscriptions for different resource types. Datta et al. [3] also provide a search engine within the discovery layer that facilitates finding of desired resources. Georgakopoulos et al. [4] present an abstract IoT model with three layers: device layer, data layer and application layer. Each of the aforementioned layers should have their own discovery service. Device layer should discover available IoT devices, data layer should discover data sources, and application layer organized or analyzed data for its applications.

Discovery solutions can be divided into centralized and distributed [3]. Jara et al. [7] proposed a solution which allows sensors to be registered into a common centralized architecture. A mobile service is developed allowing the clients to discover and access sensors. Sensors can connect to a centralized gateway via NFC, 6LowPAN, IPv6 etc. Helal et al. [6] eliminate the need of a dedicated gateway, hence proposing a distributed solution. They state that when serving large number of devices, the use of gateways might be inefficient because large number of gateways might be required, which would cost more compared to regular sensors. They propose organizing network into clusters, where cluster heads (CHs) would be responsible for discovery of new devices, and storing data about the registered ones. The authors state that their solution maximizes energy efficiency and success rate of satisfying user requests.

Discovery mechanisms can have different scopes. They can be local, when service discovery mechanism enables communication between geographically concentrated smart objects that are connected to a single gateway and are situated in the same network, or global when it enables communication between objects in different networks [2]. Wang et al. [14] propose a local solution which enables an application on mobile phone to automatically discover IoT services in its environment. Additionally, the discovery can be one-time or long standing [3]. In one-time discovery the networks discovers the devices and resources once, usually during registration, while in long standing discovery publication-subscribe mechanisms are used to notify on the changes related to the resources within the network.

Discovery protocols for constrained networks need to be resource efficient in terms of low processing and memory overhead. Additional characteristics of small devices in IoT networks are long inactivity periods, and support of small frame sizes [1]. This paper focuses on proxy layer proposed in [3]. It goes a step further in assuring energy efficiency since along with service discovery it optimizes the duty cycle of constrained IoT devices. In such a way the devices perform sensing tasks in cycles which are in line with user requirements. Duty cycling is also proposed in [6], but the authors adjust cycles according to number of smartphones (as potential data consumers) in the area, while our approach takes into account real application requirements. Service discovery and management is also addressed by standardization organizations such as oneM2M [8]. Their functional architecture specifies two components that are related to it: device management and discovery. While in device management, which can use protocols as for example OMA-DM [9], there exists the functionality of configuring device capabilities, the possibility to optimize duty cycles is not specified according to our knowledge.

3 Agent Model for Service Management in IoT Systems

Discovery and registration of smart objects is an important aspect that needs to be considered within IoT systems. When a new IoT device is ready to be connected, it should have the possibility to automatically register to the system, while IoT gateway should be able to automatically discover new devices, and enable their registration. This section presents an agent-based registration and discovery mechanism of smart objects planned to be implemented within an existing IoT system presented in [11]. The system takes advantage of a few opportunities in the area of IoT, brought by the possibilities to access sensor nodes via Internet, and by advancement of hardware components used by IoT devices. First opportunity is performing more complex algorithms, e.g. for data analysis, on IoT devices. Second opportunity is that one IoT device can have more than one sensor connected to it, and the fact that it can provide data to more than one service. The third opportunity arises from the fact that services do not need to be active all the time, but only through certain periods of time. Therefore, obtaining data from sensors should be executed only at time periods when user interest for the certain data exists. Users can express their demand for data either directly, by starting or stopping a service in their application, or by specifying rules (e.g., to turn off the heating when a user leaves the apartment). The designed solution takes advantage of the aforementioned opportunities and brings the following novelties to IoT systems with battery-powered IoT devices, as presented in our previous work [11]:

- enables a user-centric service management, based on needs or rules defined by users;
- data processing tasks can be executed on IoT devices or on IoT gateway, the decision is made according to energy efficiency of possible solutions; and
- each service has a priority level, where the goal is to extend the lifetime of the services with higher priority.

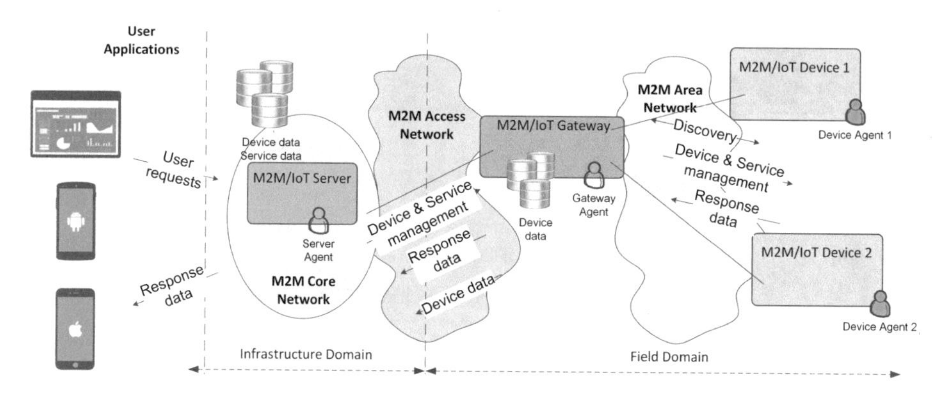

Fig. 1. Agent-based model for service discovery and management in IoT systems

Agent-based model of the system that enables energy-efficient service management is shown in Fig. 1. It is based on the oneM2M functional architecture [8]. OneM2M uses the terms Machine-to-Machine Communications (M2M) and IoT in the same context, but in standards entitles the nodes as *M2M* nodes. We will use both M2M and IoT in the designation for the nodes within the system because using the IoT nomenclature has become more widespread in literature. M2M/IoT Server receives user requests to start or stop certain services. It executes an algorithm that allocates tasks to the devices that will ensure maximum service lifetime. The algorithm takes into account which services can be executed on which M2M/IoT Devices, the current battery level of every M2M/IoT Device, and how much energy would execution of certain services consume. Server Agent located on the M2M/IoT Server is responsible for executing the algorithm and notifying the Gateway Agent situated at M2M/IoT Gateway of the tasks that need to start its execution on chosen M2M/IoT Devices. Gateway Agent then forwards the command to the corresponding Device Agents about the tasks that need to start/stop its execution. As mentioned earlier, this paper presents the upgrade of the described system by introducing a mechanism for service discovery. Gateway Agent is responsible for detecting Device Agents in the M2M Area Network (Discovery message) and in Device&Service message it configures the M2M/IoT Devices so that they are ready to start tasks execution. In the remainder of this Section the protocol for service management in IoT systems will be presented, followed by the newly introduced discovery and management protocol.

The protocol for service management responsible for allocating tasks to IoT devices is shown in Fig. 2. Device management application or a certain rule within the system initiates a service. Task allocation algorithm is executed on the M2M/IoT Server, a component in the network which runs a rule-based engine or receives commands directly from users to start or stop a service. Server Agent executes the algorithm. All task-device mappings denoted as M^x are forwarded towards the Gateway Agent which then starts tasks v allocated to itself, or forwards tasks to certain Device Agents that were chosen by the algorithm. Since

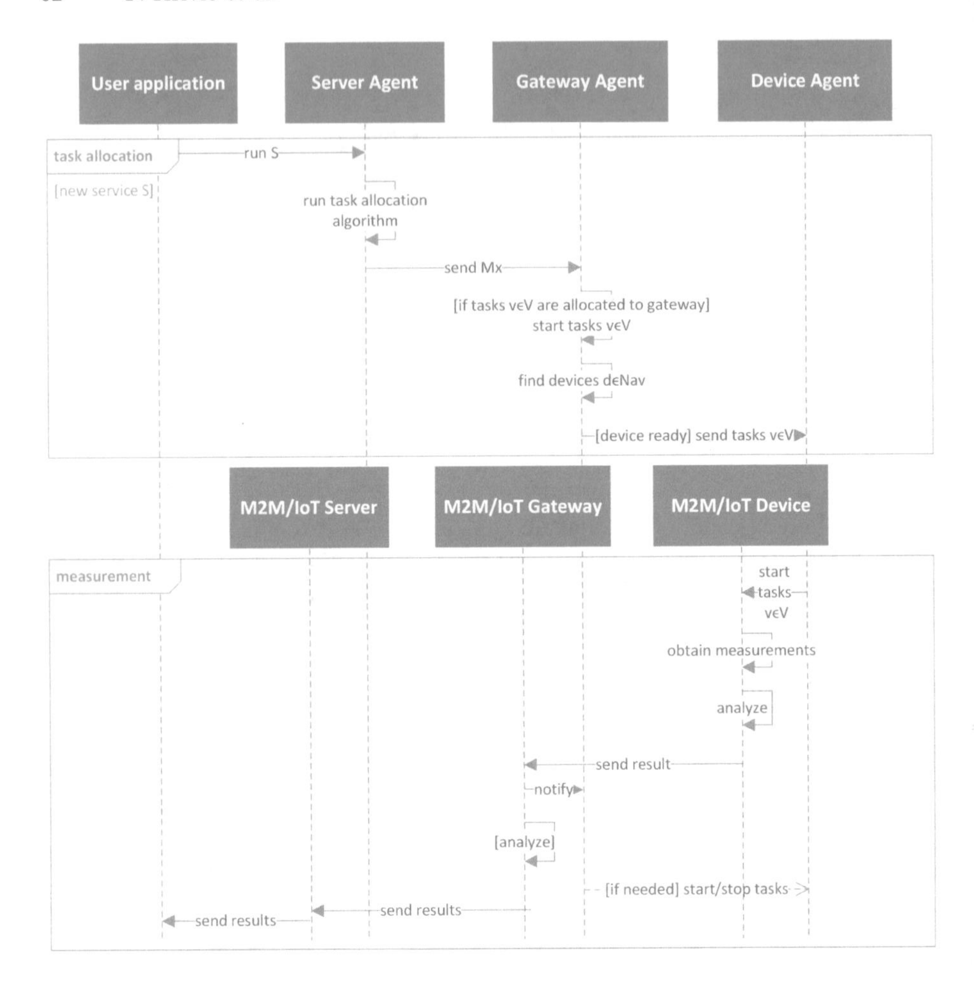

Fig. 2. Protocol for service management in IoT systems

M2M/IoT Devices spend majority of their time in low-energy mode (since they are battery-powered, in such setting it is substantial to conserve energy), the Gateway Agent should wait until they are awake to send them commands. After the Device Agent receives a new command, it notifies the M2M/IoT Device to start the execution of allocated tasks and forwarding of acquired results directly to M2M/IoT Gateway without involving the Agent. Agents do not handle the data, only control messages. Device Agents need to listen to new requests from Gateway Agent in specified intervals.

The mechanism for discovery of new smart objects introduced in this paper is shown in Fig. 3. The discovery, along with the registration process, is planned to be executed periodically or just before the process of starting/stopping a service shown in Fig. 2. The mechanism functions as follows: each M2M/IoT Device entering the system should have a previously deployed Device Agent. Further-

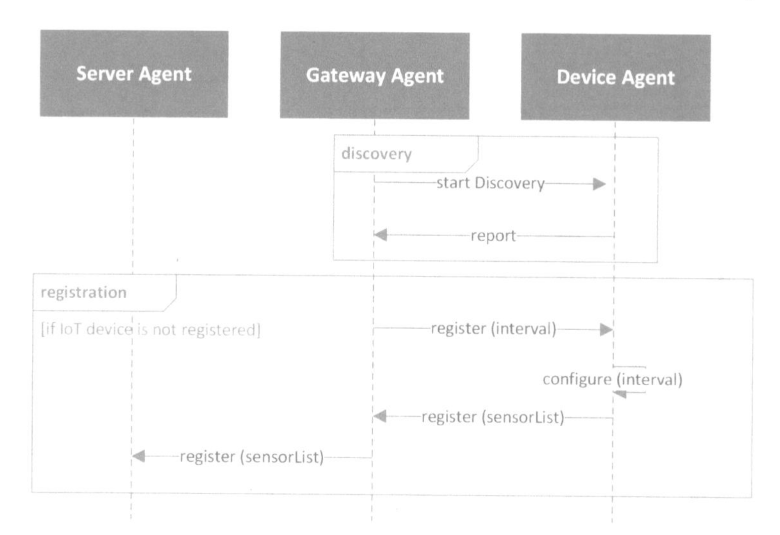

Fig. 3. Discovery and registration mechanism

more, installed software should enable the M2M/IoT Device to read the measurements from connected sensors when the request for their data is received. Before joining the system, the Device Agent should register by providing data about sensors it offers, and configure the intervals in which the Gateway Agent could reach it. The Gateway Agent should at certain time periods (e.g. before starting a new service) initiate the discovery process (start Discovery message in Fig. 3). If the process recovers unregistered devices (report message), in the next stage it should suggest the intervals in which M2M/IoT Devices should be available (register (interval) message). During the rest of the time, the M2M/IoT Device should perform the assigned operations or stay in low-energy mode in order to save energy. Furthermore, the device should send data about connected sensors (register (sensorList) messages), that should be stored by the Server Agent in the database located at the M2M/IoT Server. After the planned discovery and registration process, the tasks on the newly registered M2M/IoT Device should be managed by using the protocol presented in Fig. 2. The process of discovery and registration should work regardless of the communication technology.

4 Use Case Scenario in Smart Home Environment

This section describes how the discovery mechanism introduced in this paper works along the service management mechanism in a smart home environment. Figure 4 shows the setup with two M2M/IoT Devices which are connected to M2M/IoT Gateway by using XBee communication modules. M2M/IoT Gateway is connected to the M2M/IoT Server. User applications are also connected to the M2M/IoT Server through which they state their requests and receive data

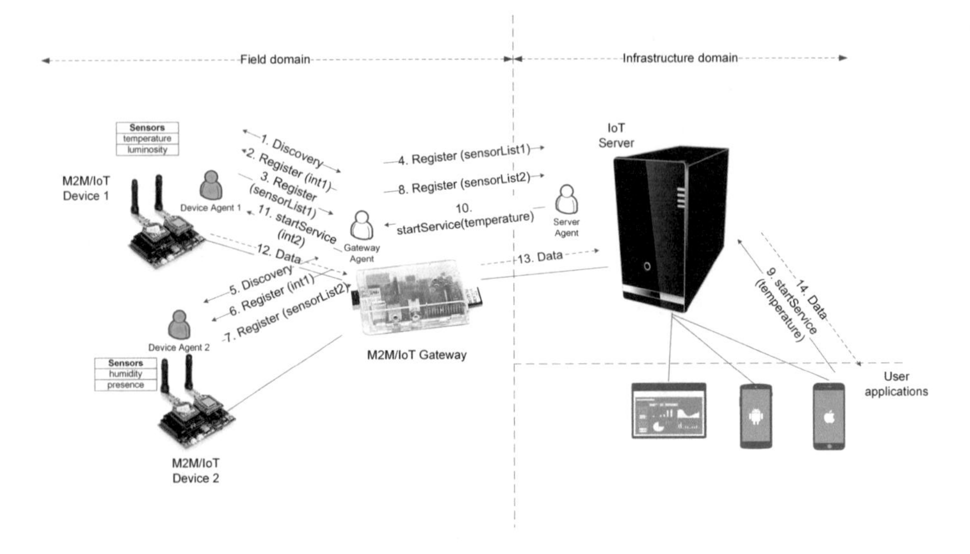

Fig. 4. Use case for the service discovery and management

from the Field domain. Each M2M/IoT Device has its Device Agent which communicates with the Gateway Agent for discovery, registration, and management. Gateway Agent receives commands from the Server Agent which is triggered by user-defined requests.

In the scenario depicted in Fig. 4 M2M/IoT Device 1 is firstly connected to the system and needs to be configured. As mentioned earlier in Sect. 3, prerequisite for this process is that the device has a running Device Agent, and deployed software that enables receiving commands from the Device Agent, and performing necessary tasks. In this example, the tasks are acquiring measurements from sensors (temperature and luminosity sensor for M2M/IoT Device 1, humidity and presence sensor for M2M/IoT Device 2), and forwarding these measurements to M2M/IoT Gateway. In the message 1 from the Fig. 4, the Gateway Agent sends a broadcast message that is received by the Device Agent 1 representing the M2M/IoT Device 1 which has just entered the system. Device Agent 1 responds to the message, after which it receives message 2 from the Gateway Agent containing the interval in which it should be available for receiving commands. Device Agent 1 responds with message 3 containing its sensor list. Gateway Agent forwards this message to the Server Agent which stores this information (message 4). The same process is executed when M2M/IoT Device 2 joins the system, as shown in messages 5–8.

After the initial discovery and registration, the devices are asleep in the duration of the interval $int1$ defined in the messages 2 and 6. This interval can change according to user requests. The request for temperature measuring service is initiated by one of the user applications in message 9. Server Agent receives that request, and allocates the temperature measurement task to M2M/IoT Device 1, since it is the only device in the system with a temperature sensor.

Request for starting this task is forwarded to Gateway Agent, and to the Device Agent 1 (messages 10 and 11). New interval is assigned in message 11 which specifies the period in which measurements from temperature sensor need to be acquired and forwarded. The data is sent directly from M2M/IoT Device 1 via M2M/IoT Gateway and M2M/IoT Server to the user application without going through the Agents, as defined in Fig. 2. The Agents receive and process only service management messages.

The main benefit of this approach is that it enables duty cycles with longer periods in the times when no data is needed by the users. In one duty cycle the device wakes up, receives messages from the Gateway Agent, and acquires measurements if necessary, and goes to low-energy mode. For instance, the interval $int1$ in the environment shown in Fig. 4 could be 10 min, and interval $int2$ requested by the service could be 1 min. In the case when no data is requested, the device would only wake up every 10 min and receive commands from Gateway Agent without acquiring the measurements. If we compare this to the case when data is always acquired and forwarded to the M2M/IoT Gateway in the interval of 1 min, the device would conserve energy necessary for 9 wake ups and 10 measurement tasks.

5 Conclusion

In this paper we presented an agent-based system which enhances autonomy in IoT environment by lowering the need for human interference in service discovery and management. Management of end-devices is in most cases a repulsive factor in wider acceptance of smart solutions, which is why we believe that more effort needs to be invested in overcoming this barrier. Because of the great amount of unnecessary data that overloads the network, M2M/IoT Devices should have an option to adapt their work based on existing user interest in their services as proposed in this system. This approach also enhances energy efficiency of M2M/IoT Devices in cases when they are battery-powered. In our opinion this is the right direction towards more autonomous and effective IoT solutions, and we believe that the use of agents could be very beneficial towards reaching complete device and protocol autonomy of IoT environments.

Acknowledgements. This work is supported by the H2020 symbIoTe project, which has received funding from the European Union's Horizon 2020 research and innovation programme under grant agreement No. 688156.

References

1. Carballido Villaverde, B., Alberola, R.D.P., Jara, A.J., Fedor, S., Das, S.K., Pesch, D.: Service discovery protocols for constrained machine-to-machine communications. IEEE Commun. Surv. Tutor. **16**(1), 41–60 (2014)
2. Cirani, S., Davoli, L., Ferrari, G., Leone, R., Medagliani, P., Picone, M., Veltri, L.: A scalable and self-configuring architecture for service discovery in the Internet of Things. IEEE Internet Things J. **1**(5), 508–521 (2014)

3. Datta, S.K., Da Costa, R.P.F., Bonnet, C.: Resource discovery in Internet of Things: current trends and future standardization aspects. In: Proceedings of the IEEE World Forum on Internet of Things, WF-IoT 2015, pp. 542–547 (2015)
4. Georgakopoulos, D., Jayaraman, P.P., Zhang, M., Ranjan, R.: Discovery-driven service oriented IoT architecture. In: Proceedings of the IEEE Conference on Collaboration and Internet Computing, CIC 2015, pp. 142–149 (2016)
5. Gubbi, J., Buyya, R., Marusic, S., Palaniswami, M.: Internet of Things (IoT): a vision, architectural elements, and future directions. Fut. Gener. Comput. Syst. **29**(7), 1645–1660 (2013)
6. Helal, R., ElMougy, A.: An energy-efficient service discovery protocol for the IoT based on a multi-tier WSN architecture. In: Proceedings of Conference on Local Computer Networks, LCN 2015, pp. 862–869, December 2015
7. Jara, A.J., Lopez, P., Fernandez, D., Castillo, J.F., Zamora, M.A., Skarmeta, A.F.: Mobile digcovery: a global service discovery for the Internet of Things. In: Proceedings of the 27th International Conference on Advanced Information Networking and Applications Workshops, WAINA 2013, pp. 1325–1330 (2013)
8. oneM2M: M2M Functional Architecture. Technical Specification (2016). http://www.onem2m.org/images/files/deliverables/Release2/TS-0001-%20Functional_Architecture-V2_10_0.pdf
9. Open Mobile Alliance (OMA): OMA Device Management Protocol. Technical Specification (2016). http://www.openmobilealliance.org/release/DM/V1_3-20160524-A/OMA-TS-DM_Protocol-V1_3-20160524-A.pdf
10. Palattella, M.R., Dohler, M., Grieco, A., Rizzo, G., Torsner, J., Engel, T., Ladid, L.: Internet of Things in the 5G era: enablers, architecture, and business models. IEEE J. Sel. Areas Commun. **34**(3), 510–527 (2016)
11. Skocir, P., Kusek, M., Jezic, G.: Energy-efficient task allocation for service provisioning in machine-to-machine systems. Concurr. Comput. Pract. Exp. **29**(23), 1–22 (2017)
12. Statista: Internet of Things (IoT) connected devices installed base worldwide from 2015 to 2025 (in billions) (2018). https://www.statista.com/statistics/471264/iot-number-of-connected-devices-worldwide/
13. Vinyals, M., Rodriguez-Aguilar, J.A., Cerquides, J.: A survey on sensor networks from a multiagent perspective. Comput. J. **54**(3), 455–470 (2011)
14. Wang, E., Chow, R.: What can i do here? IoT service discovery in smart cities. In: IEEE International Conference on Pervasive Computing and Communication Workshops, PerCom Workshops 2016 (2016)

Agent-Based Modeling and Simulation for Two-Dimensional Spatial Competition

Masashi Miura[1(✉)] and Hidetoshi Shiroishi[2]

[1] Innovation Center for Engineering Education, Tottori University, Tottori, Japan
miura@icee.tottori-u.ac.jp
[2] Faculty of Regional Sciences, Tottori University, Tottori, Japan
hshiroishi@rs.tottori-u.ac.jp

Abstract. We examine a feasibility study on agent-based modeling and simulation for spatial competition originating from Hotelling model. Hotelling model is a theory that explains the consequence of the spatial competition between two shops. This model is employed to explain little product differentiation and agglomeration of retail shops. Conventionally, the researches for spatial competition has employed the analytical approaches in the game situations. On the contrary, our research intends to introduce the agent-based approach. With the agent-based approach, we can deal with discrete and non-uniform consumer distributions which are closer to the actual situation. This paper introduces our agent-based model as extension of two-dimensional Hotelling model and the agent simulator which we've developed as a prototype.

1 Introduction

Spatial competition is an important topic on which marketing researchers have often focused in the context of location choice by retail shops and horizontal product differentiation [1–3]. The study of spatial competition is originated from a pioneering model of Hotelling [4]. According to Hotelling model, the two shops who sell homogeneous products are both located at the center of the line market where consumers are uniformly distributed. This conclusion is known as "principle of minimum differentiation." The Hotelling model can be used to explain the homogenization behavior of competing retailers who are located relatively close to each other, and competing companies who have similar products in the same product category. It can be applied not only to the market competitions but also to the similar policies set by two political parties.

Because of its high applicability, Hotelling model has attracted researchers of social sciences and many researchers attempted various extensions [5, 6]. Some of their arguments are opposite from the principle of minimum differentiation. For example, d'Aspremont et al. [7] concluded that a key calculation in Hotelling's model is incorrect. They exhibited that two shops are located at both ends in the line market respectively, by proposing the two-stage location-price game.

Although various situations are modeled in existing researches, their basic approaches are common. Usually the analytical approaches based on game theory are

G. Jezic et al. (Eds.): KES-AMSTA-18 2018, SIST 96, pp. 67–75, 2019.
https://doi.org/10.1007/978-3-319-92031-3_7

applied to those researches. On the contrary, in our research, we aim at extending Hotelling model with agent-based modeling and multi-agent simulation. In the agent-based modeling, phenomena are described as collections of "agents." Agent is the unit who makes decisions and interacts with each other according to the given rules [8, 9]. In multi-agent simulation, behaviors of multiple agents and results of their interactions are calculated with a computer.

By introducing the agent-based approach into spatial competition researches, it will be possible to describe models that are closer to the actual situations. In the existing analytical approaches, it is assumed that consumers are distributed continuously and uniformly. Actually, the consumer distribution is neither uniform nor continuous. On the other hand, with agent-based modeling, we can describe the consumers distributed discretely and non-uniformly. Our suggestion is to extend the spatial competition models using the agent-based approach and improve their applicability to the more actual situations.

This paper, for the first, introduces the spatial competition model of Tabuchi [10] which extended Hotelling model to two-dimensions. Then agent-based model as extension of Tabuchi model is introduced. Finally, the multi-agent simulator which we've developed as a prototype and the results of fundamental simulation are shown.

2 Two-Dimensional Spatial Competition

This section introduces the two-dimensional spatial competition model of Tabuchi [7]. In this model, it is assumed that consumers are distributed uniformly and continuously in the two-dimensional market space (x, y).

The timing of the game is as follows. First, shop1 is located on (x_1, y_1) and shop2 is located on (x_2, y_2) at the same time. Next, with each location given, shop1 and shop2 set prices p_1 and p_2 respectively. Then the i-th consumer located on (x_i^c, y_i^c) decide to purchase only one goods from the shop with the lower sum of the price and the movement cost as follows (1). The movement cost is determined with a positive real number t and the distance between the shop and the consumer. t represents the weight for the distance.

$$\min\left\{p_1 + t(x_i^c - x_1)^2 + t(y_i^c - y_1)^2, p_2 + t(x_i^c - x_2)^2 + t(y_i^c - y_2)^2\right\} \tag{1}$$

After all consumers choose the shop, the profits of the shops can be calculated as Eqs. (2) and (3).

$$\pi_1 = p_1 D \tag{2}$$

$$\pi_2 = p_2(1 - D) \tag{3}$$

D represents the market share of shop1 and it can be calculated as Eq. (4). In the game, each shop behaves aiming at maximizing its own profit.

$$D = \int \{(x, y) | p_1 + t(x - x_1)^2 + t(y - y_1)^2 \leq p_2 + t(x - x_2)^2 + t(y - y_2)^2\} dxdy \quad (4)$$

According to Tabuchi [7], when the market space is more elongated rectangular, shops will intend to locate far from each other in the longer axis and to minimize the deference of the location in the shorter axis. Then shops can avoid price competition.

3 Agent-Based Model

In this section, the behaviour of competing shops shown in the previous section is re-described as an agent-based model.

3.1 Agents

Our model has two competing shop agents (shop1 and 2) and N consumer agents. When the shop agents decide their location and then price, consumer agents choose one of them according to Eq. (1). In the model of Tabuchi, the market share of shop1 is calculated as the area in the market space as in Eq. (4). In the agent-based model, it should be replaced by the number of consumer agents who chose shop1. To describe the market share, the function n_i is introduced. It has the value of 0 or 1 according to the sum of the price and the movement cost for the i-th consumer as shown in Eq. (5)

$$n_i = \begin{cases} 1 \text{ if } p_1 + t(x_i^c - x_1)^2 + t(y_i^c - y_1)^2 \leq p_2 + t(x_i^c - x_2)^2 + t(y_i^c - y_2)^2 \\ 0 \text{ if } p_1 + t(x_i^c - x_1)^2 + t(y_i^c - y_1)^2 > p_2 + t(x_i^c - x_2)^2 + t(y_i^c - y_2)^2 \end{cases} \quad (5)$$

Using this function, the market share of shop1 can be described as in Eq. (6)

$$D = \sum_i n_i \quad (6)$$

Then the profit functions for the shops are described as Eqs. (7) and (8).

$$\pi_1 = p_1 D \quad (7)$$

$$\pi_2 = p_2 (N - D) \quad (8)$$

Shop agents act strategically to maximize each profit function. For shop1, the strategic variables are the location (x_1, y_1) and the price p_1. It is similarly for shop2.

The behaviour of the shop agents consists of two stages of location choice and pricing. Then we employ the backward induction method [11] which is the basic solution for two stage game.

3.2 Pricing Stage

For the first, we should consider the pricing when the locations of both shops are given. Tremendously large computation resources will be necessary for calculating the profit in all combinations of the prices. Therefore, in our model, shop1 and shop2 alternately update each price and when the price change becomes enough small is treated as the equilibrium situation.

For further saving computation resources, the candidates for the update price are narrowed down to the reservation prices for each consumer agent. By this ingenious, we can achieve to reach the equilibrium state with a small number of steps. When the price of shop2 is p_2, the reservation price to shop1 of i-th consumer agent who located at (x_i^c, y_i^c) is described as Eq. (9).

$$\bar{p}_1^i = p_2 + t(x_i^c - x_2)^2 + t(y_i^c - y_2)^2 - t(x_i^c - x_1)^2 - t(y_i^c - y_1)^2 \tag{9}$$

If the shop agent chooses some price which is not equal to the reservation price for any consumer agent, then the shop can increase the price up to the nearest reservation price without changing the number of consumers who choose him as shown Fig. 1. Therefore, the reservation prices of each consumer can be the candidate of the next price.

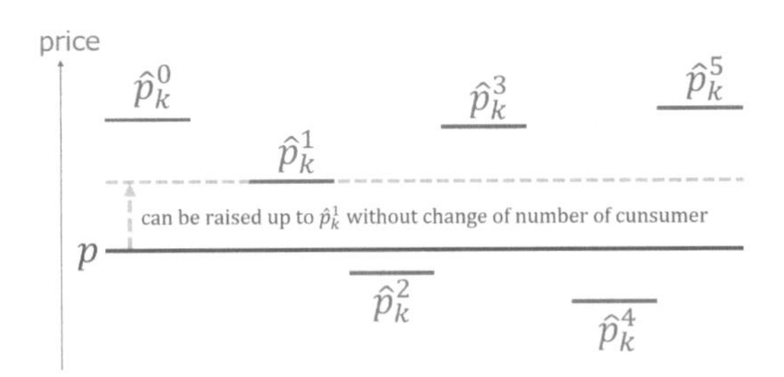

Fig. 1. Why the reservation prices can be a candidate of the next price

In the price updating process for shop1, the profit calculation by Eqs. (5)–(7) is performed for every reservation price which is calculated with Eq. (9). Then the reservation price giving the highest profit is adopted as the next price for shop1. After updating the price of shop1, shop2 update its price with the same process. It is repeated until the price change becomes to be enough small. The flow of this process is shown in Fig. 2.

3.3 Location Stage

Because it is not realistic to handle all possible location combinations, in our model, it is assumed that shop1 and shop2 repeat their location choice alternately as same with the pricing stage. And it is assumed that shop agents choose their each location from the candidate locations that are set in the market space in a lattice.

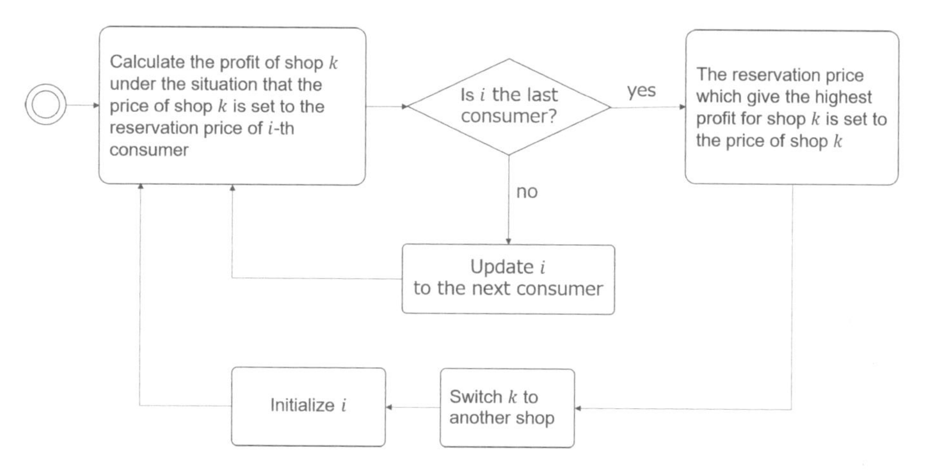

Fig. 2. The flow chart of the process of pricing stage

In the process of location choice for shop1, it chooses the location while fixing the location of shop2. For all candidate locations, the process of pricing stage explained in the previous subsection is carried out under the assumption that shop1 is located at each candidate location. Then the candidate location which gives the highest profit for shop1 is chosen for the next location. For the next, shop2 chooses the location in the same way (while fixing the location of shop1). Thus two shops repeat updating each location alternately. In the Fig. 3, the flow of the entire process of this model is shown.

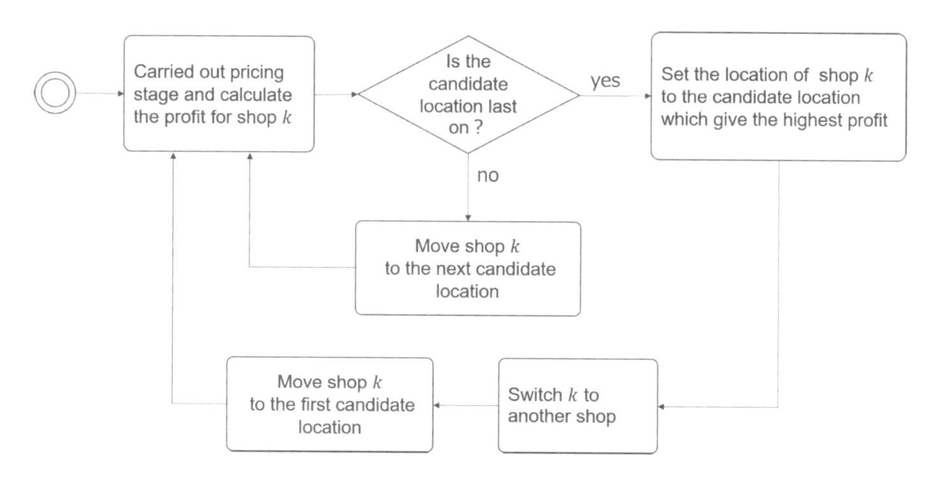

Fig. 3. The flow chart of the entire process of competition

4 Simulation

4.1 Graphical Simulator

We developed the multi-agent simulator for the agent-based model of two-dimensional spatial competition. In the simulator, the behaviors of the agents are calculated according to the model explained in the previous section. Figure 3 is corresponding the main sequence of the simulator. The simulator is developed with artisoc that is a platform for developing multi-agent simulation [12]. It has the graphical interface with which you can see results of the simulation as animation. And it has the graphical control panel to set the parameters like number of agents, initial location, initial price and consumer distribution. Figure 4 shows the capture image of our graphical simulator.

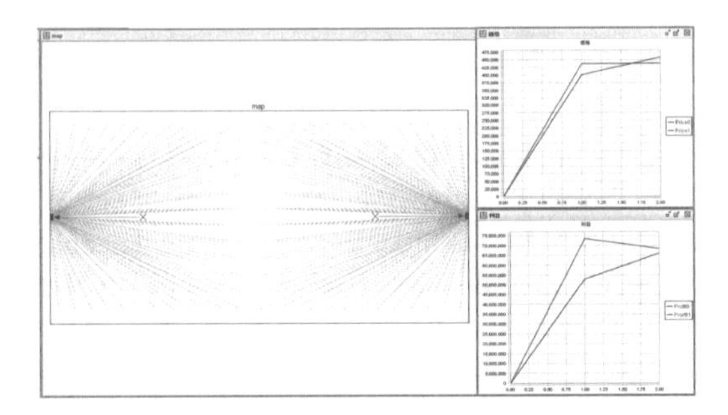

Fig. 4. The captured image of the graphical simulator

The source code of this simulator is available in our web page [13]. Although the source code can be opened with the general text editor, you need to install artisoc for running it. artisoc is distributed by KKE, inc. [14].

4.2 Results

In Figs. 5, 6, 7 and 8, the simulation results in different conditions are shown. Each figure expresses the market space. The green circular plots represent the consumer agents, the blue and red rectangles represent shops and the cross marks represent the location history of the shops.

The condition of case I (Fig. 5) in which consumers are distributed uniformly in a line is corresponding to the model of d'Aspremont et al. [7]. The simulation result of case I agrees with the discussion in [7]. And the simulation result of case II (Fig. 6) agrees with that of Tabuchi model [10]. From these results, it can be confirmed that our agent model is valid. In case III (Fig. 7) and case IV (Fig. 8), consumer agents are distributed discretely at random. Analytic approaches for spatial competition cannot be applied to such case. On the other hand, the agent-based approach can be applied to such cases. By constructing agent-based model and simulation, we can get a powerful tool to investigate the different and complex spatial competition situations (Table 1).

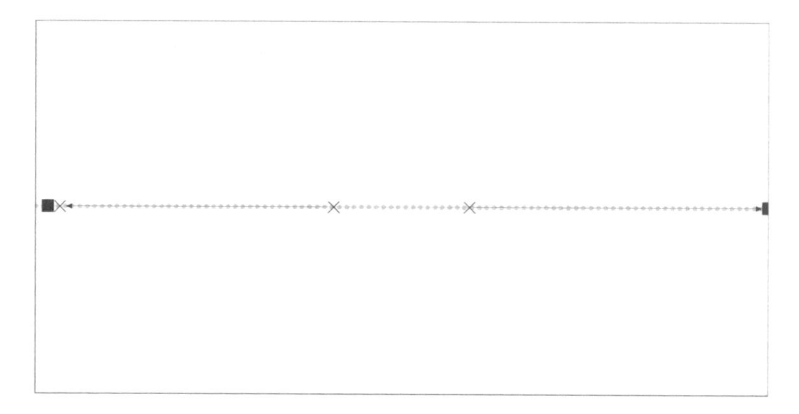

Fig. 5. The results of simulation in case I

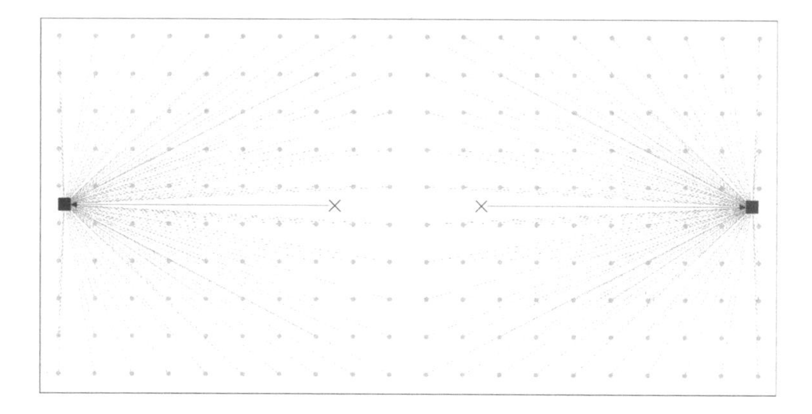

Fig. 6. The results of simulation in case II

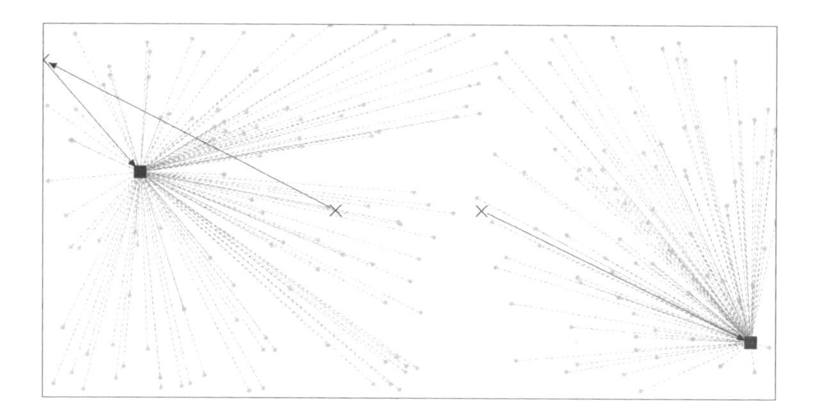

Fig. 7. The results of simulation in case III

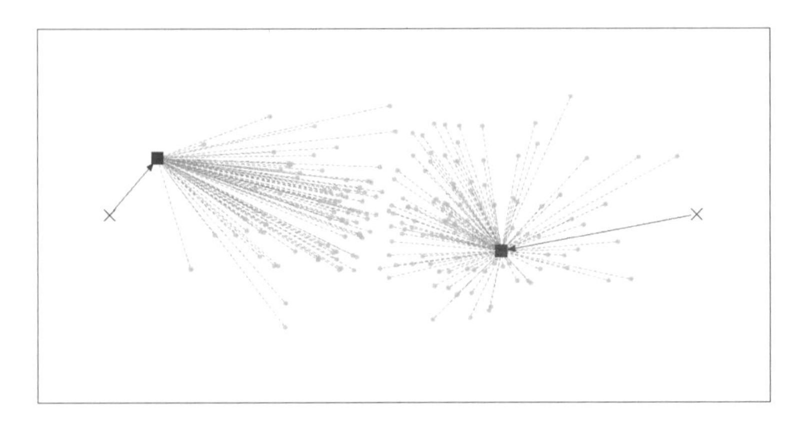

Fig. 8. The results of simulation in case IV

Table 1. The simulation conditions for each case

	Number of consumers	Consumer distribution
Case I	100	Uniformly in the line
Case II	200	In the grid pattern
Case III	200	Uniformly at random
Case IV	200	Randomly and condensed

5 Summary and Future Research

We intended to apply the agent-based approach to the researches of spatial competition and constructed the agent-based model for two-dimensional spatial competition. Our model can describe the discrete and non-uniform consumer distributions. Moreover, we developed the graphical simulator for that model and showed simulation results in some cases. The model validation was achieved by comparing the results of our simulation with those of the previous researches. And it is confirmed that the situations with the discrete and non-uniform consumer distribution could be simulated with the suggested method.

In the future, we are going to investigate spatial competition under different and complex conditions using our model and simulator. The analytical approaches have difficulties in dealing with the situations in which decision making of consumers (and shops) are affected by the results of their own previous behavior. For example, the situation with the switching cost is corresponding to it. Our agent-based model can deal with such situation. Including the above example, our research plans to utilize and extend the agent-based model obtained in this paper for improving the applicability of the spatial competition research.

References

1. Brown, S.: Retail location theory: the legacy of harold hotelling. J. Retail. **65**(4), 450–470 (1989)
2. Numan, W., Willekens, M.: An empirical test of spatial competition in the audit market. J. Account. Econ. **53**(1-2), 450–465 (2012)
3. Thomadsen, R.: Product positioning and competition: the role of location in the fast food industry. Mark. Sci. **26**(6), 792–804 (2007)
4. Hotelling, H.: Stability in competition. Econ. J. **39**(153), 41–57 (1929)
5. Anderson, S.P., Neven, D.J.: Cournot competition yields spatial agglomeration. Int. Econ. Rev. **32**(4), 793–808 (1991)
6. Yasuda, Y.: Instability in the Hotelling's non-price spatial competition model. Theor. Econ. Lett. **3**(3), 7–10 (2013)
7. d'Aspremont, C., Gabszewicz, J.J., Thisse, J.-F.: On Hotelling's stability in competition. Econometrica **47**(5), 1145–1150 (1979)
8. Bonabeau, E.: Agent-based modeling: methods and techniques for simulating human systems. Proc. Natl. Acad. Sci. United States Am. **99**(3), 7280–7287 (2002)
9. Macal, C.M., North, M.J.: Tutorial on agent-based modelling and simulation. J. Simul. **4**(3), 151–162 (2010)
10. Tabuchi, T.: Two-stage two-dimensional spatial competition between two firms. Reg. Sci. Urban Econ. **24**(2), 207–227 (1994)
11. Aumann, R.J.: Backward induction and common knowledge of rationality. Games Econ. Behav. **8**(1), 6–19 (1995)
12. Yamakage S.: Modeling and Expanding Artificial Societies: Introduction to Multi-Agent Simulation with artisoc. Kozo Keikaku Engineering Inc. (2009)
13. Miura, M.: https://m-miura.jp/mas/scmas-kes-amsta2018/. Accessed 28 Feb 2018
14. KKE, inc.: http://mas.kke.co.jp/modules/tinyd0/index.php?id=13. Accessed 28 Feb 2018

Agent Based Simulation of Network Routing: Reinforcement Learning Comparison

Krešimir Čunko[2], Marin Vuković[1], and Dragan Jevtić[1(✉)]

[1] Faculty of Electrical Engineering and Computing, University of Zagreb, Unska 3, 10000 Zagreb, Croatia
dragan.jevtic@fer.hr

[2] Erste Group Card Processor, Radnička 45, 1000 Zagreb, Croatia

Abstract. The paper considers and compares two methods applicable for self-adaptive routing in communication networks based on immobile agent. Two different reinforcement learning algorithms, *Q*-learning and *SARSA,* were employed in the simulated environment and results were gathered and compared. Since the task of routing is to find the optimal path between source and destination for every information piece of the service, the critical moment for routing in communication networks is quality of service, which includes coordination and support by many dislocated devices. These devices change their properties in time, they can appear and they fall down. Thus the task of the agents is to learn to predict new situations while continuously operates, i.e. to self-adapt its function. Our experiments show that the *SARSA* agent outperforms *Q* agent in information routing but in some situations, both agents fall. The circumstances in agent environment for which the agents are not prestigious were detected and depicted.

Keywords: Reinforcement Learning · Q and SARSA learning Agent · Routing

1 Introduction

The client-server communication paradigm enhanced by intelligent software agents is to be considered as a valuable and promising solution for the reduction of an overall network traffic and improved service quality. Thus, operating remotely, the agents could reduce overall network load and bandwidth requirements. Information routing is still challenging task where agent paradigm and self-adaptive methods could provide promising answers. The central problem of network routing is requirement to fast adaptation to traffic changes, failures, and congestions. Gathering and updating information from and to remote nodes generate additional network traffic and return information about the change of network states, but with delay. However, the methods able to learn from experience could give a new dimension for solutions of that typical problem for information routing.

Reinforcement learning (RL) [1–3] is the promising one but although applied in many fields, there are still dilemmas between proper exploration and exploitation strategies for action selection policy. The frequently applied RL algorithms are *Q*-

G. Jezic et al. (Eds.): KES-AMSTA-18 2018, SIST 96, pp. 76–86, 2019.
https://doi.org/10.1007/978-3-319-92031-3_8

learning and *SARSA* algorithms. However, they posse different properties, which could be significant difference when applied for routing inside the communication networks. The main motivation of this paper was to discover and to compare potentials of *Q*-learning and *SARSA* capabilities for information routing within the agent-based paradigm. Routing in communication network could initiate massively data transport for a single service but route always involves remotely located devices. Therefore, routing process introduces additional communication between remote devices and opens other dimensions for application of RL. There are number reports on effective applications of RL for simulation and modeling of games [4, 6], routing on chip [5] using Q-learning to improve balancing and saving the energy, furthermore the modification of *Q*-learning by combination of *Sarsa* and *Q*-learning [7] which outperform single *Q*-learning and *Sarsa* algorithms but applied to mechanical balancing system. All these applications are focused on single location without significant effect of delays and communication with dislocated devices. However, routing in communication networks is critical for the quality of service, which is coordinated and supported by many dislocated devices and which change their properties in time, and which appear and fall down.

This paper is organized as follows: firstly, the chosen model of the information system for self-adaptive control is illustrated. Secondly, reinforcement learning with its *Q* and *SARSA* algorithms are described. Thirdly, the function of an immobile agent and its location in client-server architecture are described. Fourthly, network architecture designed in the simulation environment is described, and three experiments were performed. Results of *Q* and *Sarsa* agents operating in information routing are compared and depicted on diagrams and tables. Finally, conclusion summarizes the results.

2 Model of Information System and Self-adaptive Control

Client/server paradigm is widely used architecture in a realization of telecommunication services in which communication entities are consisted as server process (typically on powerful computers) and client process (usually applications running by the users). However, client/server architecture is general model in which a numerous messages/packets are generated by both sides and forwarded to the network interfaces and through network nodes. One of the benefits of that architecture is the centralized administration of the service, i.e. any shared resource (e.g. data, files, email, CPU time, backup, etc.), however, processing is a type of distributed computing based on cooperative processing where the clients and servers cooperate in the processing of the whole service. Furthermore, clients and servers can be located on the same computer but the main benefits emerge when located remotely, where consistency of the information flow between client and server processes affects the quality of the whole service. Adaptive information routing, which follows the traffic changes, device failures, fluctuating delays, and faults is still challenging task, which affects directly to the service but also to the global network functionality. The idea was to locate immobile agents as the mediators on the server and on the client side, which continuously explore the environmental fluctuations and acts to find optimal route toward the destination.

The model of client and server location is presented in Fig. 1 where the information routing is hidden in nodes of the communication network, the Internet actually.

The service execution includes communication between processes where small pieces of information, called packets, being forwarded between nodes and using different paths between source and destination. When the packet is forwarded in the direction of an already overloaded link or node then loss or delay will appear. The packet losing produces additional increase of the network load and cause delays, jitter, interruptions, and possible degradation of the service.

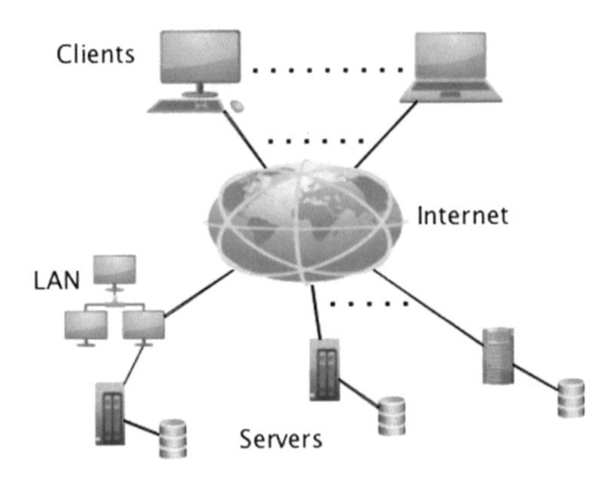

Fig. 1. General model of client and server location in the Internet.

3 Reinforcement Learning and Algorithms

Reinforcement learning (RL) surpasses important problems of the traditional programming approach [1, 2]. This is, RL that requires less programmer effort because an automatic training process does the most important work. Secondly, continuous environmental changings are followed by pieces of training, i.e. without additional programming. Thirdly, it is mathematically guaranteed that RL will converge to the optimal policy when certain assumptions are followed. Therefore, training during operation, varying environment, and guaranteed convergence to the desired policy, i.e. self-adaptation, are crucial properties of RL.

Reinforcement learning for the single agent is as follows:

- A discrete-time system ($t = 0, 1, 2, \ldots$) wherein the states and transitions depend on the actions performed by the agent.
- A new state s_{t+1} (causing a transition from the previous state s_t) after action a_t is determined with probability p (i.e. Markovian process):

$$p(s_{t+1}|s_t, a_t, s_{t-1}, a_{t-1}, \ldots) = p(s_{t+1}|s_t, a_t).$$

In this process, rewards are accumulated and can occur in specified states,

$$Q(s_o, a_o) = \lim_{N \to \infty} E\left[\sum_{t=0}^{N-1} \gamma^t g(s_t, a_t, s_{t+1})|s_o, a_o\right]$$

where Q denotes the cost estimate for a starting pair (state, action), i.e. (s_o, a_o), E stands for expectation, subsequent actions a_t's are determined by an policy, g is reward, and γ is a discount factor.

Among others, we considered two on-line learning algorithms to simulate agent based self-adaptive route selection and to evaluate effectiveness of the service quality parameters, and afterward to compare results. The both use Q value to compute cost function evaluation. The first is Q-learning algorithm (Watkins 1989) where updating is online, without explicitly using probability estimates [1–3]. The updating is based on actual state transitions,

$$Q(s_t, a_t) \leftarrow Q_t(s_t, a_t) + \alpha\left[r_{t+1} + \gamma \cdot \max_a Q_t(s_{t+1}, a) - Q_t(s_t, a_t)\right],$$

where $\alpha \in (0, 1)$ is the learning rate and $\gamma \in (0, 1)$ is discount factor. The discount factor regulates the impact of future cost functions to a current Q value, e.g. when $\gamma = 0$ then the agent is "near-sighted" - maximizes the impact of post-state. Learning rate indicates the measure to which the reward will affect the existing Q value.

The other is *SARSA* (*State-Action-Reward-State-Action*) algorithm for which the update value is given by,

$$Q(s_t, a_t) \leftarrow Q_t(s_t, a_t) + \alpha[r_{t+1} + \gamma \cdot Q_t(s_{t+1}, a_{t+1}) - Q_t(s_t, a_t)],$$

and also $\alpha \in (0, 1)$ is the learning rate and $\gamma \in (0, 1)$ is discount factor, but the main factor in updating is Q value associated to action applied on the next state $Q_t(s_{t+1}, a_{t+1})$.

The Q and SARSA learnings are similar and follow the same procedure in which the update rule is different (Fig. 2).

```
Initialize Q(s, a)
Repeat (for each episode):
    Initalize s
    Repeat (for each step):
        Choose a from s using policy derived from Q (e-greedy)
        While (s is not a terminal state):
            Take action a and observe r (s')
            Choose a' from s' applying policy derived from Q
            Compute new value for Q_t - (Q or SARSA updating)
            s ⇦ s', [also (a ⇦ a') for SARSA]
```

Fig. 2. Procedural form for Q and *SARSA* learning.

Therefore, the main difference is in cost function estimation. In Q-learning the updates are independent of the next state agent's action, while *SARSA* learning includes exploration steps to find optimal Q-value. Namely, the updating of Q-learned agent is based on greedy actions regardless on selected action, i.e. ignore actual policy (for example, exploration), however updating for *SARSA* assumes that chosen policy continues to be followed. This "small" difference reflects the global knowledge and behavior of an agent. Generally speaking, Q value associated with the single action from state s converges toward the expected sum of all rewards received by applying that action - i.e. follows optimal direction by means of state action pairs.

4 Agents and Client-Server Architecture

In the searching for self-adaptive methods, models, and solutions able to continuously forward packets selecting appropriate path section by observing traffic parameters we created a simulator in which the agent continuously controls the path during communication between client and server processes. We assumed that the agents controls requests and responses on the client and server side, respectively. The agents are immobile but able to observe network resources, to choose routing point and to select destination. This strict separation of client from the server side is not clear for some services because the end devices act as client and server simultaneously (e.g. IP and video telephony). The model assumes two independent immobile agents located on the client and server side (Fig. 3). The essential task of the agent is to move the packet from the queue toward its communicating pair by observing throughput and choosing the sections of the path.

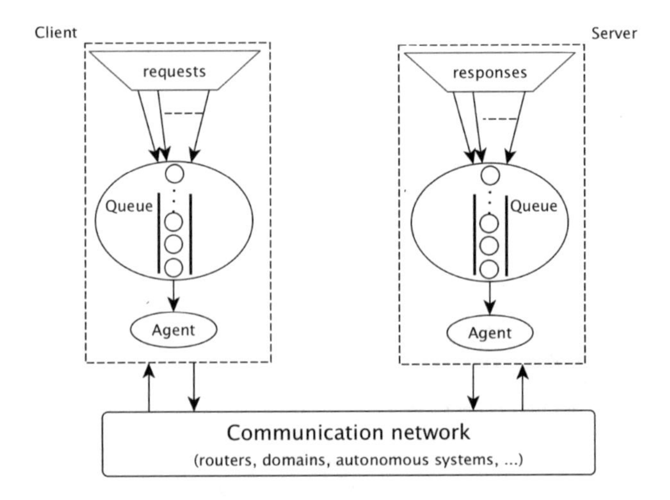

Fig. 3. Agent's operating position in a model of client server communication

We differentiate two kinds of agents on the base of applied online learning, *Q* or *SARSA*, and we call them *Q* agent and *SARSA* agent. These agents were observed in separated experiments, i.e. they do not operate concurrently. Regardless of the agent's type and location, the task of these agents is to maximize the load of the links toward its communicating pair taking packets from its queue.

Looking the difference between *Q* learning and *SARSA* we can expect advances for *SARSA* agent since in that case exploration steps enable visiting more states than *Q* learning. This could result in the quick search for optimal action when *SARSA* is used instead of *Q* agent.

5 Simulation and Results

In order to illustrate the properties of *Q* and *SARSA* agent applied for information routing we have created simulation environment using C++ (*Microsoft Visual Studio*) to test the agents operating in different network configurations. Moreover, *Round Robin* scheduling method was implemented as an additional comparative indicator of efficiency these selected self-adaptive methods. Therefore we talk about three agents: Q agent, SARSA agent and Round Robin agent.

Furthermore, we simulated the time periods of 24 and 72 h with traffic variations corresponding roughly to daily traffic load, i.e. with periods of low and high traffic load, which matching the Gaussian distribution. The results were gathered every minute, equidistantly from the start to the end of simulation time.

Instead of generating the packets with variable length, we introduced the term job, which represents single process with fixed runtime of 5 s. We introduced simplification, so the job represents a small group of successive packets passing the same route and with constant transport and execution time. The jobs were generated on client side and routed by the agent toward the selected server. The mission of the agent was to choose the router and destination server for each single job appeared in a queue, following defined policy. The routers, servers and links between them were set to be different capacities and loads, which are represented as available capacity. Furthermore, it is assumed that the devices are located in different time zones where corresponding traffic load per zone are shifted for a constant time period. The agent establishes a connecting path for single job by selecting the appropriate router and server independently of their locations in time zones. Every job is limited in life-time and if not processed in 60 s, then is canceled and counted as lost.

Q and SARSA operating parameters α, γ and ϵ*-greedy* were found by experiments as follows. Firstly, the parameters α, γ were fixed to $\gamma = 0.0001$, $\alpha = 0.01$, and parameter ϵ*-greedy* were changed gradually. Looking to routing quality parameters, e.g. lost jobs per minute, we marked best response in regard to ϵ*-greedy* (Table 1).

Table 1. The average jobs lost per minute related to ϵ*-greedy* value for *Q* and *SARSA* agents.

ϵ-*greedy*	Q	SARSA
0.8	1.04	1.04
0.9	1.0	0.89
0.95	0.96	0.85
0.97	**0.90**	**0.78**
1	1.01	0.99

Then we fixed ϵ*-greedy* to 0.97 and further looking for optimal α and γ using the same procedure, i.e. changing gradually the parameters α and γ and observing average jobs lost per minute, we found $\alpha = 0.001$ and $\gamma = 0.0001$ as optimal values for *Q* agent and $\alpha = 0.001$ and $\gamma = 0.00001$ (Tables 2 and 3).

Table 2. The average jobs lost per minute related to discount factor γ.

γ	0.5	0.1	0.05	0.01	0.005	0.001	0.0005	**0.0001**	0.00005	**0.00001**	0.000005
Q	2.43	1.76	1.63	1.27	1.33	1.21	1.00	0.92	0.95	**0.85**	0.94
SARSA	2.17	1.39	1.47	1.09	1.00	0.96	0.80	**0.79**	0.88	0.93	0.98

Table 3. The average jobs lost per minute related to learning rate α.

α	0.5	0.1	0.05	0.01	0.005	**0.001**	0.0005	0.0001
Q	1.70	1.25	1.19	0.98	0.87	**0.71**	0.87	0.85
SARSA	1.55	1.27	1.04	0.81	0.84	**0.75**	0.87	0.75

5.1 Experiment 1 - Gaussian Distribution

In order to obtain comparative measure for *Q* and *SARSA* agents acting as route selectors while state of the environment regularly changes, we created environment with 4 routers and 6 servers located in different time zones. The agents send the jobs toward the servers via selected routers using specified policy for *Q*, *Sarsa* or *Round Robin*.

The capacity of client's link is the reference value, i.e. the agents can't send more information toward the network than capacity of this connection allows. Therefore, we can represent the available capacities of all network links in relation to the client-network connection. This means, if the path/s can take all client jobs it will take at maximum 100% capacity of client/network connection. Agent's acting should maximize usage of client's link, and consequently should minimize waiting in queue and the number losing jobs. Table 4 shows the initial load of the servers and routers represented in time period (24 h) as percentage of client's link capacity.

Table 4. The initial load of routers and servers represented as part of client's link capacity.

Time intervals (hh:mm:ss)	Load [%]
00:00:00–06:59:59	10
07:00:00–12:59:59	60
13:00:00–18:59:59	80
19:00:00–23:59:59	40

As noted before, network devices could be located in different time zones, i.e. shifted distribution of the traffic load in regard to client location. The set of initial values for available links and capacities from client to all servers is depicted in Table 5 (left). For example, available capacity to route the job toward Server 4 choosing Router 1 is 100%, and choosing Router 2, 3 and 4, the capacities are 45%, 5%, and 90%, respectively.

Table 5. Link capacities between client and pairs (router - server) represented as percentage (left) and time shift of the domain where the device belongs to (right).

Routers	1	2	3	4
Server 1	15	65	25	90
Server 2	95	25	100	50
Server 3	70	30	80	50
Server 4	100	45	5	90
Server 5	55	30	90	90
Server 6	75	30	100	45

Device	Time shift [hours]
Router 1	3
Router 2	0
Router 3	-2
Router 4	12
Server 1	-7
Server 2	-4
Server 3	-7
Server 4	0
Server 5	-10
Server 6	-7

The time shift in regard to the time zone of the initial load is given in Table 5 (right) and shows how many hours are shifted the traffic load depicted in Table 4. The simulation for the previously described environment was performed during three days with maximum load of 60% of client's link capacity and the Gaussian distribution of generated jobs per every day. The results are displayed by diagrams in Fig. 4, 5, and 6. Every point depicted in diagram shows an average value of 100 measured values around that point.

Each diagram shows quicker adaptivity of *SARSA* agent in regard to *Q* agent and of course in regard to Round Robin. Fewer time the jobs waiting in a queue (Fig. 4), lesser number of jobs lost (Fig. 5) and faster execution time per job (Fig. 6) - the green line.

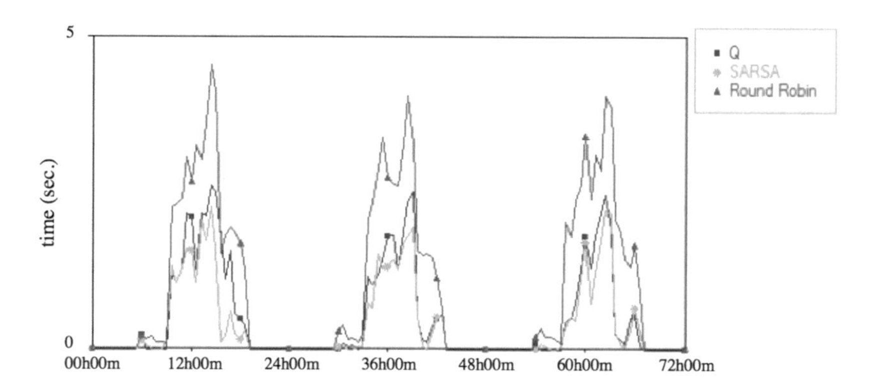

Fig. 4. Waiting time in agent's queue during simulation period.

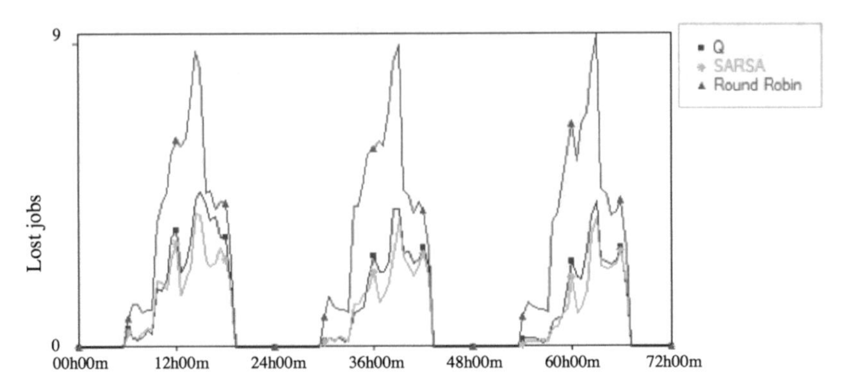

Fig. 5. Jobs lost during simulation period.

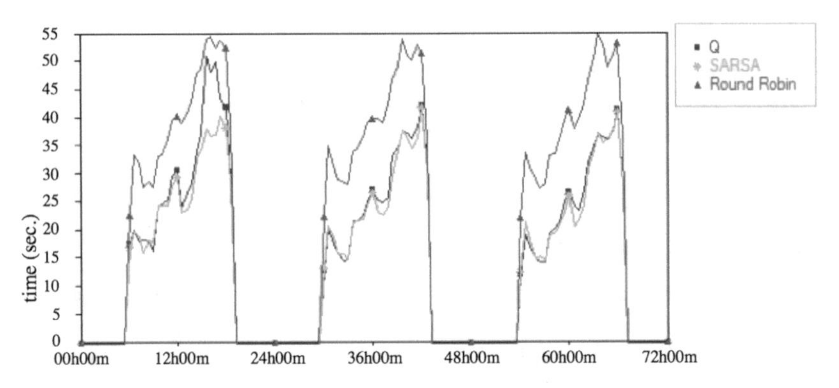

Fig. 6. Average time for job staying in the system during simulation period.

5.2 Experiment 2 - Gaussian Distribution and Large Load

The simulation from experiment 1 was modified in order to observe agent capabilities when the client requests achieved capacity of its connection link. The simulation was performed in the same environment during three days with maximum load of 100% of client's link capacity and the Gaussian distribution with standard deviation over 20 h. Time for job processing is increased from 60 to 120 s.

The diagram in Fig. 7 shows the losted jobs during period of three days. It is obvious that neither *Q* nor *SARSA* agent are not superior than *Round Robin* in all circumstances during simulation as it was the case in example 1. Still *SARSA* agent perform better or the same as *Q* agent but both of them falls down in regard to Round Robin. The points are depicted by arrow in Fig. 8. Nevertheless, the average number of the lost jobs is 7.75 for *Q* agent, 7.74 for *SARSA* and 9.20 jobs for *Round Robin*.

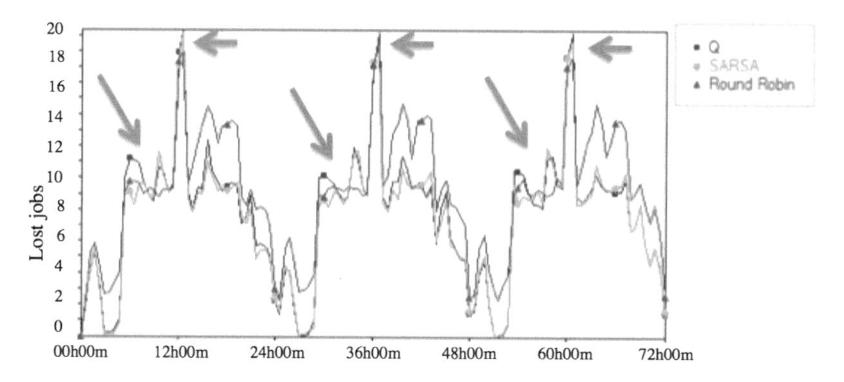

Fig. 7. Jobs lost during simulation period.

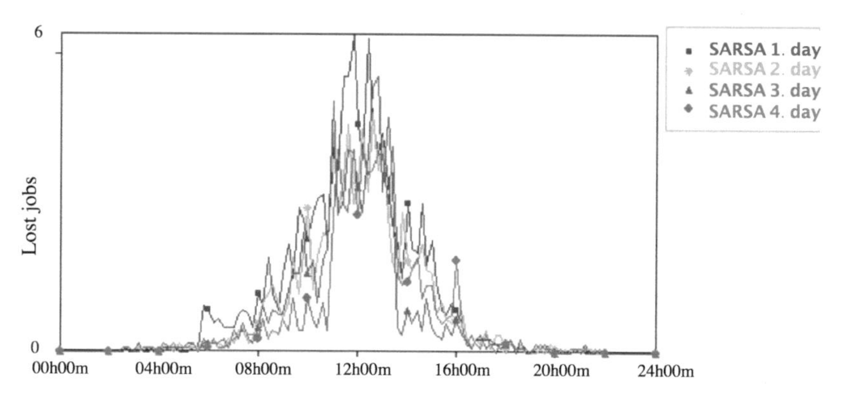

Fig. 8. Jobs lost during simulation period - SARSA agent.

5.3 Experiment 3 - SARSA Agent in Continuous Operating

In this experiment the simulation from experiment 1 was extended to period of four days. To illustrate the knowledge accumulation of single agent the same parameters from example 1 were used to observe efficiency of *SARSA* agent when repeats similar or previously seen state of the environment and collocates approximately the same number of generated jobs.

Diagram (Fig. 8) depicts lost jobs in four days for *SARSA* agent. Notice the reduction of lost jobs by improved adaptation during peaks of traffic, which was achieved by previous knowledge accumulation. The total number of canceled jobs on the first day is 1486, in the second 1146, in the third 1024, and the fourth day 933 jobs.

6 Conclusion

In order to make it easier to evaluate agent based routing in communication network whose effectiveness should be assessed a new simulator is designed and implemented. Simulator is implemented using C++ programming language. Using the simulator we have presented the results of the agent based routing for selected architectures created with *Q* and *SARSA* algorithms. The network architecture was consisted of one layer of routers and one layer of servers, each of them located in the different time zones. The agent operates by selecting the pairs router-server for every job of fixed length and limited lifetime, trying to minimize the delays in processing, i.e. retention of jobs. It was found that operating of *SARSA* agent outperform *Q* agent and faster adapt to changes. However, in some circumstances like the peak load in the network together with high load of client's requests, neither *Q* nor *SARSA* agent are not above the simple Round Robin effectiveness. Although in total *SARSA* agent shows 10–20% better performance and frequently faster adaptation, however, some results indicate the requirements for the further research.

References

1. Sutton, R.S., Barto, A.G.: Reinforcement Learning – An Introduction. MIT Press, Cambridge (1998)
2. Watkins, C.J.C.H., Dayan, P.: Q-learning. Mach. Learn. **8**, 55–68 (1992)
3. Bertsekas, D.P., Tsitsiklis, J.N.: Neuro-Dynamic Programming. Athena Scientific, MIT, Belmont (1996)
4. Takadama, K., Fujita, H.: Toward guidelines for modeling learning agents in multiagent-based simulation: implications from Q-learning and Sarsa agents. In: MABS 2004, Conference Proceedings, pp. 159–172 (2004)
5. Farahnakian, F., Ebrahimi, M., Daneshtalab, M., Liljeberg, P., Plosila, J.: Q-learning based congestion-aware routing algorithm for on-chip network. IEEE Xplore (2011)
6. Zhao, D., Wang, H.: Deep reinforcement learning with experience replay based on SARSA. IEEE Xplore (2017)
7. Wang, Y.-H., Li, T.-H.S., Lin, C.-J.: Backward Q-learning: the combination of Sarsa algorithm and Q-learning. J. Eng. Appl. Artif. Intell. **26**, 2184–2193 (2013)

Dispatching Strategies for Dynamic Vehicle Routing Problems

Besma Zeddini[1(✉)] and Mahdi Zargayouna[2]

[1] Quartz, EISTI, Avenue du Parc, 95000 Cergy Pontoise, France
bzi@eisti.eu

[2] Université Paris-Est, IFSTTAR, GRETTIA, Boulevard Newton, Champs sur Marne, 77447 Marne la Vallée Cedex 2, France
hamza-mahdi.zargayouna@ifsttar.fr

Abstract. Online vehicle routing problems are highly complex problems for which several techniques have been successfully proposed. Traditionally, the solutions concern the optimization of conventional criteria (such as the number of mobilized vehicles and the total traveled distance). However, in online systems, the optimization of the response time to the connected users becomes at least as important as the optimization of the traditional criteria. Multi-agent systems and greedy insertion heuristics are the most promising approaches to optimize this criteria. To this end, we propose a multi-agent system and we focus on the clients dispatching strategy. The strategy decides which agents perform the computation to answer the clients requests. We propose three dispatching strategies: centralized, decentralized and hybrid. We compare these three approaches based on their response time to online users. We consider two experiments configuration, a centralized configuration and a network configuration. The results show the superiority of the centralized approach in the first configuration and the superiority of the hybrid approach in the second configuration.

1 Introduction

Several real-life distribution applications, such as the good deliveries to stores, the school buses routing, the newspapers and mail distribution, etc. are instantiations of vehicle routing problems (VRP). In its original version, a VRP is a multi-vehicle traveling salesman problem. A number of nodes have to be visited only one time by a number of vehicles. The problem objective is generally to find a set of routes for the vehicles that optimize the number of mobilized vehicles and the total traveled distance. Solving these problems has high practical usefulness and they are challenging optimization problems with stimulating issues. The problem variant with time (and capacity) constraints is one of the most widely studied variants of VRP (vehicle routing problem with time windows, VRPTW henceforth) [1]. In this variant, the requests to be handled are not simple nodes, but clients who define a quantity to be transported, a node to be visited and two temporal bounds between which it has to be visited by a vehicle.

G. Jezic et al. (Eds.): KES-AMSTA-18 2018, SIST 96, pp. 87–96, 2019.
https://doi.org/10.1007/978-3-319-92031-3_9

Vehicles have limited capacities and the quantities associated with the clients in the same route must not be bigger than the capacity of the concerned vehicle.

Vehicle routing problems can be divided in two categories: static problems and dynamic problems. In the static problems, the system knows all the problem data before execution. In the dynamic problems, the problem data reveals as the optimization is being performed. The data may concern any entity of the problem, such as the traffic data or the available vehicles, but the dynamism usually refers to the clients to be served. The operational problems are never completely static and we can say that a static system cannot meet nowadays operational configurations anymore. Indeed, in real-life vehicle routing problems, and even when all the clients are known in advance (with a reservation system for instance), there always exists some element that makes the problem actually dynamic. These elements might concern no-shows, delays, breakdowns, etc. Online vehicle routing problems could be seen as an extreme case of dynamic vehicle routing problems. Indeed, not only the problem data, and specifically the clients, are not known before the optimization starts, but the clients connect in real-time to the system and expect quasi-immediate answers to their requests. The response time of the system in this configuration is then vital. If the system needs, say, two more minutes to gain one or two kilometers in its routes, it is not worth it in online problems, since the client will not wait that long to have an answer to its request.

To meet the requirement of short response times, we rely on the multi-agent paradigm for solving the online vehicle routing problems. An agent is an intelligent entity that is situated in an environment and that applies autonomous actions to satisfy its objectives [2,3]. A multi-agent modeling of the online VRP is relevant for the following reasons. On the one side, choosing a design allowing for computing distribution should provide shorter response times to clients requests. On the other side, nowadays vehicles are more and more connected, and have onboard computers. In this context, the transport system is, *de facto*, distributed and necessitates an adapted modeling to take profit of these equipments. The multi-agent system (MAS) that we propose in this paper simulates a distributed version of the so-called "insertion heuristics". These are methods that consist in inserting the clients following their appearance order in the routes of the vehicles. The vehicle chosen to insert the considered client is the one that would have the minimal additional cost to visit it (the incurred detour for instance). This is the fastest known heuristic, since there is no reconsideration of previous insertion decisions. In this context, there is still a choice to perform with respect to the dispatching of clients requests to the vehicle agents of the multi-agent systems. We propose three dispatching strategies and we compare them following their ability to provide better response times to the clients. The dispatching strategy decides which agents perform the computation to answer the clients requests. In the centralized strategy, the planner agent performs most of the computation. In the decentralized strategy, the vehicle agents perform most of the computation in a collaborative way. Finally, in the hybrid strategy the work is split between clients and vehicles.

The remainder of this paper is structured as follows. In Sect. 2, we discuss previous proposals for the dynamic VRP w.r.t our approach. The multi-agent system and the three dispatching strategies architecture of the MAS are presented in Sect. 3. We provide our experimental results in Sect. 4 and then conclude with a few remarks in Sect. 5.

2 Related Work

The majority of the proposed solution methods to vehicle routing problems are heuristic or metaheuristic methods, which provide good results in non-exponential times, and which have presented good results with benchmark problems. Generally speaking, most of the works dealing with the dynamic VRP are more or less direct adaptations of static methods. Among the static methods, insertion heuristics are the most widely adapted in a dynamic environment (e.g. [4]). Insertion heuristics are, in their original version, greedy algorithms, in the sense that the decision to insert a given client in the route of a vehicle is definitive. The advantage of using insertion heuristics is that they are intuitive and fast.

In their vast majority, multi-agent approaches of the literature rely, at least partially, on insertion heuristics. In [5], Thangiah *et al.* propose a multi-agent architecture to solve a VRP and a multi-depot VRP. In [6], Kohout and Erol propose a multi-agent architecture to solve a dial-a-ride problem. The principle of these two proposals is the same: distribute an insertion heuristic, followed by a post-optimization step. In [5], the clients are handled sequentially. They are broadcasted to all the vehicles, which in turn propose insertion offers and the best proposal is retained by the client. In the second step, the vehicles exchange clients to improve their solutions, each vehicle knowing the other agents of the system. Since vehicles are running in parallel, the authors envision to apply different heuristics for each vehicle, without changing the architecture.

For the reasons that we have given in the introduction, we choose a multi-agent modeling to solve the dynamic VRP. For their fast execution times and their adaptation to dynamic settings, we privilege a solving grounded on insertion heuristics. Thus, from a protocol and an architecture point of view, our system sticks with the multi-agent systems we have just described, since we propose a distributed version of insertion heuristics. However, in these proposals, none have focused on the response time of the system to online clients. In our previous works (e.g. [7–9], we have addressed the optimization criteria of the VRPTW. In this paper, we do not focus on the optimization problem for itself. Our focus here is on the three dispatching strategies, and our result indicate which one is the best, with respect to the chosen implementation configuration.

3 Dispatching Strategies

Each solution for a given vehicle routing problem instance is a set of vehicles with a specific route. Each vehicle's route is composed of a sequence of clients, together with their corresponding visit time. The three requests dispatching strategies that we propose in this paper are defined in the framework of a multi-agent system. Three categories of agents are defined in the system. The client agents represent users of the system (persons or goods, depending on the problem). The vehicle agents represent vehicles and the interface agents represent the interlocutor with the external world (GUI, simulator, etc.). When a user logs to the system, the interface agent create a representing client agent, representing the human user. A fourth agent type is defined for the only centralized dispatching strategy, which is the planner agent and is responsible of performing all the routing.

In online problems, the response time of the system is key, and only very fast approaches can compete in this configuration. The fastest approach, and the most popular one is the greedy insertion approach, originally proposed by Solomon [1]. The principle is to insert clients progressively in the vehicles routes. To do so, the insertion price of inserting a client in the route of a vehicle is calculated, and the vehicle with the minimal price is chosen for inserting the client. To compute this insertion price, the cost of the current itinerary (the total traveled distance) and the cost of the new itinerary are compared. The difference between the two quantities is the additional effort or insertion price for the new client's insertion. Determining the chosen vehicle consists in selecting the vehicle with the minimal insertion price.

When the solving system is a MAS, there are several alternatives regarding who handles the request. Each alternative is called a "dispatching strategy". In this section, we describe and compare three possible dispatching strategies that we have designed, implemented and compared to model the dynamic VRP: a centralized dispatching, a decentralized dispatching and a hybrid dispatching. The objective is to check which dispatching strategy is the most effective, in terms of response time to clients requests. The evaluation of the different strategies does not consider the traditional optimization criteria (number of mobilized vehicles, total traveled distance and total waiting time). Indeed, in terms of optimization, the three dispatching strategies follow the same algorithm, the only difference concerns the response time, i.e. the time that the system takes to decide about which vehicle will serve the customer.

3.1 Centralized Dispatching

In the centralized approach, all the treatments are performed by a central entity, which create vehicle plans and schedules. One of the main advantages of this approach is that it allows for central online optimization techniques. Online optimization (e.g. in [10]) allows to profit from optimization techniques, while reducing response times. The principle is to discretize the processing time into time intervals. During each interval an optimization is performed with the known clients. The new clients are kept in a queue, waiting for the next interval. The known clients that could not be served and the new clients are submitted for the new optimization round. As we said in the previous section, our objective is to compare the same solving approaches while comparing response times, our centralized approach then mimics insertion heuristics. In our proposal (see Fig. 1 (left)), all client requests are treated by the same planner agent. The planner agent has all the necessary information about each vehicle and each client and their current status. With these information, it assigns the current client in the least costly position between all the possible vehicles.

The scenario is the following. A user appears and interacts with an interface agent who creates a client agent representing him. When created, the client sends a request to the planner agent, who tries to insert it in the route of every vehicle of the system, in every feasible position. To this end, it executes sequentially, for each vehicle, a procedure computing the insertion price for the vehicle, and chooses the vehicle and the insertion position with the minimal price. If no vehicle can insert the client a new vehicle agent is created and the client is inserted in the only possible position in its route. Finally, the planner informs the client and the vehicle of the outcome of the procedure. Vehicle agents in the centralized dispatching strategy do not perform any calculation and only acknowledge the updates in their routes.

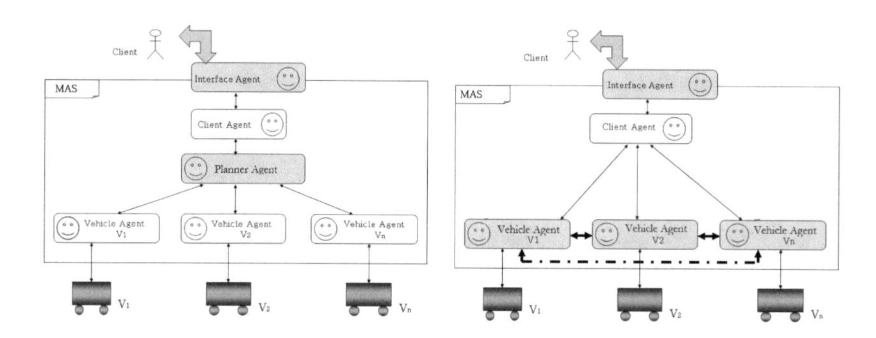

Fig. 1. Centralized architecture (left) and decentralized architecture (right)

The centralized dispatching poses two main problems. On the one side, it is not possible to distribute the execution over several hosts in order to limit the response time of the system, which is the primary concern in online systems. On the other side, the breakdown of the planner agent would result in a complete breakdown of the system. Nevertheless, the centralized dispatching strategy offers the advantage of minimizing the communications between agents, which are restricted to the notifications of the computing outcome to the clients and vehicles ($N(1 + V)$), with N the number of clients and V the number of vehicles.

3.2 Decentralized Dispatching

The decentralized dispatching is illustrated in Fig. 1 (right). Following this method, there is no bottleneck for routes calculations. Following the principle of the greedy insertion heuristics, every vehicle agent tries to insert the new client in its route, and proposes an insertion price, corresponding to the "cheapest" position where it can insert the client. The chosen vehicle will be the one having the minimal insertion price to transport the client.

In this dispatching strategy, the choice of the vehicle with the minimal price, the computation of costs, and the choice of the vehicle, all these steps are performed in a distributed way. Indeed, the scenario is the following. When a new client shows up, it broadcasts its request to all the vehicles of the system. When the request is received, every vehicle computes its insertion price. When it finishes its computation, the vehicle broadcasts a message to all the vehicles with its identifier and its price. For the processing of these messages and the inference of the winner vehicle agent, we propose the following process.

Every vehicle agent broadcasts its own computed price to the other vehicle agents. When he receives a new message containing a price that was computed by another agent, he sorts the received offers, including its own offer, following their prices. When all the other vehicle agents have proposed a price, the vehicle agent either checks if it is the best vehicle. If so, it updates its route with the new inserted client.

This dispatching strategy offers the advantage of completely distributing the processing and to be fault-tolerant. Indeed, breakdowns might occur for the agents, which would block the whole system (cf. centralized dispatching). In this approach in the contrary, for each new client request, the vehicles have to negotiate to choose which one is the most appropriate to serve the client, instead of a central entity that would decide for them. However, the number of exchanged messages might increase dramatically, which is generally the price to pay for a distribution of the processing. The number of exchanged messages between vehicles with this dispatching strategy is equal to $N \times V^2$. The overall number of messages is equal to $N(1 + V(1 + V))$.

3.3 Hybrid Dispatching

The hybrid dispatching is a compromise between the centralized approach and the decentralized approach. In the hybrid approach (cf. Fig. 2), the client agent plays the role of a dispatcher. The client agent broadcasts the client request, collects the offers of the vehicle agents and chooses the one proposing the minimal price.

The hybrid approach follows the following protocol. A new user provides the interface agent the information concerning his transport request. The interface agent creates a client agent representing him. Then, the new client agent broadcasts a message to all the vehicles. Every vehicle agent verifies if it can insert the client in its route. The vehicle agent then sends its price to the client agent. The client agent collects the answers of the vehicles and chooses the vehicle that proposes the minimal price. Once it has chosen the best vehicle that can answer the new request (if there is at least one that can insert the client), it broadcasts a new message to the vehicles informing them about its decision and asking the winner vehicle agent to insert it in its route and to serve it. When the vehicle

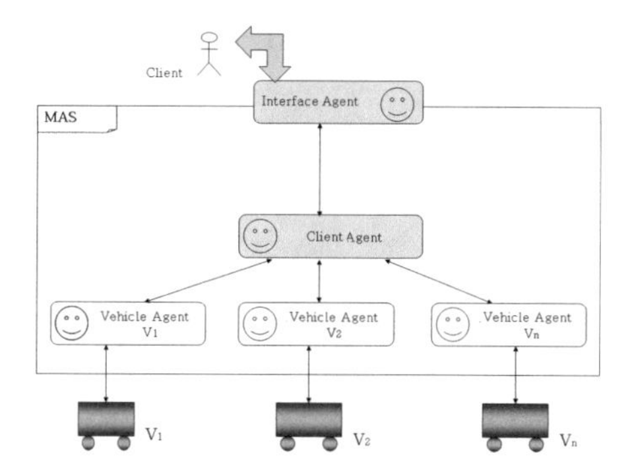

Fig. 2. Hybrid architecture

receives the message from the client informing it that it is the winner, it updates its route and inserts the client.

Thus, the objective of the hybrid approach is to relax the planner from all the calculation, and to limit the communications between vehicles. The overall number of messages in the hybrid dispatching strategy is equal to $3VN$.

4 Experiments

Our objective is to verify the impact of dispatching strategies on the response time of time-constrained online vehicle routing systems. We have generated several client files with 100, 200, 300 et 400 clients, while varying the number of vehicles between 4 and 8. The spatial environment is a plane of 50×50 and the depot is in the center of the plane. The customers are geographically uniformly distributed with time windows varying between 10 min and one hour. The service time is set to 5 min. The quantities associated with clients requests are between 5 and 20 while vehicle capacity is set to 400. The scheduling horizon is set to 10 h. For each client, we have also to define its appearance time, i.e. the moment when it becomes known by the system. We have used the Gendreau [11] method for the definition of these moments. Clients appear between 30 min and 1 min before the start of their time window. Provided the high level of randomness, we have executed each type of simulation 50 times and we report the average result values.

4.1 Centralized Experiments

We have implemented the three dispatching strategies using the multi-agent platform REPAST Simphony [12]. The simulation is made of 7200 discrete simulation ticks. Each tick corresponds to 5 s in the real world. Clients appearance times are transformed into "appearance ticks" and continuously feed the simulation at the computed ticks. We have executed our experiments on a PC with an Intel Xeon E7-4820 processor, and 50 GB of RAM. Since we use the same

Table 1. Centralized configuration

Parameters/Approaches	Nb vehicles	Nb clients	Average response time (ms)
Centralized	4	100	33
	4	200	43
	8	300	50
	8	400	62
Hybrid	4	100	38
	4	200	51
	8	300	59
	8	400	72
Decentralized	4	100	46
	4	200	63
	8	300	94
	8	400	113

deterministic algorithm for all dispatching strategies, which is a distributed version of insertion heuristics, the results for the three architectures in terms of optimization costs are the same and not reported here.

The Table 1 provides the values in terms of average response times (in milliseconds) of every dispatching strategies. The response time for a client is the difference between the moment when the client agent is created and the moment when a vehicle is chosen by the client. The centralized architecture provides the best results, followed by the hybrid architecture and the decentralized architecture. This is due to the fact that the centralized approach does not generate communications between agents and does not suppose any concurrency management. The hybrid approach provides results that are close to the centralized dispatching strategy. However, it provides results of worse quality for two reasons. On the one side, it generates more messages (linear with the number of vehicles) between the client agent and the vehicle agents. On the other side, the management of concurrent processes of the vehicles and clients, and the fact that their contexts have to be restored every time the scheduler executes them, increases the exhibited response times for the clients. Finally, the distributed approach suffers from the two drawbacks: it generates a quadratic number of messages and it uses pseudo-parallelism which slows down the processing.

However, this round of experiments being executed on a single computer, these results are not fair with the decentralized dispatching strategy, and to a lesser extent with the hybrid approach. Indeed, to use the full capacity of these strategies, we have to execute our simulations on a mini-cloud.

4.2 Network Experiments

It is possible with Repast Simphony to distribute a simulation on a network. We have deployed our three systems (one for each dispatching strategy) on a four PC network, each with the same configuration (Intel Xeon E7-4820 processor, and 50 GB of RAM). We report the new obtained results in Table 2.

Table 2. Networked configuration

Parameters/Approaches	Nb vehicles	Nb clients	Average response time (ms)
Hybrid	4	100	18
	4	200	26
	8	300	29
	8	400	32
Decentralized	4	100	23
	4	200	33
	8	300	42
	8	400	53
Centralized	4	100	35
	4	200	45
	8	300	53
	8	400	64

These results are interesting since they provide a new enlightenment concerning the most promising dispatching strategy in terms of response time to online users. Indeed, in the absence of slow-down due to single PC pseudo-parallelism, the hybrid architecture takes profit of the processing distribution, without suffering from a too big number of exchanged messages. The distributed architecture comes in the second position in terms of performances, taking profit from the distribution but suffering from their too big bandwidth consumption. The centralized architecture comes in the last position, since its gain in terms of exchanged messages does not counterbalance its sequentialization of processing. Anyway, this architecture provides results that are practically equivalent to a centralized implementation. The small difference comes from the fact that vehicle agents are executed in different hosts than the planner agent, which result in a small additional cost in terms of communication.

5 Conclusion

In this paper, we have proposed a multi-agent system with three versions, focusing on clients dispatching strategies. The dispatching strategy decides which agents perform the computation to answer the clients requests. In the centralized strategy the planner agent performs most of the computation. In the decentralized strategy, the vehicle agents perform most of the computation in a collaborative way. Finally, in the hybrid strategy the work is split between clients and vehicles. We have compared these three approaches based on their response time to online users. We have considered two experiments configuration, a centralized configuration and a network configuration. The results have shown the superiority of the centralized approach in the first configuration and the superiority of the hybrid approach in the second configuration. In our future works, we will consider more dynamic problems, in which, not only clients are not known before execution, but also traffic conditions. To this end, we will integrate our vehicle routing system inside the multimodal traffic simulator SM4T [13].

References

1. Solomon, M.: Algorithms for the vehicle routing and scheduling with time window constraints. Oper. Res. **15**, 254–265 (1987)
2. Wooldridge, M., Jennings, N.R.: Intelligent agents: theory and practice. Knowl. Eng. Rev. **10**(2), 115–152 (1995)
3. Bessghaier, N., Zargayouna, M., Balbo, F.: Management of urban parking: an agent-based approach. In: International Conference on Artificial Intelligence: Methodology, Systems, and Applications, pp. 276–285. Springer, Heidelberg (2012)
4. Diana, M.: The importance of information flows temporal attributes for the efficient scheduling of dynamic demand responsive transport services. J. Adv. Transp. **40**(1), 23–46 (2006)
5. Thangiah, S.R., Shmygelska, O., Mennell, W.: An agent architecture for vehicle routing problems. In: Proceedings of the 2001 ACM Symposium on Applied Computing, SAC 2001, pp. 517–521. ACM Press, New York (2001)
6. Kohout, R., Erol, K.: In-Time agent-based vehicle routing with a stochastic improvement heuristic. In: Proceedings of the Sixteenth National Conference on Artificial Intelligence and the Eleventh Innovative Applications of Artificial Intelligence (AAAI 1999/IAAI 1999), pp. 864–869. AAAI Press, Menlo Park (1999)
7. Zeddini, B., Temani, M., Yassine, A., Ghedira, K.: An agent-oriented approach for the dynamic vehicle routing problem. In: IWAISE 2008, pp. 70–76. IEEE (2008)
8. Zargayouna, M., Balbo, F., Scemama, G.: A multi-agent approach for the dynamic VRPTW. In: ESAW 2008 (2008)
9. Zargayouna, M., Zeddini, B.: Fleet organization models for online vehicle routing problems. In: Transactions on Computational Collective Intelligence VII, pp. 82–102. Springer, Heidelberg (2012)
10. Grootenboers, F., de Weerdt, M., Zargayouna, M.: Impact of competition on quality of service in demand responsive transit. In: Dix, J., Witteveen, C. (eds.) MATES 2010. LNCS, vol. 6251, pp. 113–124. Springer, Heidelberg (2010)
11. Gendreau, M., Guertin, F., Potvin, J.Y., Taillard, E.D.: Parallel tabu search for real-time vehicle routing and dispatching. Transp. Sci. **33**(4), 381–390 (1999)
12. North, M.J., Howe, T.R., Collier, N.T., Vos, R.J.: The repast simphony runtime system. In: Agent 2005 Conference on Generative Social Processes, Models, and Mechanisms (2005)
13. Zargayouna, M., Zeddini, B., Scemama, G., Othman, A.: Simulating the impact of future internet on multimodal mobility. In: AICCSA 2014. IEEE Computer Society (2014)

Securing Mobile Agents, Stationary Agents and Places in Mobile Agents Systems

Donies Samet[1](✉), Farah Barika Ktata[2], and Khaled Ghedira[3]

[1] ENSI, Tunis, Tunisia
donies.samet@yahoo.com
[2] ISSAT, Sousse, Tunisia
farah.ktata@gmail.com
[3] ISG Tunis, Tunis, Tunisia
khaled.ghedira@isg.rnu.tn

Abstract. A challenging problem to solve to take full advantage of the benefits of mobile agent (MA) technology is to overcome security problems. In fact, in most cases, the security aspects are taken into account on the implementation stage of software development process, which may lead to the negligence of several aspects of security. However, including security properties in the design stage could lead to the development of a more secure MA system, especially that current MA platforms do not cover all security criteria. In this paper, we present a detailed modelling of security properties related to the security of mobile agents, stationary agents and places. The proposed model allows taking into account the security aspect earlier on the software development process and also it can be a first step to help researchers ameliorating security level on MA platforms.

Keywords: Mobile Agent · Security · MA-UML profile
Agent oriented software engineering

1 Introduction

Despite the fact that mobile agent technology presents several advantages like reduction of network traffic, reduction of communication costs and disconnected operations, security issues still act as the principal factor that hinder the expansion of the use of this technology. Generally, the definition of mobile agents' security requirements is considered in the implementation stage of the development process. However, the integration of security requirements during the design stage could help towards the development of more secure systems based on mobile agents [1]. This thought is reinforced, especially that current mobile agent platforms present a lack of establishment of security aspects and of mobility related permissions [2, 20]. In this context, we propose a new security model based on accurate security properties that allow the prevention of different categories of attacks including attacks against authentication, integrity, authenticity, confidentiality, access control, availability and non-repudiation. In this work we are interested in securing mobile agent, home place, stationary agent and execution place in mobile agents system. The defined security properties represent

G. Jezic et al. (Eds.): KES-AMSTA-18 2018, SIST 96, pp. 97–109, 2019.
https://doi.org/10.1007/978-3-319-92031-3_10

new extensions to MA-UML profile. In this paper the sections are organized as follows. In Sect. 2, we provide a detailed description of security issues in mobile agent systems. In Sect. 3, we discuss some related work. Then in Sect. 4, we propose the new model as an extension of MA-UML profile in order to ensure the security of mobile agents, stationary agents and places. In Sect. 5 we will describe the implementation of a medical transfer application based on mobile agents using the new proposed security model. Finally, Sect. 6 concludes the paper and offers directions for future work.

2 Security in Mobile Agent Systems

Security performance of mobile agent system is identified by its level of obedience to the security criteria. Security criteria in mobile agent systems are [2]: confidentiality, integrity, access control, availability, authentication, authenticity and non-repudiation. Based on the literature [3–5], there are three categories of attacks in the mobile agents based systems. Attacks of an Agent to another present the first category and it includes: attack against authentication (masquerade), attack against confidentiality (unauthorized access), attack against availability (denial of service DoS and distributed denial of service DDoS) and repudiation attack. The second category is Attacks of a Place on an Agent, it includes: attack against authentication (masquerade and cloning), attack against confidentiality: (eavesdropping, stealing and reverse engineering), attack against availability (denial of service, delay of service and refusal of transmission), attack against integrity (alteration, incorrect execution of the code) and repudiation attack. The third category is Attacks of a Place/Agent on a Place, it comprises: attack against authentication (masquerade), attack against availability (denial of service and distributed denial of service), attack against access control (unauthorized access) and repudiation attack.

3 Related Work

Recently, related work has been interested in integrating security concepts at the design stage of the software development process. The various approaches may be roughly grouped as formal approaches and semi-formal approaches. Formal approaches such as the works proposed in [6–8], the most common shortcomings of these approaches are that they intervene only on the implementation stage and do not address all the security criteria related to mobile agents system. Semi-formal approaches include three classes of contributions. The first class presents approaches such as SecureTropos [9] and Nemo [10] which provide agent oriented methodologies. The second class presents object oriented approaches which introduce extensions to UML, such as UMLSec [11] and SecureUML [12] which is based on the role based access control model RBAC [13]. The third class corresponds to extensions of UML which are agent oriented approaches. In this category, the works of [14–16] present several security aspects, but they do not treat concepts which are specific to the security of mobile agent systems. The approaches [5, 17], which present extensions to MA-UML profile [18], deal with the security aspects in mobile agents system. Authors in [5] proposed some security properties to ameliorate security of the place and they are limited to protecting it from malicious agent in the context of individual attacks. The MA-UML extension proposed by [17] presented security properties that protect only stationary agents during their communications with visitor mobile agents. Although the two works proposing MA-UML extensions present interesting contributions, the proposed properties are not precise enough to take into account certain scenarios of denial of service and cooperatives attacks that can be generated due to unsolicited traffic on or toward a target entity (agent, place, resource). In addition, these extensions do not introduce a detailed description of authentication properties and how the authentication mechanism is established. The extensions of MA-UML do not address the integrity, the confidentiality and the authenticity criteria. Also, the definition of permissions is required whenever an agent joins the system which is not ideal especially for open and large-scale systems.

4 Proposed Extensions to MA-UML Profile

In order to ameliorate the security level in mobile agents systems we propose to integrate new security properties that treat different types of individual and cooperative attacks. We propose to extend the MA-UML environment diagram, agent diagram and itinerary diagram to design secure places and secure agents in systems of mobile agents. The proposed security properties are based on role based access control RBAC by including role definition, agent role assignment and permissions. The fact that permissions are based on this combination of concepts allows the definition and application of an easy way to configure security policy for places and for agents. In the following sub-sections, we present the proposed new security properties.

4.1 Extensions of MA-UML Environment Diagram

Places in mobile agents systems must use security properties that allow establishing solutions against threats that jeopardize the performance of the place. In order to protect the place, the mobile agents, the stationary agents and resources against individual and cooperative attacks, we propose security properties that define permissions to entities (the mobile agents and the other places) according to the roles.

Extension 1: Preventing attacks executed by a place against the availability of another place. An attacker place or places may choose to send too much simultaneous attacker agents to flood the place and weakening its performance. Consequently, defensive security properties against this attack must be defined for monitoring and avoiding it. In this context, we propose to define new extensions to the MA-UML environment diagram, to avoid a cooperative attack executed by agents from the same origin against the availability of a place. We define a new stereotyped reflexive association-class, called «PlacePlacePermission» on the stereotyped class «Place» (Fig. 1). This association-class must contain a set of properties that sets the privileges assigned for each place which is a sender of agents regarding the receiver place. As properties that must contain *«PlacePlacePermission»*, we define:

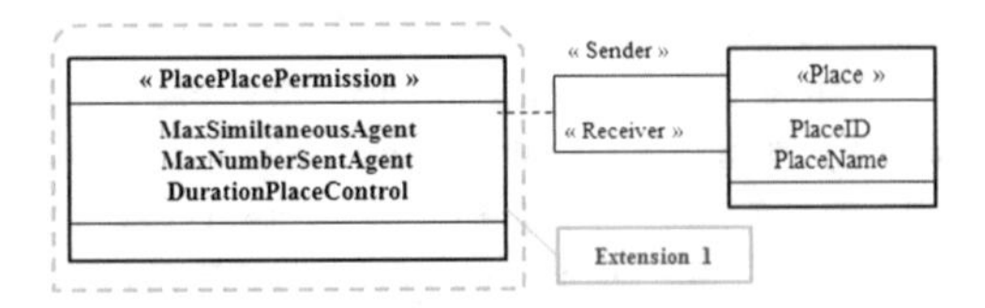

Fig. 1. Extension of the environment diagram to prevent the attack by malicious cooperative agents belonging to the same origin against the availability of a place

MaxSimiltaneousAgent: This property sets the maximum number of agents from a sender place that can exist simultaneously on a receiver place. MaxNumberSentAgent: This property sets for a sender place the maximum number of all sent agents toward a receiver place. DurationPlaceControl: This property sets the duration for which the MaxSimiltaneousAgent and MaxNumberSentAgent are defined.

Extension 2: Including RBAC model for permissions specification. To make the configuration of the security policy easier we rely on the role-based access control (RBAC) notation to integrate the notion of role in the definition of permissions. By doing so, we avoid the need to define permissions for each new agent joining the system. For this we define a new stereotyped class, called «Role» which is related by relationships of inheritance to the new stereotyped classes «MobileRole» and «StationaryRole» (Fig. 2). As properties of «Role», we define RoleID (sets the identity of the role) and RoleName (sets the role name). It is to be noted that the use of the notion of role is very useful to help set the values to be assigned to the security properties defined in the presented model in order to guarantee the good functioning of the system while maintaining the security; so that the values of the properties depend on the scenario of the application. They will be defined according to maximum values that an entity with a given role may need. Beyond these values, the behavior of the concerned entity is identified as a deviant behavior.

Extension 3: Preventing attacks executed by a mobile agent against the availability of a stationary agent and against the availability of a place during a communication. In a place, a stationary agent is exposed also to a denial of service attack in the case when a malicious mobile agent makes unsolicited communication toward it. Authors in [17] address this type of attack but the defined properties are not detailed enough to avoid all possible attacks in this category. For example, we can mention the cases when a malicious mobile agent sends a big number of messages, very long messages or successive messages. In order to prevent such attack we define a new stereotyped association-class, called «MobileStationaryComPermission» (Fig. 2). This association-class must contain a set of properties that sets the privileges assigned for each mobile agent when communicating with a stationary agent. As properties that must contain the association-class «MobileStationaryComPermission», we define: MSDurationMessagingControl: This property sets the duration for which the maximum number of messages sent by a mobile agent to a stationary agent is defined. MSMaxNumberMessages: This property sets the maximum number of messages allowed to be sent by a mobile agent to a stationary agent. MSMinDelayTwoMessages: This property sets the minimum delay allowed between two messages sent by a mobile

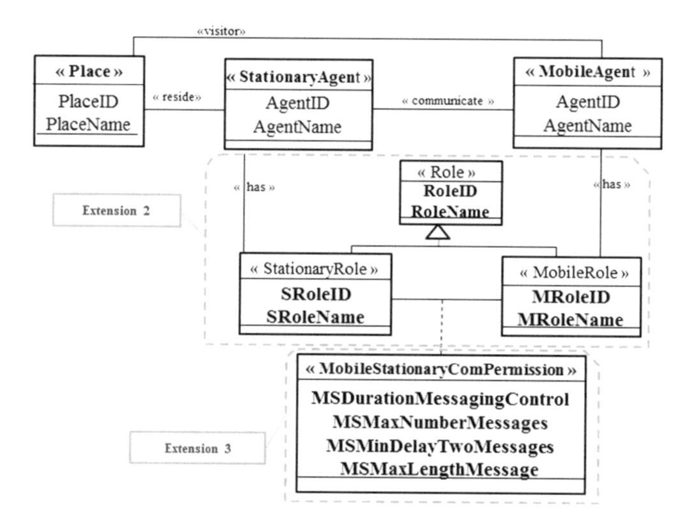

Fig. 2. Extension of environment diagram to prevent Dos attack caused by unsolicited communications towards a stationary agent

agent to a stationary agent. MSMaxLengthMessage: This property sets the maximum length allowed of a message sent by a mobile agent to a stationary agent.

Extension 4: Preventing attacks executed by a mobile agent against the availability of a place. In order to protect places, it is necessary to establish a set of privileges that must be respected by visitor mobile agent. Malicious mobile agents can cause denial of service attack by visiting a place successively or for a visit duration that is longer than required or known for a given role. For a sophisticated cloning attack it is not enough to just set the number of clone allowed during a period of time such as what is proposed in [5], but permissions regarding cloning must include properties that set the invocation number of cloning method, the number of clones that are allowed to exist simultaneously and the time to live for a clone. In this context, we define a new stereotyped association-class (Fig. 3), called «PlaceRolePermission». This association-class introduces the following security properties: MaxNumberCloningInvocation: This property sets the maximum number of cloning invocations allowed to a mobile role in a given place. DurationCloningControl: This property sets the duration for which the maximum number of cloning invocations «MaxNumberCloningInvocation» is defined. MaxSimultaneousClones: This property sets the maximum number of clones which can exist simultaneously in a place for a given mobile role.

Extension 5: Preventing attacks executed by a mobile agent against access control and availability of a Resource. Permissions regarding access to resources must be defined to avoid attacks executed by malicious mobile agents against access control and denial of service. Authors in [5] treat this attack but the defined properties cannot avoid sophisticated attacks on a resource which are executed by consuming repeatedly a resource in a short window of time. In this context and in order to avoid such attacks we extend the MA-UML environment diagram with the new stereotyped

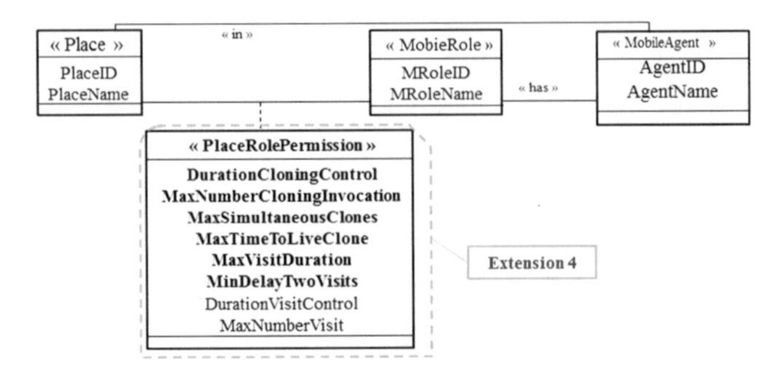

Fig. 3. Extension of the environment diagram to prevent attack of a mobile agent against the availability of a place

association-class, called «ResourceRolePermission» which introduces the following security properties (Fig. 4): MaxNumberofAccess: This property sets the maximum number of access defined for a mobile role. DurationAccessControl: This property sets the duration for which the maximum number of access «MaxNumberofAccess» is defined for a mobile role. MinDelayTwoResourceAccess: This property sets the minimum delay allowed between two resource accesses for a mobile role.

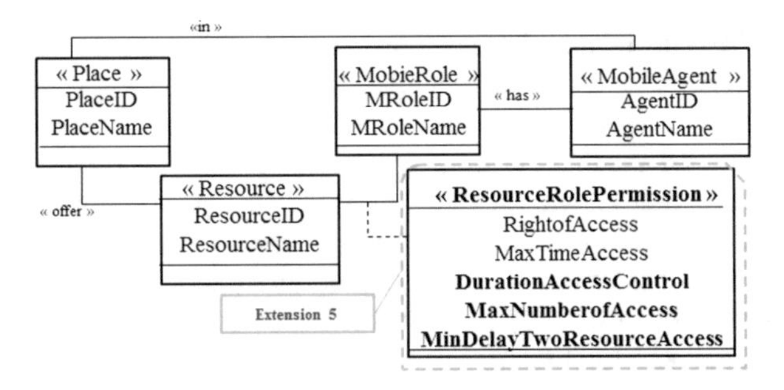

Fig. 4. Extension of the environment diagram to prevent attack of a mobile agent against access control and the availability of a resource

Extensions 6, 7 and 8: Preventing attacks against authentication, integrity, authenticity and confidentiality. Places in mobile agents systems must use efficient cryptographic mechanism to verify authentication, confidentiality, integrity and authenticity criteria. In this context we propose new extensions to MA-UML environment diagram that allow the use of keys in a PKI infrastructure to secure agent and places in terms of security criteria listed above. We adopt the use of PKI infrastructure as cryptographic mechanism because of its advantages, knowing that by using it there is no administrative burden to make a key distribution and consequently the place is less susceptible to attacks such as masquerade. The new extensions (Fig. 5) define necessary security properties to use an authentication based on the verification of the agent digital signature (signature using private key), this allow the receiver of an agent code, data or a message to know that only the sender could have encrypted it (verification using the public key). Also, the proposed security properties, allow ensuring the confidentiality (encryption using public key and decryption using private key) and integrity of agent code, messages and data which is not going to be modified after its initialization (the digital signature is a guarantee of the integrity of the signed information or agent: in case of any alteration attack the digital signature will not be valid). By verifying the integrity and the authentication we verify as a result the authenticity. Extension 6 presents the stereotyped class «KeyStore» which include security properties that define the public key and private key of a place. Extension 7 is the stereotyped class «TrustStore» which include security properties: ForeignPlaceID: this property defines the identities of trusted places for a place. ForeignPlacePublicKey: this property defines the corresponding public key for each trusted place. Extension 8 is HomePlaceSignature: this property keeps trace of the agent code signature by its home place. These properties allow the use of signature and cryptographic mechanisms on an agent code, messages and data; in order to permit its verification and then preventing attacks against confidentiality, authentication, integrity and authenticity.

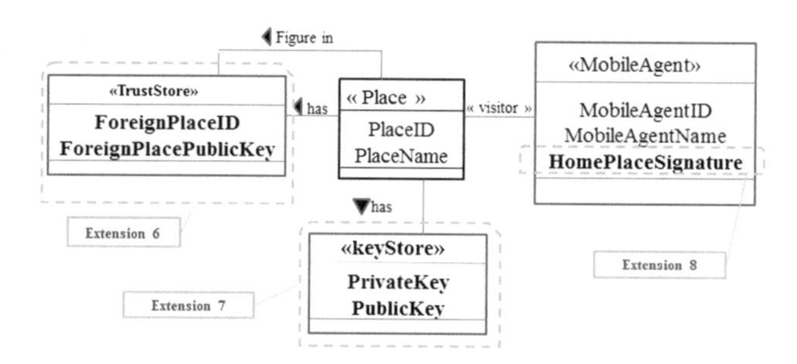

Fig. 5. Extension of the environment diagram to avoid attacks against authentication, integrity, authenticity and confidentiality

4.2 Extensions of MA-UML Agent Diagram

During the course of its itinerary a mobile agent may be subject to several attacks, hence the need to include in its internal structure and its characteristics security properties that set the permissions of the other entities of the mobile agent system towards the mobile agent.

Extension 9 and 10: Preventing attacks against the availability of the mobile agent. In order to protect the mobile agent, it is necessary to establish a set of permissions that must be respected during the communication with foreign agents either mobile or stationary, because they can be malicious and cause denial-of service by communicating excessively with the mobile agent. Thus, this can disrupt and degrade the performances of the mobile agent and it can also lead to harm its entire mission. These attacks can be executed cooperatively by agents from the same place. To avoid these attacks, it's necessary to control the communication properties toward mobile agents. In order to satisfy this constraint, we define two new stereotyped association-class, called «RoleMobCommunicationPermission» and «PlaceMobCommunicationPermission». «RoleMobCommunicationPermission» (extension 9) is a stereotyped association-class that contains a set of properties to sets the permissions assigned for each foreign agent having a given role when communicating with a mobile agent. The defined association-class has the following properties (Fig. 6): RMMaxTimeToWaitAnswear: This property sets the maximum time a mobile agent waits for an answer from a foreign agent having a given role. RMMaxNumberMessages: This property sets the maximum number of messages allowed to send by a foreign agent having a given role to the mobile agent. RMDurationMessagingControl: This property sets the duration for which the maximum number of messages sent by a foreign agent «RMMaxNumberMessages» is defined. RMMinDelayBetweenTwoMessages: This property sets minimum delay allowed between two messages allowed to send by a foreign agent having a given role to the mobile agent. RMMaxLengthReceivedMessage: This property sets the maximum length allowed of a message sent by a foreign agent having a given role to the mobile agent. «PlaceMobCommunication Permission» (extension 10) is a stereotyped association-class that contains a set of properties to set the permissions assigned for a place about communications with the agent. The defined association-class «PlaceMobCommunicationPermission» has the following properties (Fig. 6): PMMaxNumberReceivedMessages: This property sets the maximum number of messages allowed to send by the set of foreign agents which belong to the same origin place. PMDurationMessagingControl: This property sets the duration for which the maximum number of messages sent by the set of foreign agents «PMMaxNumberReceivedMessages» is defined.

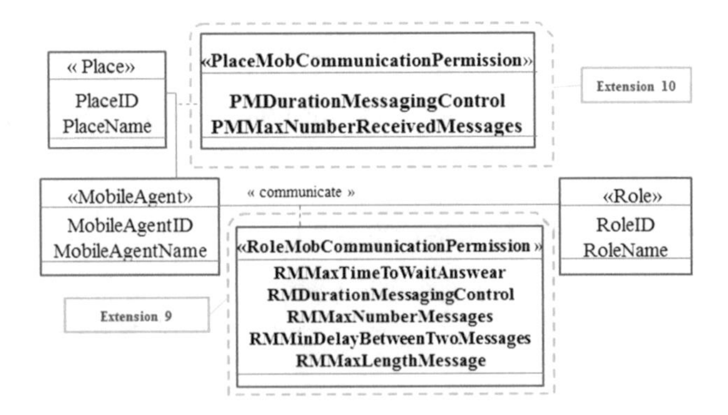

Fig. 6. Extension of agent diagram to prevent Dos and DDos attack caused by unsolicited communications towards a mobile agent

4.3 Extensions of MA-UML Itinerary Diagram

Extension11: Detecting attacks against the integrity of collected results and the non-repudiation of the results providers. Intermediate results provided by visited places to the mobile agent can be vulnerable to alteration attacks. In the other hand a malicious place can deny providing some results. So a substantial requirement here is to avoid these two attacks, knowing the importance of the intermediate results on the entire lifecycle of a mobile agent. In MA-UML mobile itinerary diagram, the itinerary of a mobile agent has a stereotyped class «Result» that describes the collected intermediate results. In order to avoid the above attacks, we propose to add to this class, the following properties (Fig. 7): IdProviderPlace, IdProviderAgent, TimeStampResult and ProviderPlaceSignature that represents the signature of the visited place which provides the result.

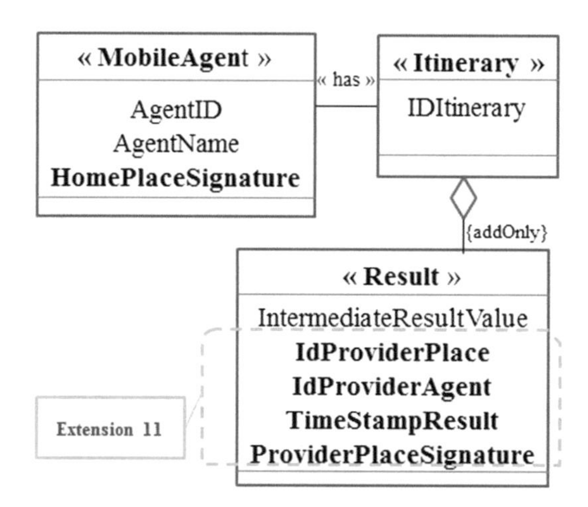

Fig. 7. Extension of the itinerary diagram to prevent alteration and repudiation attacks

5 Application Modeling and Implementation

5.1 Application Description

In order to illustrate the use and to validate the new defined extensions; we propose to apply these new extensions for modeling and implementing some scenarios of medical transfer application based on mobile agents. Our proposed medical transfer application based on mobile agents is composed of: Requestor Center: represents the place of a requestor that treats the case of a patient who needs a medical transfer because of medical conditions. Transfer Center: represents a medical center that is capable to provide the requested medical conditions. Patient Agent: is a mobile agent, which migrates from a place of requestor center to a transfer center over the network in order to search for the adequate response to the transfer request. Authentication Agent: is a stationary agent responsible for authenticating the visitor patient agent. Attending Agent: a stationary agent that represents a hospital center. A patient agent can communicate to it to be informed about the available medical resources on the hospital center. Resource Database: represent a resource which provided by the transfer center. The operation of the system can be summarized in the following points (Fig. 8): (1) The requestor fills out a form that contains the data needed for the required medical transfer and sending the Patient Agent to the hospital centers. (2) Upon receiving the patient, the authentication agent will execute the authentication process. (3) The Attending Agent will be created. (4) The Patient Agent presents all relevant information about the case of the patient and the required resources. (5) The Attending Agent treats the request and communicates the patient agent to find an accord. (6) In case of accord between Patient Agent and Attending agent, Patient Agent proceeds to manipulate resources database. (7) Patient Agent migrates to the next Transfer Center if needed; otherwise it returns to the requestor center with the results of the mission.

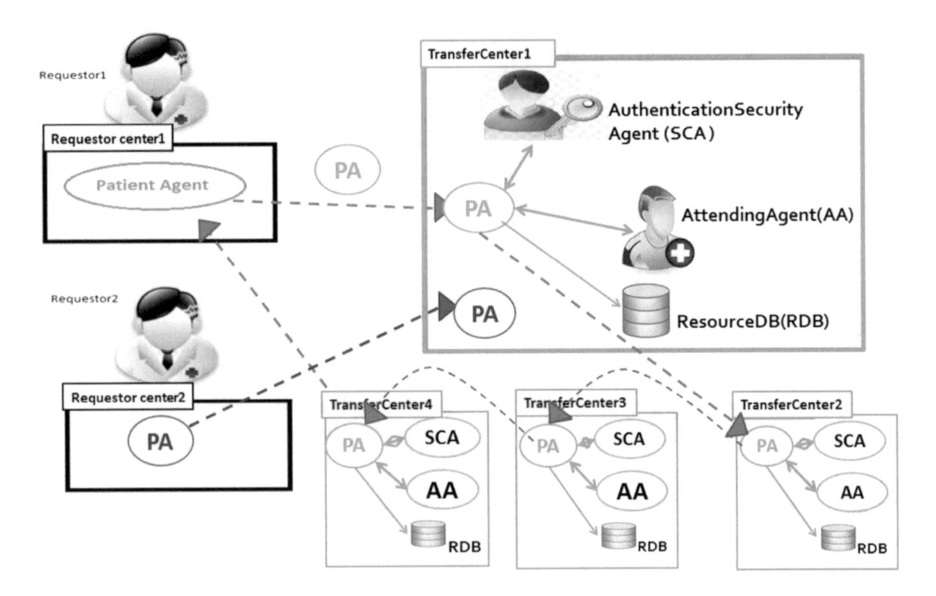

Fig. 8. The architecture of the medical transfer application

5.2 Application Modeling

We propose to model some scenarios of the medical transfer application. This modeling is based on the standard UML, MA-UML and the new proposed extensions. This modeling specifies the security aspect regarding mobile agents, stationary agents, places and resources. Figure 9 show a part of the extended environment diagram to model the secure medical transfer system based on mobile agents.

5.3 Application's Scenarios Implementation

For the implementation, we have used the Jade platform [19] and especially the Jade Security add-on [20] and jade PKI add-on [21]. This choice is not arbitrary; it is made based on the comparison study of mobile agent platforms established in [2]. We have implemented scenarios to show how the added properties allow the detection and inhibition of deviant and malicious behaviors in the medical transfer system. The first scenario (Fig. 10) is implemented to detect attacks against the integrity of collected results, it show the aim of the defined properties on the itinerary diagram: IdProviderPlace, IdProviderAgent, TimeStampResult and ProviderPlaceSignature which

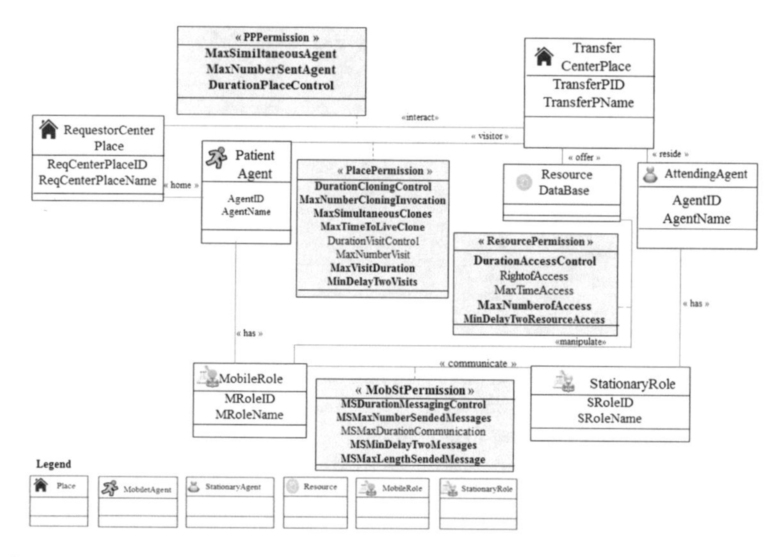

Fig. 9. A part of the extended environment diagram for a secure medical transfer system

Fig. 10. Detection of the modification of an intermediate result

ensure the integrity and non-repudiation criteria. As an example, we modify the provided result by the transfer center 1. When the patient agent comes to the next transfer center place, the authentication security agent consults the key store of the place and detects a modification of the result as the signature is not valid according to the public key of the place provider of the result. In the second scenario (Fig. 11) we use the property MaxSimiltaneousAgent to help preventing a distributed denial of service attack against a transfer center place1 by overwhelming it with agents. The medical transfer place1 prohibits the requestor place1 to have more than permitted agents which can exist simultaneously on the transfer place. In the third scenario (Fig. 11) we use the property MaxNumberofAccess to help preventing a denial of service attack against a transfer center place by repeatedly access to the resource database. The medical transfer place1 prohibits a visitor agent which has the role of patient agent to access more than permitted to the resource database. In the last scenario (Fig. 11) we use the property MaxVisitDuration to sets the maximum visit duration allowed for a mobile patient agent. A mobile agent which has the role of patient agent and visits the medical transfer center 1 is forbidden to stay for a period longer than permitted.

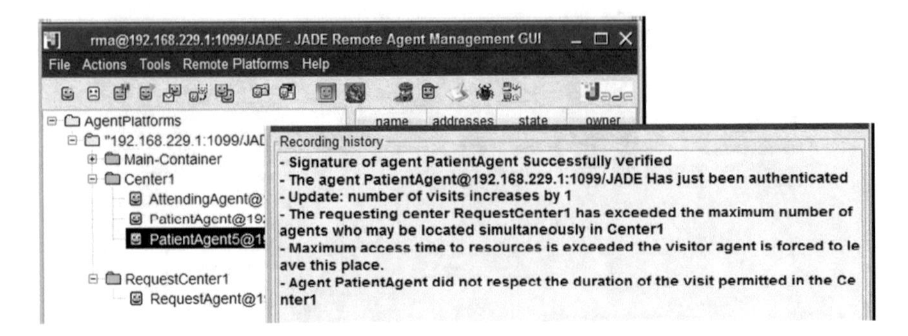

Fig. 11. Prohibition of the exceed of the maximum allowed number of agents belonging to the same requestor center place, the prohibition of the exceed of allowed resources access and the prohibition of a visit that it is longer than permitted

6 Conclusion

In this paper, we are interested in the design of security aspects for mobile agents, stationary agents, places and resources. The new proposed model helps the insurance of security criteria: authentication, integrity, authenticity, availability, access control, confidentiality and non-repudiation. The definition of permissions is easier by means of the use of role notion. The proposed extensions allow the establishment of a security policy in which values of the security properties are assigned according to the application scenario. This work opens several perspectives that we are working on it such as the integration of the proposed model for the amelioration of security level on mobile agent platforms by adopting it for the establishment of a complete security infrastructure. Finally, other extensions can be defined to MA-UML in order to take into account more detailed access control parameters and other extensions can be added for the prevention and detection of other possible sophisticated cooperatives attacks.

References

1. Mouratidis, H., Giorgini, P., Manson, G.: Modelling secure multiagent systems. In: AAMAS 2003 Proceedings of the Second International Joint Conference on Autonomous Agents and Multiagent Systems, Melbourne, Australia, 14–18 July 2003
2. Samet, D., Ktata, F.B., Ghedira, K.: Security and trust on mobile agent platforms: a survey. In: KES International Symposium on Agent and Multi-Agent Systems: Technologies and Applications, Vilamoura, Portugal (2017)
3. Jansen, W., Karygiannis, T.: Mobile Agent Security. NiST Special Publication 800-19 (2000)
4. Alfalayleh, M., Brankovic, L.: An overview of security issues and techniques in mobile agents. In: Conference on Communications and Multimedia Security. University of Newcastle, Australia (2004)
5. Hachicha, H., Samet, D., Ghedira, K.: A conceptual approach to place security in systems of mobile agents. In: 13th German Conference, MATES 2015, Cottbus, Germany, 28–30 September 2015

6. Loulou, M., Jmaiel, M., Hadj Kacem, A., Mosbah, M.: A conceptual model for secure mobile agent systems. In: International Conference on Computational Intelligence and Security, Guangzhou, China (2006)
7. Ma, L., Tsai, J.: Formal modeling and analysis of a secure mobile-agent system. IEEE Trans. Syst. Man Cybern. Part A Syst. Hum. **38**(1), 180–196 (2008)
8. Rekik, M., Kallel, S., Loulou, M., Hadj Kacem, A.: Modeling secure mobile agent systems. In: KES International Symposium on Agent and Multi-Agent Systems: Technologies and Applications, Dubrovnik, Croatia (2012)
9. Mouratidis, H., Giorgini, P.: Secure tropos: a security-oriented extension of the tropos methodology. Int. J. Softw. Eng. Knowl. Eng. **17**(2), 285–309 (2007)
10. Huget, M.-P.: Nemo: an agent-oriented software engineering methodology. In: OOPSLA Workshop on Agent Oriented Methodologies, Seattle (2002)
11. Jürjens, J.: UMLsec: extending UML for secure systems development. In: International Conference on the Unified Modeling Language, Dresden, Germany (2002)
12. Lodderstedt, T., Basin, D., Doser, J.: SecureUML: a UML-based modeling language for model-driven security. In: International Conference on the Unified Modeling Language, Dresden, Germany (2002)
13. Sandhu, R.S., Coyne, E.J., Feinstein, H.L., Youman, C.E.: Role-based access control models. IEEE J. Mag. **29**(2), 38–47 (1996)
14. Xiao, L., Peet, A., Lewis, P., Dashmapatra, S., Saez, C., Croitoru, M., Vicente, J., Gonzalez-Velez, H., Ariet, M.L.I.: An adaptive security model for multi-agent systems and application to a clinical trials environment. In: IEEE Annual International Computer Software and Applications Conference Volume II, pp. 261–268. IEEE Computer Society (2007)
15. Beydoun, G., Low, G.C., Mouratidis, H., Henderson-Sellers, B.: Modelling MAS_specific security features. In: International Conference on Advanced Information Systems Engineering, Trondheim, Norway (2007)
16. Beydoun, G., Low, G.C., Mouratidis, H., Henderson-Sellers, B.: A security-aware metamodel for multi-agent systems (MAS). Inf. Softw. Technol. **51**(5), 832–845 (2009)
17. Zrari, C., Hachicha, H., Ghedira, K.: Agent's security during communication in mobile agents system. In: KES 2015, Singapore (2015)
18. Hachicha, H., Loukil, A., Ghedira, K.: MA-UML: a conceptual approach for mobile agents' modelling. Int. J. Agent-Oriented Softw. Eng. **3**(2/3), 277–305 (2009)
19. Bellifemine, F., Caire, G., Poggi, A., Rimassa, G.: JADE: a software framework for developing multi-agent applications. Lessons learned. Inf. Softw. Technol. **50**(1-2), 10–21 (2008)
20. Jade Security Guide (2005). http://jade.cselt.it/doc/tutorials/JADE_Security.pdf. Accessed 05 Dec 2017
21. http://jade.tilab.com/doc/tutorials/PKI_Guide.pdf. Accessed 05 Dec 2017

How Research Achievements Can Influence Delivering of a Course - Siebog Agent Middleware

Milan Vidaković[1(✉)], Mirjana Ivanović[2], Dejan Stantić[3], and Jovana Vidaković[2]

[1] Department of Computing and Control, Faculty of Technical Sciences, University of Novi Sad, Novi Sad, Serbia
minja@uns.ac.rs

[2] Department of Mathematics and Informatics, Faculty of Sciences, University of Novi Sad, Novi Sad, Serbia
{mira, jovana}@dmi.uns.ac.rs

[3] School of Information and Communication Technology, Griffith University, Gold Coast, QLD, Australia
d.stantic@griffith.edu.au

Abstract. Computer science is one of the fastest changing area and we constantly witness novelties and therefore it is extremely important to also introduce new approaches for computer science education and introduce significant change into curricula. This paper presents the experience in using agent middleware, which has been initially developed as part of the research project in agent oriented graduate-level course. We present the course structure, the tools, teachers' and students' experiences with regard to the course delivery. Experiences with the introduction of the teamwork in the course, and characteristics particularly related to the specific methodology have been elaborated.

Keywords: Agent middleware · Teaching · Courses

1 Introduction

Computer science education is under constant reevaluation and development. This is especially important if we look at the opinions presented in [1]. That is why it is extremely important to impose and suggest different approaches for revitalizing computer science education and introduce significant change to the Information Communication Technology (ICT) curricula. According to the rapid development of ICT and new trends in software development, we are facing a world of computing where the majority of software is distributed. So there is an obvious need to innovate and change educational practice. Until recently undergraduate computer science and informatics curricula only had one course in Artificial Intelligence (AI) where general topics and themes from AI have been covered. Rarely such courses presented distributed systems and agent technology in more details. However, it is evident that the field of "intelligent agents and multi-agent systems" sufficiently matured and it is time to face the challenge and introduce those topics to undergraduate students as well. This paper elaborates on

G. Jezic et al. (Eds.): KES-AMSTA-18 2018, SIST 96, pp. 110–120, 2019.
https://doi.org/10.1007/978-3-319-92031-3_11

some challenges and initial experience of the delivery of elective courses on agents and multi-agent systems, at the University of Novi Sad (UNS), Serbia.

Unfortunately, as it is probably the case at other universities, we faced a lot of obstacles in our ambition to introduce such innovative and demanding courses. First of all, there is a great discrepancy between our students' previous knowledge and ambitions. The majority of students lack motivation for studying and gaining a higher level of knowledge and skills in any course of the curricula. In fact, they want to finish their studies with minimal efforts, so generally, they try to avoid demanding courses. On the other hand, a minority of students have shown to be highly motivated and prefer to be challenged with new, interesting and innovative courses. While introducing multi-agent systems (MAS) to undergraduate students, we have also faced some general obstacles such as the lack of a choice of appropriate textbooks and teaching materials as well as the lack of a common set of software tools to support key concepts [15]. It is obvious that a lot of existing textbooks on agents and MASs provide a foundational contribution to the field [2], however, different courses consider different and usually a unique set of requirements.

This course we introduced is an elective in the program and motivations to enroll in it are very diverse. Some of the students will choose this course with the intention to learn a lot of important agent concepts in depth while others prefer to be introduced with very basic concepts of agent technology and pass the course with limited effort. So we came to the conclusion that we could not rely on any of the available textbooks, but to write a detailed set of class notes on key concepts using different sources [15]. Having the need for appropriate software tool for the practical part of the course, we also concluded that use of the in-house developed system would obtain easier maintenance of students practical and laboratory activities. So we have decided to use the Siebog, a multi-agent middleware developed at UNS University [3, 4].

The rest of paper is organized as follows. Section 2 covers the related work. Section 3 presents detailed structure and examples of exercises of the Agent Technology course. Section 4 describes lessons learned by both students and teachers while Sect. 5 concludes the paper with some closing remarks.

2 Related Work

The rapid development of ICT and other supportive technologies in the last decade influenced the appearance of different approaches for revitalizing computer science education. One of the important trends and areas in software development is oriented towards distributed, pervasive networks and Internet of things where agents and agent technologies play an important role. The wide range of heterogeneous software and hardware platforms exists, different computing devices are interconnected, and they exchange information via heterogeneous network communication channels. So far future jobs and engagements within ICT companies, students have to be informed to learn about new and challenging environments and implementations of software, like ubiquitous and pervasive computing, mobile computing, sensor networks, high-performance computing, cloud computing and the Internet of things. Therefore, the rigorous design, integration, and harmonization of various topics of distributed systems and agent technologies into Computer Science and Informatics curricula presents a quasi-permanent challenge taking into account the various constraints of time, resources, effort, and expertise of educators and students [5].

Also, an interesting approach for revitalizing introductory undergraduate or high school computer science curricula was through the deep integration of agent-based modeling (ABM) and multi-agent systems (MAS) perspectives is presented in [6]. Authors intensively used their own system i.e. MAgICS (Multi-Agent Introduction to Computer Science) framework. By introducing different topics (as searching and sorting, machine learning, networks and security) with a focus on parallel, distributed, and stochastic methods, they can make traditionally upper-level topics both motivating and accessible to introductory-level students. The primary learning goal of the course was that students would be able to use decentralized thinking and to understand the trade-offs between decentralized and centralized approaches. Another being an advanced goal in that students are able to consider issues of parallelism and distribution also from programming, but also from the conceptual design of systems point of view.

Extensive experiences of teaching several undergraduate courses on agents and multi-agent frameworks, over the three years at two different universities are presented in [3]. Authors paid attention to three key issues.

The first intention of the course was to support students' intuitive understanding of essential concepts of agent technology and multi-agent systems. Using "science fiction materials" and appropriate games, they offered to students some kind of excitement and additional motivation for studying multi-agent systems. The second important thing is the selection of right material, so authors proposed several criteria that have been useful for selecting such material. The third aspect of this approach is that teachers are oriented to the philosophical, ethical and social issues connected to agents and multi-agent systems. An important part of the paper is devoted to lessons learned and feedback from students in order to try to improve the course.

In different papers that present experiences in delivering agent-oriented courses, teachers have reported a variety of environments they use in order to increase active learning. In paper [7] the authors suggest that NetLogo presents an excellent platform for the teaching purposes. This platform possess number of interesting features for educational purposes: a simple, expressive programming language with a small learning curve, rapid GUI creation and custom visualizations, easiness of modeling of complex environments and agents, etc. In the first several years of course delivery the students reported that they enjoyed the course in spite that it was too theoretical. They expressed interest to acquire hands-on experience. It is widely recognized that some practical program development within agent and multi-agent courses is needed.

Our intention was to prepare a smooth introduction in agent technology and MASs using as much as possible our previously developed teaching resources and self-developed software tools for creation MASs [4, 5, 15]. The main advantage of our approach is that we use in-house agent middleware for teaching. So we are able to offer deep knowledge of our system to the students. Also, it is possible to make corrections and extend the system, because we do not depend on external programmers. Students perform different tasks with the system and use it to implement simple MAS and during programming and testing their solutions they test on the framework.

In the future, we plan to introduce another kind of task that will be connected with the implementation of new components and functionalities of the framework. In this way, we will be able to constantly innovate, develop and extend our system. The system is continuously being developed and it is placed on the public Git repository [3] and it is easily accessible for students but also for other parties interested in it.

3 Ease of Use

With respect to the common set of conceptual tools, the basic idea within the course is to motivate the students, but more importantly, it is to help bridge the gap in students' understanding of multi-agent systems. The course on Agent Technologies is held during the eighth semester at the Faculty of Technical Sciences, UNS but using some experiences from [4]. The course appeared as a sequel of other artificial intelligence related courses, like "Artificial Intelligence" and "Business Intelligence" which give students essential knowledge on characteristic AI issues such as genetic algorithms, fuzzy rule-based systems, planning, knowledge-based systems, and machine learning. By completing this course, students are awarded 4 ECTS points.

The course is designed to give an introduction to software agents, agent frameworks (middleware), protocols, languages and technologies needed to program and work with agents. The course prerequisites define that students must be familiar with the theory and practice of Computer Programming (using the Java and JavaScript programming languages), Operating Systems and Computer Networks.

The course structure includes lectures, lab exercises, and project implementation. The course takes 14 weeks, 3 h/week lectures and laboratory work 3 h/week. 10 to 15 students take this course every year. The grading of the course is based on the final exam and lab assignments. The final exam counts 70% of the final mark and comprises 30% based on a questionnaire for assessing gained theoretical knowledge and 40% based on the project completed which involves the design of a small-scale distributed agent middleware. The lab grading counts 30% of the final mark and it is based on a set of smaller lab assignments distributed over the semester.

During the semester, as a part of the lab activity, the students are also introduced to a number of tutorials of necessary frameworks, libraries, and application servers required for carrying out their lab/project assignments. The aims of our course are to:

- Make an introduction to the concept of intelligent agents and provide an introductory study of the various types of agents, communication protocols, and supporting architecture for their operation.
- Provide an introduction to the multi-agent systems and the various issues involved in agent lifecycle management, communication, and interaction.
- Discuss possible applications of the agent technology in various fields (knowledge management, data harvesting, heavy computing, etc.).
- Discuss the advantages of the agent-based approach for solving complex engineering problems.

3.1 Role of the Siebog Agent Middleware

Students are introduced to the agent technology using the Siebog agent middleware [3] in both lectures and lab exercises. Siebog is the in-house multiagent middleware built by integrating two, also in-house, developed components. XJAF [8] and Radigost [9] are middlewares developed as separate components. They are integrated into a single framework Siebog, on Fig. 1. That way, the Siebog middleware provides support for both server-side and client-side agents. Server-side agents are supported by the XJAF-part of the Siebog framework and are written in Java programming language.

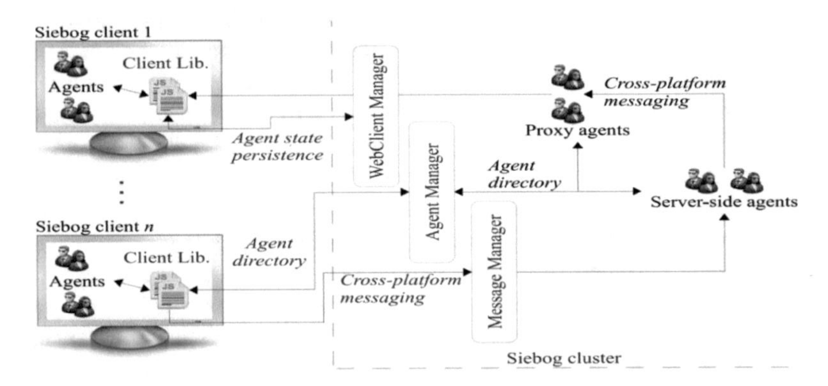

Fig. 1. Siebog architecture

Agents written for the Siebog framework can be executed in the distributed, clustered environment, having the ability to be load-balanced and safely transferred to another node if their node in the cluster fails. Load-balancing and safe fail-over have been implemented using existing industrial-ready solutions, instead of programming the proprietary solutions. Results published in [4, 10] proved that use of proven technologies instead of developing proprietary was the right way.

Client-side agents (originated from the Radigost) are written in JavaScript and can migrate to the server-side, if necessary. While being deployed as client-side agents, they exist as JavaScript objects, loaded within the web page. So, they can execute in a wide range of devices as desktops, laptops, tablets, smartphones, smart TVs, etc. Since the state of a client-side agent can be destroyed when the page is leaved (or browser closed), it was necessary to develop a way of persisting the state of an agent to the server, so it could continue with the execution when the page becomes reloaded. Persisting agent state is done transparently from the agent's point of view. The other key feature of the client-side agents is their ability to communicate to both client and server-side agents (by employing the Web Workers and Sockets technologies).

3.2 Organization of the Course

Theoretical lectures and practical exercises are tightly interconnected and are organized in a manner of step by step increasing complexity. We start with the basic elements of agenthood and simple practical tasks and solutions. After that, we introduce more complex concepts and illustrate them with appropriate examples. Practical exercises are adjusted to the topics presented during the theoretical part of the course. They follow the path defined by the lectures and cover increasingly complex problems. All the teaching materials (for lectures and exercises) are available to students, as well as the software, which is open-source and free. Materials are prepared and produced based on our experience in the field and on some available papers from the past 15 years.

In the beginning, during the lectures, students learn about the fundamentals of the agent technology, organization of agent code, agent lifecycle, communication and distribution of the code. FIPA protocols and FIPA ACL [11] are also covered. Students are expanding their technical knowledge with all the necessary skills for the agent middleware development, including JavaEE, aspect-oriented programming, and advanced JavaScript. All the examples in lectures and lab are demonstrated using our in-house developed framework [3]. Lab exercises begin with illustration of the simple communication between two agents, on a single computer. Students are trained to create agents, deploy them on the Siebog, and run them. During the lab exercises, students are divided into small teams (3 or 4 members). Teams create agents capable of communicating with several different types of agents on the same computer, employ FIPA communication protocols, and finally perform a communication between multiple agents on several computers, realized as the FIPA Contract Net protocol [12]. Listing 1 shows the essential parts of the Initiator Agent code in the Contract Net implementation on the Siebog platform, which students need to create.

```
@Stateful // agents are stateful or stateless EJB beans
@Remote(Agent.class)
/**We have developed an abstract class for the ContractNet
 * initiator participant. Developer just needs to extend this. */
public class InitiatorExample extends Initiator {
  ...
  /**
   * Iterates over a list of proposals and returns the best one.
   * This function is used to implement the criteria for choosing the
   * best proposal. @return The best proposal. */
  @Override
  public Proposal getOptimalProposal(List<Proposal> proposals) {
    int bestValue = 10000;
    Proposal bestProposal = null;
    // Iterate over the list of proposals
    for (Proposal p : proposals) {
      if (Integer.parseInt(p.getContent()) < bestValue) {
        bestProposal = p;
        bestValue = Integer.parseInt(p.getContent()); } }
    return bestProposal; }
  /**
   * Used to create a Call for Proposal. @return CfP, containing:
   * AID of the Initiator, timeout for the reply, and finishing time for
   * the job. */
  @Override
  public CallForProposal createCfp() {
    CallForProposal cfp = new CallForProposal(myAid);
    cfp.setReplyBy(System.currentTimeMillis() + 10 * 1000);
    needsToBeFinished = System.currentTimeMillis() + 30 * 1000;
    return cfp; }
... }
```

Listing 1. Contract Net Initiator Agent code part

Students develop the agent code in the Eclipse IDE [13], and deploy the solution into the JBoss application server. Siebog middleware has the built-in web console for the agent management. This console also dumps all the logged messages, as shown in the Fig. 2. This console greatly improves debugging and analysis of the code since it gives students the interactive way of monitoring the Siebog operations.

After the first part of the course is finished, students are further introduced to the clustering of application servers, and realization of agents in clustered environments. The Siebog is also used for these exercises as it supports/allows such functionalities.

The next step is to introduce technologies needed for the realization of agent middlewares. During lectures, students complete their knowledge of JavaEE, specifically: Enterprise Java Beans – EJB, Java Naming and Directory Interface – JNDI, Java

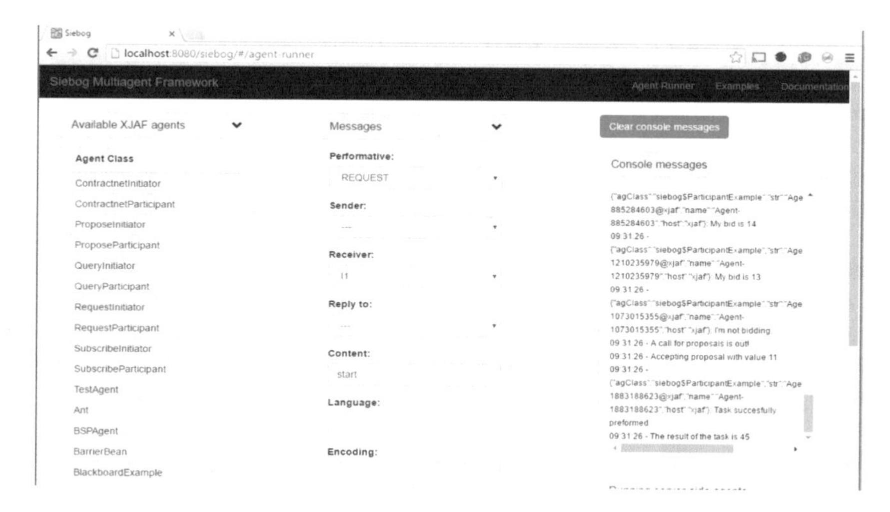

Fig. 2. Siebog web console for the contract net example

Messaging System – JMS, Message Driven Beans – MDB, JAX-RS specification for the REST services, and Web Sockets. Students also complete their knowledge about reflection, class loading, security managers, and advanced JavaScript.

During the fourth part of the course, students are building their own agent middleware in the lab, having the Siebog as a model. They have to implement only a subset of all Siebog features and are required to demonstrate agent functionality by implementing chosen FIPA agent communication protocol on a real-life problem. One of the examples is devoted to the implementation of trading agents and stock exchange. The key feature of the student-built middleware is the interoperability. They must implement their middleware following the same API. That way, no matter how the students implement an agent middleware, all instances are able to communicate to each other. This means that agents developed by different teams, and deployed on different platforms can communicate to each other without some additional requirements or implementation efforts. The use of the REST services provides this interoperability.

By the end of the course we expect a student to acquire some agent programming skills, gain appropriate knowledge and achieve general course learning outcomes:

- Understand the concept of agents and agent middleware, the agent lifecycle, message exchange, mobility and other related concepts.
- Understand the diversity between specificities of agents and other software entities.
- Be able to design and implement agent-based solutions.
- Be able to solve problems being part of the team.
- Critically analyze the expected benefits of using agent technology.

4 Feedback and Lessons Learned

According to the University quality assurance policy, students are required to fill the questionnaires assessing all courses they have had during the particular semester. Students' feedback on the Agent Technologies course has been encouraging. Regarding our course, students were satisfied to be able to learn advanced JavaEE and JavaScript technologies and they also liked the opportunity to work with distributed systems.

Additionally, students were very satisfied with the use of the Siebog agent middleware, because they were able to get instant help regarding any problem that appears during the lab exercise since the Siebog system is being developed at our university, with the professor and assistant deeply understand implementation, functionalities and functioning of the system. Both professor and assistant were able to solve all the problems immediately, instead of searching on the Internet, or solutions on different forums.

As a result of this support, during the 2014/2015 school year, out of 12 students attending this course, 4 proceeded with the work in the field of agent technology by taking upon (and finishing on time) a bachelor thesis related to agent frameworks. The same trend continued at the 2015/2016 year: 5 students out of 20 students, advanced to doing a bachelor thesis related to agent frameworks. During the 2016/2017 year, 6 students out of 29 students, took bachelor thesis work related to agent frameworks.

Another important methodological approach that we have adopted in our agent course was the organization of practical activities as team work. Similar to the real-life assignments, usually in IT companies, students were organized in groups and all teams had to implement the same API given by the teachers.

That way, all the teams had to produce interoperable code, which would be compatible with the solution from any other team. Another advantage of having team work is in giving more complex assignments to the students. Both students and teachers were satisfied with the results achieved from the performing tasks within the teams since all the solutions were produced on time and the code was interoperable. There is a potential drawback of team work for performing practical tasks and projects within different courses. In this approach, it is quite usual that not all students contribute equally to solving joint tasks and problems. That is the general problem in having teams and the solution can be having good project management, which is in our course equally done by the teachers and students. Student polls indicate that they are reasonably satisfied with the load balancing between team members.

Finally, very important issue necessary to be considered is to try to raise the number of students that will attend this course. During the 2015/16 school year, we have had more students than in previous years (20 of them). Furthermore, during the 2016/17 school year, 29 students attended our course. This is valuable information as it would seem that more and more students are becoming aware of the necessity and application of agent technology. They realized that such a course would give them the opportunity to improve on distributed programming skills necessary for their future careers.

5 Conclusion

This paper elaborates on our experience and insights into the Agent Technologies course which has been offered for the past several years at the Faculty of Technical Sciences, UNS, Serbia inspired by [4] and subsequently improved [5, 15]. The course is attended by the students who already have some basic knowledge in Object-Oriented and Network programming. Students have had classes and exercises performed using the Siebog agent middleware which is being actively developed at the UNS.

Our experience showed that the in-house solution used for teaching such advanced courses is very welcomed by students since they can obtain instant help regarding problems, or understanding complex concepts. The team work provided students with the possibility to develop distributed applications the way it is usually done in IT companies, which prepares them for the real-life working situations in their future jobs.

Based on previous experiences in delivering this course we are aware of the fact that there are still a wide range of opportunities for the expansions and introduction of innovations in the course structure and organization. We plan to use ALAS agent programming language [14], which is also being developed by the Siebog team. Introduction of such in-house developed agent programming language will allow students to focus more on the agent-based problem solving, especially in the clustered environment.

Generally speaking, some of the topics, either the same or similar to these presented in our Agent Technology course, could be included in specific courses within different undergraduate computer science and ICT curricula. For example, typical courses in which agent technologies could be introduced are courses on Artificial Intelligence, Distributed Computing [16], or Distributed Network Application Development [5].

References

1. Rashid, R.: Image Crisis: Inspiring a new generation of computer scientists. Commun. ACM **51**(7), 33–34 (2008)
2. Bowring, E., Tambe, M.: Introducing multiagent systems to undergraduates through games and chocolate. In: Beer, M., Fasli, M., Richards, D. (eds.) Multi-Agent Systems for Education and Interactive Entertainment: Design, Use and Experience, pp. 101–114. Hershey, Information Science Reference (2011)
3. Siebog Home. https://github.com/milanvidakovic/siebog. Accessed 13 Nov 2017
4. Mitrović, D., Ivanović, M., Vidaković, M., Budimac, Z.: A scalable distributed architecture for web-based software agents, In: Proceedings of the 7th International Conference on Computational Collective Intelligence (ICCCI), Madrid, Spain. LNAI, vol. 9329, pp. 67–76. Springer (2015)
5. Bădică, C., Ilie, S., Ivanović, M., Mitrović, D.: Role of agent middleware in teaching distributed network application development. In: Proceedings of the 8th International Conference KES-AMSTA 2014 Chania, Greece, Agent and Multi-Agent Systems: Technologies and Applications. AISC, vol. 296. Pp. 267–276. Springer International Publishing (2014)

6. Stonedahl, F., Wilkerson-Jerde, M., Wilensky, U.: MAgICS: toward a multi-agent introduction to computer science. In: Beer, M., Fasli, M., Richards, D. (eds.) Multi-Agent Systems for Education and Interactive Entertainment: Design, Use and Experience, pp. 1–25. Information Science Reference, Hershey (2011)
7. Sakellariou, I., Kefalas, P., Stamatopoulou, I.: An intelligent agents and multi-agent systems course involving NetLogo. In: Beer, M., Fasli, M., Richards, D. (eds.) Multi-Agent Systems for Education and Interactive Entertainment: Design, Use and Experience, pp. 26–50. Information Science Reference, Hershey (2011)
8. Vidaković, M., Ivanović, M., Mitrović, D., Budimac, Z.: Extensible Java EE-based agent framework – past, present, future. In: Ganzha, M., Jain, L.C. (eds.) Multiagent Systems and Applications. ISR Library, vol. 45, pp. 55–88. Springer (2013)
9. Mitrović, D., Ivanović, M., Budimac, Z., Vidaković, M.: Radigost: Interoperable web-based multi-agent platform. J. Syst. Softw. **90**, 167–178 (2014)
10. Mitrović, D., Ivanović, M., Vidaković, M., Budimac, Z.: Extensible Java EE-based agent framework in clustered environments. LNCS, vol. 8732, pp. 202–215. Springer (2014)
11. FIPA ACL Message Structure Specification. http://www.fipa.org/specs/fipa00061/SC00061G.html. Accessed 13 Nov 2017
12. Contract Net implementation on Siebog. https://github.com/milanvidakovic/siebog/tree/master/siebog/src/siebog/interaction/contractnet. Accessed 13 Nov 2017
13. Eclipse homepage. http://www.eclipse.org. Accessed 13 Nov 2017
14. Mitrović, D., Ivanović, M., Budimac, Z., Vidaković, M.: Supporting heterogeneous agent mobility with ALAS. ComSIS J. **9**(3), 1203–1229 (2012)
15. Ivanović, M., Putnik, Z., Mitrovic, D., Stantic, B.: Possible routes on a highway of elearning - promising architecture for elearning systems. In: KES-AMSTA, pp. 289–302 (2014)
16. Computer Science Department, Griffith University. https://www2.griffith.edu.au/study/engineering-it/computer-science. Accessed 13 Nov 2017

Agent Communication and Social Networks

Sending Messages in Social Networks

Matteo Cristani[1](✉), Francesco Olivieri[2], Claudio Tomazzoli[1], and Guido Governatori[2]

[1] Dipartimento di Informatica, Università di Verona, Strada Le Grazie 15, Verona, Italy
{matteo.cristani,claudio.tomazzoli}@univr.it

[2] Data61, 70-72 Bowen Street, Spring Hill, Brisbane, QLD 4000, Australia
{francesco.olivieri,guido.governatori}@data61.csiro.au

Abstract. Since the birth of digital social networks, management research focused upon the opportunities of social media marketing. A marketing campaign has the best success when it reaches the largest number of potential customers. It is, however, difficult to forecast in a precise way the number of contacts that you can reach with such an initiative.

We propose a representation of social networks that captures both the probability of forecasting a message to different agents, and the time span during which the message is sent out.

We study reachiability and coverage from the computational complexity viewpoint and show that they can be solved polynomially on deterministic machines.

1 Introduction

Web-based social networking services make it possible to connect a vast pool of heterogeneous people in several cases many millions. *Gift economy* of information can be seen as one the factor which most contributed to the success of social networks. According to recent Facebook surveys, more than 60% of the users of the most common social networks consider these networks to be their main source of news, and this factor led many enterprises to extensively deploy social media marketing as the main tool for their marketing campaigns.

The activity of sending messages through a social network is completely different from the one of sending the same message on a mailing list. In the latter, the sending process is entirely under control, In the former, instead, we need the members of the network to cooperate in the process itself by pro-actively re-transmitting the message to their contacts.

This is the so-called *word of mouth* effect, the result of enhancing the likelihood of message forwarding in those people receiving a message from a friend (or a reliable contact of theirs, such as a web influencer, an authority or an institution) [8]. The observation of the above mentioned phenomena has given rise to the idea of *social media marketing*, namely the strategy of message delivery informing about products and services by means of social media, such as

G. Jezic et al. (Eds.): KES-AMSTA-18 2018, SIST 96, pp. 123–133, 2019.
https://doi.org/10.1007/978-3-319-92031-3_12

Facebook, VKontakte, Twitter, Pinterest, Instagram [1]. More in detail, these studies have brought to the formulation of a spread-out model that is inspired by epidemiological researches (see for instance [6]).

In particular, consider a social network in which individuals have a certain number of contacts in the network itself, suppose a minimum of n being m the number of these individuals in the network. If we assume that every member of the network can be reached by every other member by a chain of contacts, and if the message content and the selected members of the network are good enough to guarantee that every member of the network sends out the message to every contact of hers, we can reach the m members by selecting a subset of the members (called *seeds*) that is $\log_n(m)$.

The current investigations on social media marketing try to find methods to forecast this number without trying to determine precisely the members involved. The most common approach to this selection process is the usage of *social network analysis* (see [10] for a basic introduction and an analysis of the state of the art, still rather up-to-date), sometimes with a semantic flavour (see for instance [4]).

In this paper, we propose a representation of social networks that captures a notion of *activation threshold* as related to the connections, and the notions of *minimum* and *maximum latency time* to describe the fact that individuals in the network typically wait for a minimum temporal span before starting to send out the message, and when a maximum time span is reached, they quit sending out the message.

The activation threshold seizes the notion that different agents have different levels of interest with respect to different topics. Moreover, such an interest is typically known by the other agents of the network, and such information can be used during the send-out process to know which users can be interested in which topic. For instance, Joe knows that Ed is very passionate about technology, while Sarah is not. When Joe receives a new promotion on the latest smart phone, it is very likely that he will pass this information to Ed, and less probable that he will forward it to Sarah.

Between the minimum and the maximum latency time, we may have two possible behavioural models: (1) the message is sent out once, and (2) the message is sent out a number of times that is only limited by the number of available time instants in the interval. Multiple send-outs might appear unnatural, but we it is more adequate to model socio-technical systems. The sender has no warranty that the receiver got her message, and even the automated notification of receipt (as in Whatsapp) does not ensure that the message has been provoking the desired effects, for instance, the activation. Nonetheless, we might say that, a part from those who are directly involved in the campaign – like promoters, sellers, franchisees – a typical agent who decides to send the message out to a contact of hers does so once.

When we consider probabilistic models (activation thresholds not equal to 1), we essentially hypothesise some sort of intermediate decision process. When the activation thresholds are all equal to 1, we have an automated response configuration.

On the model described above, we formulate two computational problems: the *Coverage Problem* and the *Reachability Problem*. The first is the problem of determining the probability of fully covering the graph within a given time span, whilst the second one is the problem of determining the minimum time span within which the whole network is covered with a certain probability. A rather immediate application of the above sketched representation paradigm is to viral marketing.

Example 1. A marketing service is in charge of organising a social media campaign for a customer. The client is a company that sells high-tech, in particular smart watches, and has decided to advertise exclusively on the web to remark in the campaign itself the nature of the products. The marketing service has discovered that the entire set of prospect customers possesses a profile on Facebook, and aim at using that network to deliver the campaign.

They know that: Ed and Sarah follow Joe; Sarah follows Ed; Mary follows Sarah; Joe follows Mary. Once the marketing service has chosen the seeds (Ed and Joe are returning customers, and very active on social networks, so they chose them), they wonder whther the campaign will be successful. They want to know whether the message will be delivered to all the members of the network within a given time, settled for the campaign.

The plan is as follows. We start with Sect. 2 by defining some basic notions to be used in the rest of the paper. We then provide a formalisation of the problems and define two algorithms that solve the above mentioned problems in Sect. 3. In Sect. 4 we discuss the relevant literature of the field. Finally, Sect. 5 takes some conclusions and sketches further work.

2 Preliminary Definitions

In this section we introduce a formalisation of the notion of social network (henceforth $\mathcal{SN}$) as adopted in this paper.

Definition 1 (Social network). *A* basic social network *($\mathcal{BSN}$) is a graph $S = \langle V, E \rangle$, where V is a finite set of* members *(the vertices) and E is a relation on V (the edges), whose elements are* communication channels. *An edge connects the source vertex to the target vertex and the intended meaning is that the target vertex receives messages from the source vertex by the communication channel established by the edge.*

A communication channel is intended to represent both the models of communication of message re-send, and of the social activities of sharing.

We consider a labelling of the above defined graph with one label λ_r^v on each vertex and three distinct labels λ_a^e, λ_{min}^e and λ_{max}^e on each edge, where:

- λ_r^v is the *reception time* for a vertex v, which is either *null* (when the vertex has not yet been reached by the message) or a positive integer, and represents the instant when the message arrives to the member. The label 0 is reserved to represent the situation in which a vertex is a *seed*;

- λ_a^e is the *activation threshold* on edge e, a real number such that $0 \leq \lambda_a^e \leq 1$, used to denote the probability that the member sends the message she has received on communication channel e, at every admissible instant of time;
- λ_{min}^e is the *minimum latency time* on edge e, a positive integer representing the time span passed before the message can be send out;
- λ_{max}^e is the *maximum latency time* on edge e, a positive integer representing the time span after which the message cannot be sent out anymore.

Vertex v will try to send out the message on edge e in between the interval $[\lambda_r^v + \lambda_{min}^e, \lambda_r^v + \lambda_{max}^e]$, repeatedly. Not every labelling of the vertices is coherent with the labels on the edges. Determining this coherence is a trivial task, therefore, without loss of generality, we assume it. The parameters can be determined by direct experimental testing (namely by measuring the sendout probabilities and the latency times on a sample).

When all the activation thresholds are fixed to 1 and, for all the edges, minimum and maximum latency times coincide, we name the network *deterministic* ($\mathcal{DSN}$).

For an edge (v, v') we call v the *source vertex* and v' the *target vertex.*

A *seeding* P of a $\mathcal{SN}$ is a partial labelling of the vertices with initial temporal instants, namely 0. A seeding consists in the selection of a set of agents aimed at delivering the message first.

We use the term *configuration* of a $\mathcal{SN}$ to denote the labelling of vertices with temporal labels that are coherent with the seeding.

Definition 2 (t-precedence). *Given a $\mathcal{SN}$ S and the current (global) instant time T, a configuration $C_1 = \langle S', (t-1)\rangle$* t-precedes *a configuration $C_2 = \langle S', t\rangle$ on a given $\mathcal{SN}$, when the activation threshold of vertices is generated coherently with the constraints as expressed in the definition of configurations, passing from the label in C_1 to the label in C_2 in the admissible ways:*

- *When $\lambda_r^v = null$ in C_2 (of a vertex v), then $\lambda_r^v = null$ in C_1 as well;*
- *When $\lambda_r^v = T$ in C_2 (of a vertex v), then $\lambda_r^v = null$ in C_1. Moreover, there exists u, a vertex connected to v, such that $\lambda_r^u < T$ and $(T-1) \in [\lambda_{min}^{(u,v)} + \lambda_r^u, \lambda_{max}^{(u,v)} + \lambda_r^u]$ in C_1.*

A sequence of configurations $C_1, C_2, \ldots, C_n$ is correct iff every pair $(C_i, C_{(i+1)})$ with $1 \leq i \leq (n-1)$ is such that C_i t-precedes $C_{(i+1)}$. A configuration C is *admissible* on a $\mathcal{SN}$ S with a given seeding, if there is a finite sequence of configurations on S that starts with the seeding and ends with C. A configuration is *terminal* when it does not t-precede any other configurations.

Computing all the configurations, while determining their probabilities, permits to establish the probability of reaching coverage configurations, and can therefore be used to solve the coverage problem. The configurations are the ways in which we can leave vertices with empty labels, that means, at worst, the subsets of the vertices. Since two configurations differ both from the viewpoint of vertices involved, and in the labelings we may compute the number of total configurations as $O(n^2 \cdot k^2)$ where k is the number of temporal labels to be employed.

In Sect. 3 we provide an analysis of two cases: *deterministic* and *probabilistic* $\mathcal{SN}$. The combinatorial approach adopted in the above analysis can be used to solve the problems of coverage and reachability in all the configurations of the problems, and represents the actual evolution of a network, namely the process of starting from an initial configuration and reaching another one. We can however provide polynomial solutions to both deterministic and probabilistic cases.

3 Algorithms

Definition 3. *Given a* $\mathcal{SN}$ S*, a seeding* V' *on* S*, and a probability threshold* τ*, we define* Reachability Problem *the problem of determining the time span* T *needed to guarantee that* S *evolves to a coverage configuration within* T *instants from the seeding configuration* V' *with a probability* $\pi \geq \tau$*.*

Definition 4. *Given a* $\mathcal{SN}$ S*, a seeding* V' *on* S *and a time span* T*, we define* Coverage Problem *the problem of determining the probability that* S *evolves to a coverage configuration within* T *instants from the seeding configuration* V'*.*

At first, we shall consider the case of a deterministic social network (which was name $\mathcal{DSN}$). In this scenario, we can find a solution to the coverage problem by reasoning as follows.

Given a $\mathcal{DSN}$ S, consider an evolution where a vertex has been reached twice. It does not make any sense to consider the paths including a cycle. The vertex which has been reached twice remains inactive after the first reception. Therefore, we can state that a $\mathcal{DSN}$ S is actually equivalent to an acyclic graph, that simply considers the cycles as impossible to pass through.

Johnson's approach on cycle detection can be used to cycle elimination with the following procedure: we eliminate cycles, by just detecting them, and then adding a new vertex (that is the copy of one vertex in the cycle) where the cycle itself is divided in two subpaths. The first path goes from the vertex x in word onto the last vertex of the cycle (the one that precedes x in the cycle), and one subpath formed by one edge connecting the last vertex in the cycle to the new vertex x', copy of x. From the vertex x' no new edge is added. This task is performed in $O(|V| + |E|)$.

Once the cycles have been determined and eliminated, we can compute the time required to reach a vertex from the seed vertices. This can be performed by using all-pairs-shortest path method, also known as Floyd-Warshall method. Now consider, for every vertex, the length of the shortest path between that vertex and each seed, and compare it against the input time span. When the length of the path results less than the time span with at least one of the seeds, then the vertex is reached within the interval. Thanks to this step we can answer to the question for $\mathcal{DSN}$: do we cover the whole network with the given seeding in the given time span? The proposed approach is described in Algorithm 1.

With $d(x, y)$ we intend the label on the edge from vertices x to y after Floyd-Warshall's application. Conversely, if we consider the deterministic reachability problem we can apply a slightly modified version of Algorithm **d-Coverage**, consisting in computing the maximum of the length on the edges, after the first two steps. This works only when the graph is connected. The same reasoning can apply to each connected component of the network. Algorithm 2 solves the deterministic reachability.

Algorithm 1. [d-Coverage] Three-steps coverage method for $\mathcal{DSN}$.

```
Data: A DSN S, with vertices V and edges E, a time span T, a seeding V' ⊆ V.
Result: YES if S can be covered within T while starting with V', NO otherwise.

Execute Johnson's modified method to eliminate cycles;
Execute Floyd-Warshall' method all-pairs-shortest-paths;
Flag-v=YES;
for y ∈ V do
    Flag-s := NO;
    for x ∈ V' do
        if d(x, y) ≤ T then
            Flag-s=YES;
            Break;
    if Flag-s = NO then
        Flag-v := NO;
        Break;
Return Flag-v
```

For the probabilistic version of social networks we can propagate the probabilities through the network itself for as many instants of time as have been specified in the input of the coverage problem.

The start is given onto the initial instant 0, when the seeding is performed. After that, at each step, when a time instant t is considered passed, we evaluate all the vertices $V(t)$ that can be reached from the set of vertices reached by the message in the time instant before, that we kept traced in the set $V(t-1)$. $V(0)$ is the set of seeds. During the propagation, we mark the vertices of the network by λ_r^v, the reception time for v.

A vertex is in the set $V(t)$ when the following conditions hold: (1) there is a vertex $v' \in V(t-1)$, (2) the network contains an edge e connecting v' with v, and (3) the instant of time we look at is admissible for send-out (if $T = (t-1) - \lambda_r^v$, we have $\lambda_{min}^e \leq T \leq \lambda_{max}^e$).

We hence consider the elements of $V(t-1)$ as the vertices reached by the message at time instant $(t-1)$, that have not yet delivered the message forward. If a vertex v is in $V(t)$ there can be two possible reasons: either the message has flown from a vertex v' in $V(t-1)$ through an edge connecting v' to v, or the message has reached the vertex v at time $(t-1)$ but, due to the minimum or

Algorithm 2. [**d-Reachability**] Three-steps reachability method for $\mathcal{DSN}$.

Data: A $\mathcal{DSN}$ S, with vertices V and edges E, a seeding $V' \subset V$.
Result: The minimum time span T needed to reach the vertices in S while starting with V'. When one vertex in V is not reachable starting from V', ∞.

```
Execute Johnson's modified method to eliminate cycles;
Execute Floyd-Warshall' method all-pairs-shortest-paths;
T := 0;
for y ∈ V do
    Flag-v := NO;
    for x ∈ V' do
        if (x, y) ∈ E then
            Flag-v := YES;
            if d(x, y) > T then
                T := d(x, y);
    if Flag-v=NO then
        Return ∞
Return T
```

the maximum latency time, in relationship with the time of synchronisation (t), the vertex did not send the message out.

At the first step, we consider the event of the message delivery to the seed vertices *definite*, and mark those vertices with probability value 1 every seed vertex, and with 0 the probability of receipt in any other vertex. After one instant of time, the edges connecting the seeding vertices with the rest of the network contain a label each indicating the probability of send-out.

After the minimum latency time on the edge, the message is delivered through that edge with the probability on the edge itself. If that edge is the only one connecting one seed to a new vertex in the network, the probability that the latter vertex is reached is the product of the former with the probability on the edge. On the other hand, if the edges are more than one, we consider the opposite event, that being *'in no edge the message has been sent out'*. This can be computed by the probabilistic complement of the products of the probabilistic complements on the edges: $1 - \prod(1 - p_i)$, where p_i is the probability that the vertex is reached from the vertex v_i.

Consider now the case in which the edges reach a target vertex that is not marked with 0. The update consists in considering, again, the opposite event.

In Eq. 1, we refer by $p(v, t)$ the probability that the vertex v is reached by the message not later than time instant t, by $E(v, t)$ the set of edges that connect the vertices $v' \in V$ to the vertex v, that are *active* at time instant t, namely such that the time passed since the message reached firstly v', at temporal instant t, is included between the minimum latency time and the maximum latency time, on the edge connecting v' to v. By $p(s(e))$ we denote the probability label on the source edge of the edge e and by $p(e)$ we denote the activation threshold on the edge e.

$$p(v,t) = p(v,(t-1)) + (1 - p(v,(t-1))) \cdot \prod_{e \in E(v,t)} [1 - (p(s(e)) \cdot p(e))] \quad (1)$$

Algorithm 3. [Compute-probabilities]

Data: A $\mathcal{SN}$ S, with vertices V and edges E, a time span T, a seeding $V' \subseteq V$.
Result: The graph is labelled in every vertex with the probability that the network is covered within the time span T starting with the seeding V'.

```
for x ∈ V do
    if x ∈ V' then p(x) := 1; t(x) := 0
    else p(x) := 0
for t = 1...T do
    for x ∈ V(t − 1) do
        for y ∈ V do
            if (x, y) ∈ E and λ_max^(x,y) + t(x) ≤ t then
                Add y to V(t);
                t(x) := t(x) + 1;
    for v ∈ V(t) do
        for e ∈ E(v, t) do
            Update p(v) with p(v, t) as in Eq. 1;
```

A slight change to Algorithm 4 will provide the solution to reachability. This consists simply in running the same internal cycle to re-label the vertices. Within the cycle we check the probability of the input against the computed overall products. The output -1 indicates that the time needed to reach the vertices is impossible to compute. Proofs are omitted for the sake of space, and will be made available in a further extended version of this study.

Algorithm 4. [Coverage] Coverage for $\mathcal{SN}$

Data: A $\mathcal{SN}$ S, with vertices V and edges E, a time span T, a seeding $V' \subseteq V$.
Result: The probability of reaching a coverage configuration within T.

```
Execute Compute-probabilities(S, t);
α := 1;
for v ∈ V do
    α := α · p(v);
Return α
```

Theorem 1. *Algorithm 1 correctly computes the solution of the d-Coverage problem $\langle S, \tau \rangle$ in $O(\epsilon \cdot n)$ with n number of vertices and ϵ number of edges in S. Algorithm 2 correctly computes the solution of the d-Reachability problem in $O(\epsilon \cdot n)$ with n number of vertices and ϵ number of edges in S.*

Algorithm 5. **[Reachability]** Reachability for $\mathcal{SN}$

Data: A $\mathcal{SN}$ S, with vertices V and edges E, a probability threshold τ, a seeding of $V' \subseteq V$.

Result: The minimum time needed to reach a coverage configuration.

```
α := 1; α' := 0; t := 1;
while α ≠ α' do
    Execute Compute-probabilities(S, t);
    α' := α; α := 1 ;
    for v ∈ V do
        α := α · p(v);
    if α ≥ τ then
        Return t;
    t := t + 1;
Return −1
```

Theorem 2. *Algorithm 4 correctly computes the solution of the Coverage problem $\langle S, \tau \rangle$ in $O(\tau \cdot (\epsilon + n))$ with n number of vertices, ϵ number of edges in S, and τ number of instants in the time span.*

Algorithm 5 correctly computes the solution of the Reachability problem in $\langle S, \tau, \alpha \rangle$ in $O(\tau \cdot (\epsilon + n))$ with n number of vertices, τ number of instants in the time span, and ϵ number of edges in S.

4 Related Work

Numerous investigations have been carried out in different communities regarding the ways in which messages are delivered through networks, and some recent studies have also found ways to combine these efforts [9]. In the investigations about transmission protocols (see [7] for the comparison of the most common network models), the notion of *transmission delay* has been investigated in many different and diverse contexts [13].

A wide literature is devoted to network broadcast methods, in particular the so-called ad-hoc networks, that have been often modelled with probabilistic behaviours, and studied with specific attention the flooding phenomenon, consisting in broadcasting messages with the *collaboration* of the network routers [12].

A special attention has been posed on social networks by the community of marketing research [5], as they formulate interest in the ability of the networks to deliver messages by using *epidemic* models [6].

Some recent efforts have been carried out on how can extend social network analysis to consider virality from the very beginning [2] and consequently being able to forecast the behaviour of the network itself in relation to virality [4]. Further analogous studies have been taken further on virality, such as [3,11].

5 Conclusions and Further Work

We investigated a model of social network message propagation that represents the uncertainty on send-out of messages, whilst consider reception as deterministic. This model can cautiously be employed to forecast the probability of success for a social media marketing campaign, and to forecast the time span needed to guarantee a minimum probability of success.

There are several different ways to take this research further. First of all, it is possible to investigate other models of send-out and reception of messages. For instance we could model send-out without repetition, or allow multiple reception. All these studies are worth but require a very specific effort.

An important limit of this investigation that we aim at filling with further studies is that we do not have yet a model of how to provide the measures used to make the forecast discussed in this paper. We are devising an experimental trial whose goal is to provide a clear picture of the features used by humans to make the above mentioned forecast, and use the indicators obtained in this way to validate the model.

References

1. Bampo, M., Ewing, M.T., Mather, D.R., Stewart, D., Wallace, M.: The effects of the social structure of digital networks on viral marketing performance. Inf. Syst. Res. **19**(3), 273–290 (2008)
2. Cristani, M., Fogoroasi, D., Tomazzoli, C.: Measuring homophily. In: CEUR Workshop - Proceedings, vol. 1748 (2016)
3. Cristani, M., Olivieri, F., Tomazzoli, C.: Viral experiments, vol. 1959 (2017)
4. Cristani, M., Tomazzoli, C., Olivieri, F.: Semantic social network analysis foresees message flows. In: ICAART - Proceedings, vol. 1, pp. 296–303 (2016)
5. Esmaeilpour, M., Aram, F.: Investigating the impact of viral message appeal and message credibility on consumer attitude toward the brand. Manag. Mark. **11**(2), 470–483 (2016)
6. Gonsalves, J.N.C., Rodrigues, H.S., Monteiro, M.T.T.: A contribution of dynamical systems theory and epidemiological modeling to a viral marketing campaign. Adv. Intell. Syst. Comput. **557**, 974–983 (2017)
7. Inaltekin, H., Chiang, M., Poor, H.V.: Average message delivery time for small-world networks in the continuum limit. IEEE Trans. Inf. Theory **56**(9), 4447–4470 (2010)
8. Khan, S.K.A., Mondragon, R.J., Tokarchuk, L.N.: Lobby influence: opportunistic forwarding algorithm based on human social relationship patterns. In: PERCOM Workshops, pp. 211–216 (2012)
9. Lu, Z., Sagduyu, Y., Shi, Y.: Friendships in the air: integrating social links into wireless network modeling, routing, and analysis. In: INFOCOM - Proceedings, vol. 2016, pp. 322–327, September 2016
10. Scott, J.: Social network analysis: developments, advances, and prospects. Soc. Netw. Anal. Min. **1**(1), 21–26 (2011)
11. Tomazzoli, C., Storti, S.F., Galazzo, I.B., Cristani, M., Menegaz, G.: The brain is a social network, vol. 1959 (2017)

12. Tseng, Y.-C., Ni, S.-Y., Chen, Y.-S., Sheu, J.-P.: The broadcast storm problem in a mobile ad hoc network. Wirel. Netw. **8**(2–3), 153–167 (2002)
13. Zhu, Y., Zhang, H., Ji, Q.: How much delay has to be tolerated in a mobile social network? Int. J. Distrib. Sens. Netw. **2013**, 1–8 (2013)

ER-Agent Communication Languages and Protocol for Large-Scale Emergency Responses

Mohd Khairul Azmi Hassan(✉) and Yun-Heh Chen-Burger

Department of Computer Science,
Heriot-Watt University, Edinburgh EH14 4AS, UK
{mh42,y.j.chenburger}@hw.ac.uk

Abstract. In this paper, we introduce a new agent communication language (ER-ACL) and a corresponding protocol (ER-ACP) to be used in multi-agent systems (MAS) to assist large-scale emergency responses as a part of an Emergency Response Communication Framework. In the previous study of ACL, we found them lack the necessary richness to support communication during a large-scale disaster, inc. structure, semantics and user models. This inspired us to create a new ER-ACL to fulfil this gap. Four types of agents are supported in ER-ACL: victims, carers (medical & social workers), families & friends, and ER-rescuers & helpers (members of the public, NGOs, government agencies, etc.). The advantages of ER-ACL and ER-ACP are that they provide a well-defined foundation to connect victims with potential helpers, thereby enabling crowdsourcing via effective communication based on precise semantics. The ER-ACL represents a significant extension and specialisation of the FIPA ACL for applications in emergency response scenarios now that great technical advances have been made in telecommunication (including image and video reporting). We have also added many new message constructs from the Common Alerting Protocol. In today's uncertain world, we believe a well-managed and personalised communication system is vital to organise unstructured/opportunistic resources to save lives. Not having found one in existence to-date, we hope our efforts can help close this gap.

Keywords: Agent communication language and protocol
Emergency response · Mobile agents · Large-scale disaster rescue

1 Introduction

Communication is key to effective emergency response, especially in large-scale disaster events. Effective communication allows volunteers and rescuers to find victims quickly and accurately, allowing them to plan and carry out rescue tasks using suitable methods in a timely fashion. Communication is essential to keep families, friends, rescuers and carers informed, thereby providing effective support ASAP [1].

Multi-agent systems [2, 3] are distributed systems that encompass many autonomous self-directional and actionable agents. Such systems are ideally placed to model and support Emergency Response Scenarios. Engineering such a multi-agent system requires rigorous specification, homogenization, standardization and a suitable

G. Jezic et al. (Eds.): KES-AMSTA-18 2018, SIST 96, pp. 134–143, 2019.
https://doi.org/10.1007/978-3-319-92031-3_13

foundation to support a good level of richness in conversations in the communication language and interaction protocols among agents.

FIPA-ACL is a widely used standard Agent Communication Language [4]. One of the motivations behind the development of FIPA-ACL was the need to address the challenges faced by the Knowledge Query and Manipulation Language (KQML) [5]. However, in this research, we found significant gaps still exist in FIPA-ACL when we tried to apply it to support emergency response scenarios.

For example, there is a lack of richness in the different types of message, which are thereby unable to support specific different agent interaction models. Examples of these are announcements, live updates, broadcast appeals, forwarded appeals and complex collaboration and planning types of conversations. For instance, announcements and live updates do not normally require a reply, but a broadcast appeal does - planning and collaboration would require back-and-forth discussion and confirmation.

FIPA-ACL also lacks a mechanism for the storage of emergency-related information, e.g. the changing status of a disaster and its impact, event and victim locations, dynamic personal health statuses, including injury type, severity and urgency, and hospital capacity. Nor does it support modern mobile technology that would allow voice, image and video file attachments to communications. Also lacking is any means of defining groups of users in order to support group-specific communication more rigorously. To address all of the above gaps, this paper discusses the Emergency Response Agent Communication Language (ER-ACL) and its corresponding protocol (ER-ACP) that we have designed to support communication in large-scale disaster emergency response.

2 Motivating Scenario

The inherent complexity and dynamism of large-scale disasters make the implementation of timely, effective, well-informed and organised emergency responses a far from trivial task, made even more complex when a large number of victims are involved. In order for a response to be effective, a broad range of information needs to be readily available and directed to the right people regarding, for instance, the changing status of the disaster itself, of locations and conditions of survivors, up-to-date shelter logistics, and communication between victims and rescuers, carers, family and friends.

Search and rescue may be framed as an agent-based problem for which the development of a suitable Agent-based Communication Language (ACL) is urgently needed. This ACL will be used via a mobile communication mechanism, such as a mobile app, that can store personal information (sharable before the emergency event) and be personalised to suit individual users' needs and their ways of communicating with others according to a set of pre-defined user groups using well-defined protocols.

To address these aims, based on the existing FIPA-ACL we have developed a new Emergency Response Agent Communication Language (ER-ACL) and a corresponding protocol (ER-ACP) in a new mobile app, Mobile Kit Assistant (MKA). This allows different information sources created by different people in different places to be connected and used together in meaningful ways based on an ontological backbone that we have created in [1].

3 Agent Communication Language and Protocol Design

When developing ER-ACL and ER-ACP, several issues have been taken into account to ensure the language is appropriate and usable. The following were considered.

3.1 Design Philosophy

To added new message constructs from the Common Alerting Protocol [6], important considerations for designing the ER-ACL are: **Interoperability** – ER-ACL should provide a well-define structure and semantics, so that messages can be understood correctly in different systems; **Completeness** – The ER-ACL should support all of the possible communication information and methods, e.g. (typical) communicated information and its formats, e.g. voice, images and video messages and an indication of their retrieval method. **Simple implementation** – The ACL should be as simple as possible to use and implement; **Flexibilities** –The constructs should remain sufficiently abstract, while being rich, to be adaptable and extendable to other coding schemes; **Multi-use format** – the same message format may be used by different message types issued by different user groups; **Familiarity** – The data elements and code values should be meaningful to originators and non-expert recipients alike; **Interdisciplinary and international utility** – The design should allow a broad range of applications in public safety and emergency management and allied applications and should be applicable worldwide.

3.2 Requirements for Design

The ER-ACL should (1) Provide a specification for a simple, extensible format for digital representation of warning messages and notifications; (2) Enable integration of diverse sensor, inc. multi-gesture signals on mobile phones; (3) Support multiple transmission systems, including Wi-Fi Direct Peer to Peer (P2P), this is needed, as standard telecommunication networks are often down or congested that alternative communication channels are much needed; (4) Provide a unique identifier (e.g., Message ID) for each warning message and for each message originator; (5) Support multiple message types and sender roles; (6) Support suitable pre-defined content (key words); (7) Referencing supplement information/files external to the message; (8) Following established standard data representation; (9) Can sustain real-world cross-platform testing and evaluation; (10) Support emergency response scenarios and promote public safety.

3.3 Emergence Response User Scenarios

In our study, there are several scenarios that can take place during and after large scale of disasters. We provide such an example in Fig. 1 This situation indicates that the victim broadcasts an ask-help message to everyone near his location in the hope to find

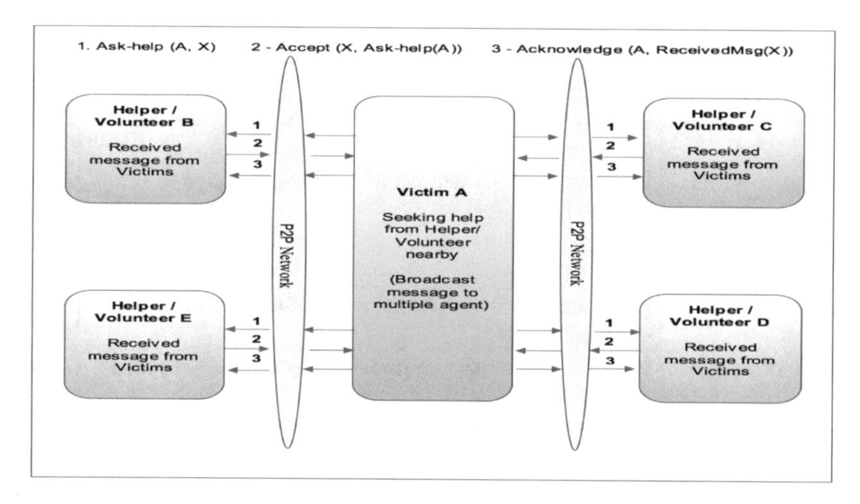

Fig. 1. Victim agent asking for help (broadcast mode)

a volunteer/rescuer that is close to the victim. The literature showed [7] that the ideal distance for wireless connectivity for smart phones [8] is maximum 100 m. Our focus is to alert nearby helpers and to reduce the congestion of telecommunication network, thus help the victims quickly after a large scale of disaster. The timing of how messages are sent is as follows: (1) Send Ask-help messages from victim to (nearby) volunteers; (2) Send Accept- message by volunteers to victims, if helping (refuse-messages are not send to reduce network congestion); (3) send Acknowledge-messages from victim to helpers.

3.4 Developing ER-ACL

Two documents have been used as main references to develop our ER-ACL and its protocol ER-ACP: the FIPA ACL [9] and Common Alerting Protocol [6]. These documents provide fundament concepts and structure. Here we present ER-ACL and the part of FIPA ACL performatives that we would normally use in emergency scenarios.

3.5 Performatives in ER-ACL

Table 1 shows the combination of Performances in our new ER-ACL and FIPA-ACL as in [10] (ER-ACL performatives are shown in bold) has been used in our study. With these extensions, we are able to support common emergency response scenarios.

Table 1. List of performative (Bold are new performatives used in ER-ACL)

Performative	Description	Status
Ask-help	Used by sender (victim) to send help message to receiver (volunteer)	New
Ask-help-for-others	Used by sender (volunteer) to send help message to receiver (volunteer)	New
Offer-help	Used by sender (helper) to send offer of help message to receiver (victim)	New
Accept	Used to accept message (and reply with current situation of sender agent)	New
Forward-Message	Used to forward message from agent (victim) from sender to another receiver	New
Acknowledge	Used to acknowledge message received from sender	New
Send	Used to send normal messaging between or among agents	New
Reply-to	Used to reply in normal messaging between or among agents	New
Reply-with	Used to reply-with normal messaging between or among agents	New
Status-report	Used to report status between or among agents	New
Channel	The connection method used for data transferring	New
Refuse	Used to refuse to perform a given action, explaining the reason for the refusal	Existing

4 ER-ACL Communication Protocol

Through different scenarios of Fig. 1 what we may call two-way complex communication may exist among three main agents such as family/friend, volunteer 1 and volunteer 2. Figure 2 shows communication taking place among agents after a large scale of disaster, beginning with the victim asking for help from family/friend and then they ask help from volunteer 1 (we assume they are nearby the victim).

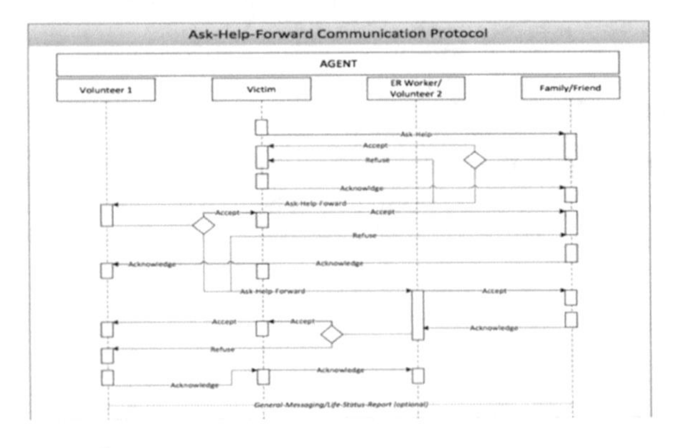

Fig. 2. Complex 3rd party ask-help and forward communication protocol in ER-ACL

However, the outcome is that volunteer 1 cannot help because they are managing another victim nearby. So volunteer 1 refuses the request, and then sends the information to another volunteer (volunteer 2). If volunteer 2 is able to help, they will accept the request and the 'accept' message will be sent to volunteer 1 and the family/friend, informing every one of the situation. The acknowledge message will be sent to the sender (family/friend) and Victim to ensure the information has been received, and the victim has only to wait for volunteer 2 to come.

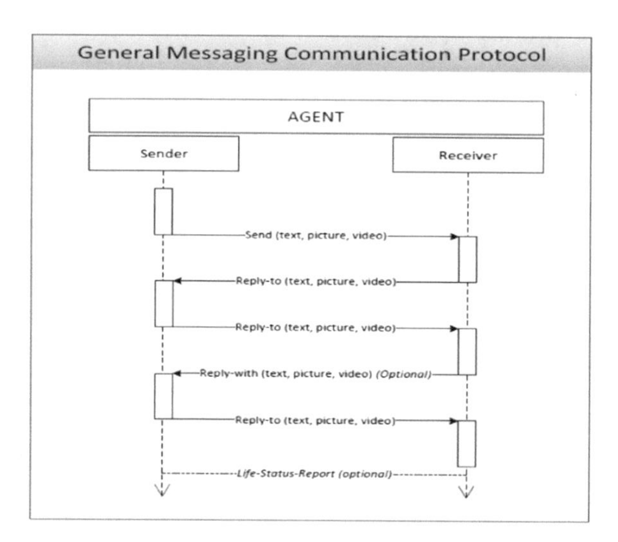

Fig. 3. Two ways general messaging communication protocol in ER-ACL

The situation shown in Fig. 3 is a protocol diagram for general messaging communication for situations that occur when two personal agents exchange information. Even the existing FIPA-ACL consist of reject-proposal, request, request-when and request-whenever performative, it is much difference with our propose performative in ER-ACL. The differences of those performative shown in Table 2 below:

Table 2. Performative differences between FIPA-ACL and ER-ACL

FIPA-ACL	ER-ACL
Reject-proposal The action of rejecting a proposal to perform some action during a negotiation	**Refuse** The action of refusing to perform a given action and explaining the reason for the refusal
Request The sender requests the receiver to perform some action One important class of uses of the request act is to request the receiver to perform another communicative act	**Ask-Help** The action of sending information for getting help by victim (sender agent) to volunteer (receiver agent) or by family/friend (sender agent) to volunteer (receiver agent). There is no action perform needed by the receiver

(*continued*)

Table 2. (*continued*)

FIPA-ACL	ER-ACL
Request-when The sender wants the receiver to perform some action when some given proposition becomes true	
Request-whenever The sender wants the receiver to perform some action as soon as some proposition becomes true and thereafter each time the proposition becomes true again	
Inform The sender informs the receiver that a given proposition is true	

5 ER-ACL Conversation Tree

Figure 4 shows a conversation tree where a family/friend asked for help on-behalf of the victim. Ask-help (C1) is the help request message sent by the family/friend to a volunteer 1. The second level applies whether volunteer 1 accepts or refuses the request. If volunteer 1 refuses to help, he/she may choose to forward the request in a new Ask-help-forward request (C1-1) to another volunteer 2 (and ride of the responsibility). An Accept message is sent to the family/friend by volunteer 2, only if help is offered by volunteer 2. The message ID, C1-1, records the trail of forwarded message of C1. This helps one to eliminate duplicated messages, if receives more than once.

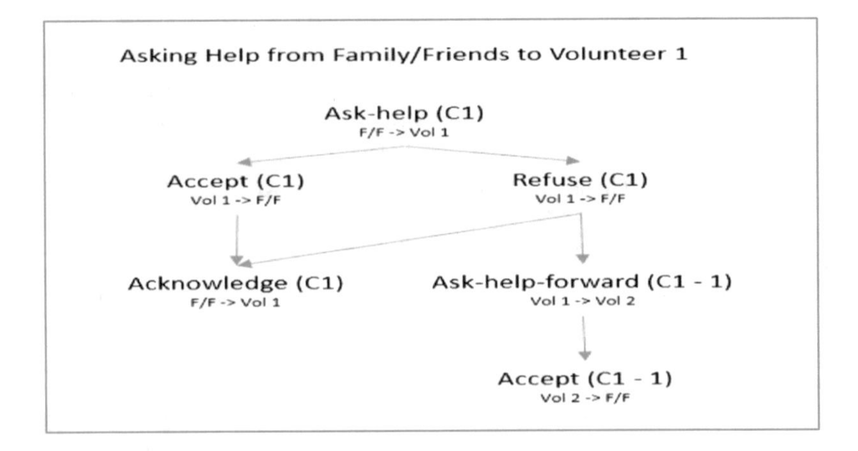

Fig. 4. Asking help from family and friends to volunteer

Figure 5, above, shows the ask-help message that is sent by volunteer 1 to volunteer 2. The difference between ask-help and ask-help-forward, as shown above, is

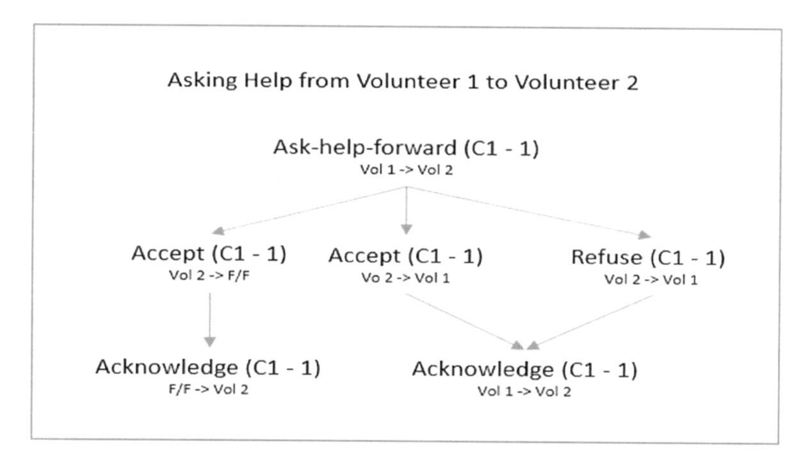

Fig. 5. Asking help from family and volunteer to volunteer

that a message sent via ask-help-forward will include the initial family/friend message sent to volunteer 1. This will help volunteer 2 glean important information such as location, time and message content, which is needed to help the victim.

6 ER-ACL Interaction Model

In this section, we describe two of our communication trees as in Figs. 4 and 5. From this communication tree, we, therefore, develop an interaction model in Tables 3 and 4. The models have explained the use of ER-ACL performative and parameters per the ACL document as follows:

Performative <parameters>

6.1 Complex Two Ways Communication Model Set

To see a more detailed structure, i.e. parameters used for complex two-way communication situations, we list all of the performative parameters used in Table 3. The acronyms of them are as follows:

Table 3. Complex two ways ask and reply communication model

Performative	Parameter
Ask-Help	$AH = <M_{id}, T_s, S, R, E_c, E_t, S_e, E_d, R_m, C_m \{C_{id}, T_m, L_s, U_m, P_{id}, VD_{id}, VC_{id}\}, L_l, C_l, B_s, M_s, C_n, P_o, O_m>$
Ask-Help-for-Others	$AHF = <M_{id}, T_s, S, R, E_c, E_t, S_e, E_d, R_m, C_m \{C_{id}, T_m, \{F_m\}, L_s, U_m, P_{id}, VD_{id}, VC_{id}\}, L_l, C_l, B_s, M_s, C_n, P_o, O_m>$
Accept	$A = <M_{id}, T_s, S, R, R_m, C_m \{C_{id}, T_m, L_s, U_m, P_{id}, VD_{id}, VC_{id}\}, L_l, C_l, B_s, M_s, C_n, P_o, O_m>$
Refuse	$RE = <M_{id}, T_s, S, R, R_m, C_m \{C_{id}, T_m, \{F_m\}\}, M_s, C_n, P_o, O_m>$
Acknowledge	$ACK = <M_{id}, T_s, S, R, C_m \{C_{id}, T_m\}, M_s, C_n, P_o, O_m>$

Table 4. Two ways general communication model

Performative	Parameter
Send	S_n = <M_{id}, T_s, S, R, C_m {C_{id}, T_m, P_{id}, VD_{id}, VC_{id}}, M_s, C_n, P_o, O_m>
Reply-to	R_t = <M_{id}, T_s, S, R, C_m {C_{id}, T_m, P_{id}, VD_{id}, VC_{id}}, M_s, C_n, P_o, O_m>
Reply-with	R_w = <M_{id}, T_s, S, R, C_m {C_{id}, T_m, P_{id}, VD_{id}, VC_{id}}, M_s, C_n, P_o, O_m>

AH – Ask Help
T_s – Time Stamp
RE – Refuse
R_m – Myrole
L_s – Life Status
L_l – Last Location
M_s – Message Status
F_m – Forward Message
S_e – Severity
VD_{id} – Video Message
M_{id} – Message Id
A – Accept
R – Receiver
C_{id} – Content Id
U_m – Urgency
C_l – Current Location
P_o – Protocol
E_c – Event Category
E_d – Expiration Date
VC_{id} – Voice Message
AHF – Ask Help Forward
S – Sender
ACK – Acknowledge
T_m – Text Message
P_{id} – Picture Message
B_s – Battery Status
O_m – Ontology
E_t – Event Type
C_n – Channel

6.2 Two Ways General Messaging Model Set

The Two Ways General Messaging Model is less complex. This model depicts a direct communication between the sender and receiver to exchange information. The messaging sequence is send, Reply-to, followed by Reply-with or Reply-to. With the Reply-with performative, the communicator can generate a messaging sub-thread; where as Reply-to would follow the same message thread. Given a Reply-with message, the following messages can either be a Reply-to or Reply-with (to generate a new sub-thread). Sub-threads are recorded via message IDs. Table 4 gives performative parameters used for two-way general messages. The acronyms are: S_n – Send; R_w – Reply-with; and R_t – Reply-to.

7 Conclusion and Future Work

This paper explains how important it is to improve the existing FIPA-ACL to suit emergency response needs. We have therefore created ER-ACL as a foundation for mobile app developers. To explain what information is needed and when communication between the victim and the rescuer should occur, we also built ER-ACP, and have provided the corresponding syntax, conversation tree and interaction models. However, the new ER-ACL has not been implemented and tested in any real emergency response system. For future work, we plan to build a distributed multi-agent communication and tracking mobile apps based on ER-ACL, ER-ACP and their underlying ontologies to understand usability issues as the mobile apps are developed. Testing and evaluation of the usability, simulations and trials of the system involving real users will be carried out based on real-world emergency response scenarios to test the robustness and effectiveness of our proposed solution. We trust that this will improve protocols as well as similar apps in the future.

References

1. Hassan, M.K.A., Chen-Burger,Y.-H.: Communication and tracking ontology development for civilians earthquake disaster assistance. In: Proceedings of ISCRAM 2016 Conference – Rio Janeiro, Brazil, May 2016 Tapia, May 2016
2. Sharmeen, Z., Martinez-Enriquez, A.M., Aslam, M., Syed, A.Z., Waheed, T.: Multi agent system based interface for natural disaster. In: Ślęzak, D., Schaefer, G., Vuong, Son T., Kim, Y.-S. (eds.) AMT 2014. LNCS, vol. 8610, pp. 299–310. Springer, Cham (2014). https://doi.org/10.1007/978-3-319-09912-5_25
3. Costin, B., Scafes, M., Ilie, S., Badica, A., Muscar, A.: Dynamic Negotiations in Multi-Agent Systems. ICT Educ. Res. (2011)
4. Juneja, D., Jagga, A., Singh, A.: A review of FIPA standardized agent communication language and interaction protocols. J. Netw. Commun. Emerg. Technol. **5**(2), 179–191 (2015)
5. Chopra, A., Singh, M.P.: Agent Communication. Multiagent Syst. Mod. Approach Distrib. Artif. Intell. 101–141 (2013)
6. Westfall, J.: Common Alerting Protocol Version 1.2, pp. 1–47, July 2010
7. Nishiyama, H., Ito, M., Kato, N.: Relay-by-smartphone: realizing multihop device-to-device communications. IEEE Commun. Mag. **52**(4), 56–65 (2014)
8. Sheikh, A.A., Ganai, P.T., Malik, N.A., Dar, K.A.: Smartphone: ANDROID vs IOS. SIJ Trans. Comput. Sci. Eng. Appl. **1**(4), 141–148 (2013)
9. Foundation for Intelligent Physical Agent: FIPA ACL message structure specification. IEEE Comput. Soc. p. 1 (2002)
10. Agent Communication. http://jmvidal.cse.sc.edu/talks/agentcommunication/

Towards a Logical Framework for Diagnostic Reasoning

Matteo Cristani[1(✉)], Francesco Olivieri[2], Claudio Tomazzoli[1], and Margherita Zorzi[1]

[1] Department of Computer Science, University of Verona, Verona, Italy
matteo.cristani@univr.it
[2] Data61, CSIRO, Canberra, Australia

Abstract. Diagnosis is widely used in many different disciplines to identify the nature and cause of a certain phenomenon. We present $t\mathsf{L}$, a new logical framework able to formalise *diagnostic reasoning*, i.e., an hybrid learning technique based both on *deduction* and *experiments*. In this paper we introduce tL, a *Labeled Modal Logic*, garnishing with temporal and statistical information and a basic propositional language.

After proposing examples on how tL effectively works, we sketch the main ideas about the full deduction system *à la* Prawitz we are currently developing.

Keywords: Hybrid reasoning · Labelled logic · Temporal logic

1 Introduction

Diagnostic reasoning can be defined as the process of evaluating the results of some operations (questions or practical actions), to establish which specific conditions hold on an individual or, generically, on a sample. This class of operations are named *tests*. A number of scientific fields exploit test-based knowledge acquisition: engineering, earth sciences (such as geology), computer science, biology and, of course, medicine. A test principally *reveals* a property, with a margin of error, and so provides information about the causes; as a consequence, by establishing 'cause and effect' relationships, tests can provide information about solutions. Consider a medical diagnosis: specific symptoms suggest which tests are to be done, the result of which helps in identifying which disease is currently on, and, ultimately, which drugs are needed.

In other terms, tests reveal truth on tested conditions in a provisional way. A question arises: How do we reason on those conditions? And, is it possible to formalise (and mechanise) diagnostic knowledge acquisition, possibly integrating it in other knowledge acquisition styles?

Logical systems have been widely used in Computer Science literature as means to represent reasoning mechanisation, and to give account to processing pieces of knowledge that, in turn, represent the ways in which humans provide

G. Jezic et al. (Eds.): KES-AMSTA-18 2018, SIST 96, pp. 144–155, 2019.
https://doi.org/10.1007/978-3-319-92031-3_14

language evidence. Although many investigations have focused upon deduction systems, and dealt with the general problem of developing formalisms able to incorporate these pieces of knowledge, *learning* of formulae has yet been dealt only as a method for incorporating truth onto knowledge bases, and general rules to manipulate those pieces of knowledge have been considered generic rules.

Generally speaking, we can say that there are three ways in which we can derive one specific piece of knowledge: (a) *empirically*, basing the knowledge received by means of the *senses*, (b) *experimentally*, by means of controlled experiences, (c) *deductively*, by deriving a statement from other statement in a consistent way. Recently, scholars have spent significant efforts in investigating the methods that bring knowledge out of controlled experiences in an automated way, including, in particular, inductive methods and sophisticated synthesis methods used to obtain verified knowledge, as happens, for instance, in data mining techniques.

Although many investigations have focused upon the problem of deriving knowledge by means of learning techniques, *the diagnostic process as a logical derivation process*, or, in other terms, *as a reasoning task*, has received too little attention in the community of Artificial Intelligence so far.

In this paper, we start to discuss a long-term approach that aims at going beyond the limits of classical logic in order to provide a framework for reasoning on diagnostic issues in a more adequate way with respect to the way in which this kind of reasoning is performed by humans.

We introduce $t\mathsf{L}$, a logical framework that implements an hybrid learning technique based both on deduction and *experiments*. We use the expressiveness of *Labeled Modal Logic* [20], garnishing a basic propositional language with temporal and statistical information (meta-linguistic additional knowledge imported by labels). Experiments are modelled in terms of *tests* ($t\mathsf{L}$ labels), viewed as bayesian classifier (see Sect. 2), which reveal one or more properties of a sample ($t\mathsf{L}$ formulae). We fully define syntax of formulae and relational rules between labels. We finally sketch main ideas about the full the deduction systems *à la* Prawitz under development and we propose examples on how $t\mathsf{L}$ effectively works.

The paper has the following agenda. Section 2 offers a brief overview about tests and binary classification concepts. In Sect. 3, we introduce the $t\mathsf{L}$ logic, defining basic alphabets, syntax for formulae, test labels (procedures applied during an experiment), and labeled formulae. Section 4 is to formalised central notions of the diagnostic based reasoning, and focuses also on relation between tests. In Sect. 5, we sketch main ideas about work in progress; particular focus is placed on the labeled deduction system currently studied and under development. Finally, Sect. 6 reviews some reference literature.

2 Experiments, Probabilistic Classifiers, and Errors

The $t\mathsf{L}$ logic is able to deal with approximate reasoning. As formally introduced afterwards in Sect. 3, a $t\mathsf{L}$ (well-formed) formula represents a property of a sample (or individual) that can be revealed, with a margin of error, by a suitable experiment, built out from a sequence of tests. Information about tests are represented by labels, metalinguistic logical objects that "adorn" the pure syntactical level of formulae.

Experiments and experimental based reasoning represent the focus of the paper. Before introducing the formal system, we briefly recall some basic notions from statistical information retrieval and learning [12] and we offer a concrete example of diagnostic procedure.

From a mathematical perspective, a test can be interpreted as a statistical classifier, i.e., a function f that, fed with an input a, is able to predict a probability distribution over a set of classes. Oversimplifying, f assigns to a a label y, that represents the answer the classification of a depends on. For example, if f encodes the problem 'Does x enjoys property P?', the output '*yes*' classifies a as an element of the set of objects that enjoys P. In literature, a large taxonomy of probabilistic classifiers has been developed. In this paper, we focus on the simplest type of classifiers, called *(Naive) Bayes (or bayesian) classifiers*, that exploit some strong statistical assumptions [18]. Despite their simplicity, bayesian classifiers work finely in many complex real-world situations.

Classifiers are prone to error. In this context, errors are described either as *false positive* results, or *false negative* results[1]. In the remainder, we omit the word *result(s)* when clear. We also speak of *true positive* and *true negative* for those answers that coincides with the answers given by a logical formula ϕ.

The intrinsic value of the scientific research of this area is to reduce errors in bayesian classifiers, obtaining therefore better methods to derive knowledge from experiments. The following example both provides a concrete instance of diagnostic reasoning and permits us to gently introduce the notions of error, and error taxonomization.

Example 1. Western-Blot is a technique used in biology to confirm the existence of antibodies against a particular pathogenetic factor. This is determined by the application of the test in a manner that can be considered without false negatives[2]. Western-Blot, however, has a number of false positives.

On the other hand, the Elisa test (simply Elisa) is analogously lacking false negatives, but it exhibits a larger number of false positives than Western-Blot when applied to the same pathogenetic factor.

The sequence of tests depends upon their *cost* more than on their reliability. For instance, Elisa is a cheaper procedure than Western-Blot, and thus Elisa is typically applied before than Western-Blot.

Assume that Elisa answers positively on a given sample. We cannot conclude with certainty that the pathogenetic factor is present in the tested organism, due to the high number of Elisa's false positives. We then apply the Western-Blot test to confirm the validity of the Elisa result. We now derive a negative answer.

[1] Roughly speaking, a false positive, commonly called a 'false alarm', is a result that indicates a given condition exists, when it does not, and on the reverse way for false negative.

[2] In principle, tests can be *considered without false negatives* when the number of false negative results is irrelevant to the decision process.

Being that Western-Blot has been assumed without false negatives, we conclude that the pathogenetic factor is not present in the organism, against the provided evidence of the Elisa test.

Example 1 shows a way of deriving truth from tests that is common in those systems. It is straightforward to see that tests with no false negatives that give a negative answer, as well as tests with no false positives that gives a positive answer, are *always* truthful.

To describe a test we use a 2×2 matrix. The four elements represent the true positives, false positives, true negatives, and false negatives. The number of *positive results* is the sum of true and false positives. Symmetrically, the number of *negative results* is the sum of true and false negatives. This representation can be made independent from the sample by substituting the results (true positives, false positives, true negatives, false negatives) with their percentage, computed by the ratio formed by the values over the total size of the sample. The cross sum of true positives and false negative expresses the *actual positives*, whilst the cross sum of true negatives and false positives expresses the *actual negatives*.

	T	**F**
P	δ_{TP}	δ_{FP}
N	δ_{TN}	δ_{FN}

Fig. 1. Confusion Matrix.

	T	**F**
P	$\frac{\delta_{TP}}{T+F}$	$\frac{\delta_{FP}}{T+F}$
N	$\frac{\delta_{TN}}{T+F}$	$\frac{\delta_{FN}}{T+F}$

Fig. 2. Probabilistic Version.

In Fig. 1 we show the called the confusion matrix of a test. In Fig. 2 we show the probabilistic version of the confusion matrix.

The explained concepts represent the statistical foundation of this reasoning framework. Error taxonomy provides a classification of tests and allows us to define (partial) order relation between tests w.r.t. their accuracy in revealing properties (see Sect. 4). In the following sections, we will provide a logical foundation of the notion described above, focusing on test-based learning. In particular, we will formalise main properties of tests and phenomena occurring during the knowledge acquisition, taming the intriguing interplay between different tests.

3 The Logic $t\mathsf{L}$

The alphabet of $t\mathsf{L}$ is built out of the variable symbol x, a denumerable sets of symbols for constants (denoted by c), d (possibly indexed) representing individuals, and a denumerable set of unary predicates, denoted by $P, Q, \ldots$, possibly indexed. Predicates represent properties, i.e., functions that, applied to a individual constant, returns as output an element of a given domain, e.g. a binary or multivalued evaluation.

Properties are *revealed* by *tests*, introduced below in the syntax of labels and not included into the syntax of formulae.

An *atomic formula* is a predicate P. A *ground atomic formula* (*ground formula* hereafter) is an atomic formula of the form $P(c)$, where c is a constant. We dub with gF the set of ground formulae.

We equip the language of $t\mathsf{L}$ with the usual logical connectives: $\rightarrow, \bot, \wedge, \neg$. Formulae are built out of the set of atomic formulae by means of logical connectives. Formally, the set $a\mathcal{F}$ of *well-formed assertion formulae* (only *formulae* or *assertions* in the following, ranged by $A, B, C, \ldots$, possibly indexed) is the smallest set Y such that: (i) $\mathsf{gF} \subseteq Y$, (ii) if $A, B \in Y$ then $(A \wedge B) \in Y$, (iii) if $A, B \in Y$ then $(A \rightarrow B) \in Y$, (iv) $\bot \in Y$, and (v) if $A \in Y$ then $\neg A \in Y$. Literals are formed by letters or negations of letters, applied to constants, i.e., $P(c)$ and $\neg P(c)$.

In the perspective to formalise a labeled natural deduction system (see Sect. 5), we follow the tradition of labeled logical systems [13,14,20], and we extend the above devised syntax by introducing a class of *labels*. Labels represent *experiments*, i.e., instants of time in which tests of properties are performed on a sample, under some environment conditions.

Labels are built out from a set, dubbed $\mathcal{R}$, of symbols for tests ranging on variables τ and ρ, possibly indexed. Tests in label symbols carry on information about *execution time* (the instant in which the test is performed) and *experimental condition* (*condition* hereafter, i.e., the history of actions performed during the experiment). This reflects the fact that a particular test can be *conditioned* by a specific situation (like the environment, a medical condition, etc.). For instance, when you conduct a carrot extraction, in geology, if the terrain is very humid, the stratification can be larger than usual, changing the forecast of the position of water underground.

To formalise these ideas, we introduce the set $\mathcal{T}$ of symbols for time instants (ranging on variable t, possibly indexed) and the set $\mathcal{A}$ of experimental conditions. Symbols for conditions range on variable ϕ and ψ, possibly indexed. In this first investigation, we define $\mathcal{A}$ simply as the set of finite compositional sequence of tests $\tau_1 \ldots \tau_k$, where $\tau_{i+1} \in \mathcal{R}$ has been applied after $\tau_i \in \mathcal{R}$ on the same sample. One can have $\phi = \emptyset$.

Lab_T is a fixed, denumerable sets of labels of the form $\tau^{(t,\phi)}$, where τ is a test able to reveal one or more properties, t represents a time instant (of a given timeline), and ϕ is the experimental condition. Labels range over variable symbols l, r, possibly indexed.

A *labeled (well-formed) formula* is then a formula of the form $\tau^{(t,\phi)} : A$, where $A \in \mathsf{gF}$ is a ground formula. $\tau^{(t,\phi)} : P(c)$ denotes the assertion 'τ reveals P at time t on the sample c, under conditions ϕ'. For instance, if we execute the Elisa test on a sample on monday, having the patient John with fever, to reveal the existence of an infection of Ebola we write $Elisa^{(Monday,Fever)} : Ebola(John)$.

Most fundamental in the evolution of such a system is the existence of a specific way to denote the fact that a formula *is not revealed* by a test. This is not the result of stating that a test reveals the negation of a formula.

We thus introduce a second kind of negation. The epistemic negation '$\sim$' ranges on labeled formula. Note that the two operators $\neg$ and $\sim$ are distinct. In particular, skolemization *does not holds*, i.e., $(\sim (\tau^{(t,\phi)} : A) \nrightarrow \tau^{(t,\phi)} : \neg A)$. In Sect. 4 we introduce partial orders between tests that are consequential to the above definitions.

4 Focusing on Experimental Knowledge: Orders and Relation for Tests and Observable Properties

We now reason on the mechanisation of the experimental reasoning. In other words, we try to understand how to provide a logical foundation of test-based knowledge. To this end, in this paper, we mainly focus on test labels, and on the delicate phenomena occurring during a procedure that aims to extract experimental knowledge from some resources (typically a sample).

In the previous section, we have introduced the class Lab_T. We want to stress out that a test label is something more expressive than a test symbol. A test label represents a test τ put into a context, i.e., equipped with additional information such as its time (when τ is applied) and the history of the experiment, i.e., the trace of previously applied tests (in the same experiment).

We can define a partial order between two test labels, both related to temporal information and on statistically measures for test performances.

We start by defining some temporal orders between labels. We need some auxiliary definition. With $t_1 < t_2$ we denote the usual temporal order between time instants. With $\phi_1 \lhd \phi_2$, we denote the order between conditions and we state that $\phi_1 \lhd \phi_2$ iff ϕ_1 is a prefix of ϕ_2.

Following the labeled logical systems tradition [13,14,20], we state the *relational formulae* below, lifting orders $<$, and $\lhd$ to labels.

Definition 1 (Temporal Relational Formulae)

- $\tau_1^{(t_1,\phi_1)} \ll \tau_2^{(t_2,\phi_2)}$ *iff* $t_1 < t_2$ *and* $\phi_1 \lhd \phi_2$
- $\tau_1^{(t_1,\phi_1)} \mapsto \tau_2^{(t_2,\phi_2)}$ *iff* $t_1 < t_2$, $\phi_2 = \phi \cdot \tau_1$ *and does not exist* $\bar{t}$ *such that* $t_1 < \bar{t} < t_2$.

We are modelling the notion of temporal composition of tests. In particular, whenever represents a general temporal application sequence, the relation $\tau_1 \mapsto \tau_2$ represents the execution of the test τ_2 *immediately* after the execution of the test τ_1. The above introduced notion necessitates the introduction of a system with branching future time (see Sect. 5).

In the following, with a little abuse of the notation, we denote $\tau_1 \mapsto \tau_2$ the test obtained by composing τ_1 and τ_2. We treat $\tau_1 \mapsto \tau_2$ as a symbol in $\mathcal{R}$, and we then use it as a label.

To discuss the possible operators that qualify and relate tests, we need to define some properties on bayesian classifiers. First, for every bayesian classifier τ revealing a formula A, we use the following semantic operator to qualify it:

- We call *precision* $\delta_\tau^+(A)$, which is the ratio of true positives over the number of elements in Δ.
- $\delta_\tau^-(A)$ is the ratio of true negative elements over the number of elements in Δ. The probabilistic complement of this measure is named *recall.*
- $\overline{\delta_\tau^+(A)}$ is the ratio of false positive elements over the number of elements in Δ.
- $\overline{\delta_\tau^-(A)}$ is the ratio of false negative elements over the number of elements in Δ.

Evidently, $\delta_\tau^+(A) + \delta_\tau^-(A) + \overline{\delta_\tau^+(A)} + \overline{\delta_\tau^-(A)}$ equals 1. We name *accuracy* of a test τ the ratio between true values and the number of elements in Δ, thus $a(\tau, A) = \delta_\tau^+(A) + \delta_\tau^-(A)$. When a test has precision 1, it has no false negatives, whilst when it has recall 1, it has no false positives.

The base of empirical reasoning about tests is the deduction of truth on tests that are *correct* (with no false positives), or *complete* (with no false negatives)[3]. We denote by $\Box_+$ the fact that a test has no false positives, and by $\Box_-$ that it has no false negatives.

We now introduce three orders based on test metrics for elements in Lab_T.

Definition 2 (Metric based Relational Formulae)

- *We write* $(\tau_1^{(t_1,\phi_1)} >_a \tau_2^{(t_2,\phi_2)})[A]$ *if* τ_1 *at time* t_1 *and under condition* ϕ_1 *is more accurate than* τ_2 *in revealing* A *at time* t_2 *under condition* ϕ_2.
- *We write* $(\tau_1^{(t,\phi)} >_p \tau_2^{(t,\phi)})[A]$ *if* τ_1 *at time* t_1 *and under condition* ϕ_1 *is more precise than* τ_2 *in revealing* A *at time* t_2 *under condition* ϕ_2.
- *We write* $(\tau_1^{(t,\phi)} >_r \tau_2^{(t,\phi)})[A]$ *if* τ_1 *at time* t_1 *and under condition* ϕ_1 *has greater recall than* τ_2 *in revealing* A *at time* t_2 *under condition* ϕ_2.

We can take $t_1 = t_2$ and $\phi_1 = \phi_2$, and compare test performances at the same step of the same experiment. Definition 2 allows a more refined comparison. Taking into account the timeline and the history of the experiment provides a general case that will be exploited in the following, when we shall model interference between different tests.

When using tests for revealing properties employed in empirical sciences, a given test can interfere in the result of other tests. For instance, when certain therapeutical tests (such as the attempt at solving a dangerous potential batteric infection by the prophylaxis with antibiotics) can make the results of other tests unreliable. Generally speaking, we say that a certain test τ_1 *obfuscates* test τ_2, if performing τ_1 on a sample before τ_2 diminishes τ_2 ability to reveal a given property. On the other hand, τ_1 *gifts* a property, as its application extends the ability τ_2 test to reveal the property itself.

Obfuscation and gift have the flavor of exclusive or and material implication, but they differ insofar as obfuscation and gift involve branching time.

[3] If a test was both correct and complete, then the test would be the property it reveals.

We postpone the discussion of obsfuscation and gift potential mapping onto logical operators to Sect. 5.

Base of the above reasoning is the application of tests in sequence, and therefore, the reason why we have introduced a notion of time. Time is discrete, and that tests are executed at a given instant of time. We introduce a notion of *absolute time* and associate directly temporal instants to test execution only.

Partial obfuscation and *partial gift* can be intuitively described as follows:

- We say that a test τ_1 (for a property A) *a-obfuscates* ($\uparrow_a$) the test τ_2 of a property B if, when τ_1 is executed before τ_2, then the accuracy of $\tau_2 : B$ is less than it would have been if the test τ_1 on A was not executed.
- We say that that a test τ_1 (for a property A) *a-gifts* ($\downarrow_a$) a property B if $\tau_1 : B$, when, contrary to *a-obfuscation*, the accuracy of the test for B increases.

Analogous concepts inspire definitions for p-obfuscation, p-gift, r-obfuscation and r-gift, as referred to obfuscation and gift for precision and recall, instead of accuracy.

More formally, we can provide the following relation, exploiting metric based relational formulae (Definition 2).

Definition 3 (Obfuscation and Gift)

- $(\tau_1^{(t_1,\phi_1)} \uparrow_a \tau_2^{(t_2,\phi_2)})[B]$ iff $t_1 < t_2$ *and* $\tau_2^{(\bar{t},\overline{\phi})} >_a \tau_2^{(t_2,\phi_2)})$ *for* $\bar{t} < t_1$ *or for* $\overline{\phi}$ *s.t.* $\tau_1 \notin \overline{\phi}$.
- $(\tau_1^{(t_1,\phi_1)} \downarrow_a \tau_2^{(t_2,\phi_2)})[B]$ iff $t_1 < t_2$ *and* $\tau_2^{(t_2,\phi_2)}) >_a \tau_2^{(\bar{t},\overline{\phi})}$ *for* $\bar{t} < t_1$ *or for* $\overline{\phi}$ *s.t.* $\tau_1 \notin \overline{\phi}$.

Similar rules for recall and precision can be obtained by replacing relations $\uparrow_a$ and $\downarrow_a$ with the counterparts $\uparrow_p$, $\uparrow_r$, $\downarrow_p$, and $\downarrow_r$.

An interference between tests τ_1 and τ_2 may occur. We write $\tau_1 \bowtie \tau_2$ if τ_1 and τ_2 are *non-interfering*, i.e., if they do not obfuscate, or gift, each other in neither directions.

5 Discussion

In this paper, we proposed some initial step towards the definition of $t\mathsf{L}$, a logical framework able to formalise diagnostic reasoning. We introduced essential notions about experiment-based deduction, following a perspective oriented to reasoning mechanisation. The challenge we are addressing is to couple deductive and test reasoning, dealing both with temporal and approximate reasoning.

The natural interpretation for $t\mathsf{L}$ is strongly related to interpretations of the branching time logic UB [1]. A labeled formula of $t\mathsf{L}$ enjoys a plan semantic in terms of state based interpretation. This is related to at least two reasons. The first one is the temporal based nature of the system, as observed above. The second one deals with the *approximate knowledge* labeled formulae try to model, that couples with a possible worlds interpretation. For the sake of conciseness, this temporal/modal flavour has not been reflected at the syntactical level yet.

We plan the definition of $t\mathsf{L}$ proof theory. The most suitable style is Prawitz' natural deduction. Following [21,22], we are developing a labelled, non-monotonic natural deduction system.

The key point in labeled framework is the interplay between labels and formula as in [22].

In Sect. 4, we have introduced modalities $\Box_+$, $\Box_-$ to express correctness and completeness of test labels. Modals $\Box_+$ and $\Box_-$ relate to accuracy, precision and recall. In fact, when two tests are differently accurate, and both lack false positives, then they are also ordered in the same way by precision. Analogously, when they lack false negatives with respect to recall. This can be formalised as follow:

$$\frac{\tau_1^{(t,\phi)} <_a \tau_2^{(t,\phi)} \quad \Box_+\tau_1 \quad \Box_+\tau_2}{\tau_1^{(t,\phi)} <_p \tau_2^{(t,\phi)}}\ MAP \qquad \frac{\tau_1^{(t,\phi)} <_a \tau_2^{(t,\phi)} \quad \Box_-\tau_1 \quad \Box_-\tau_2}{\tau_1^{(t,\phi)} <_r \tau_2^{(t,\phi)}}\ MAR$$

Interference between $(\tau_1^{(t,\phi)} >_a \tau_2^{(t,\phi)})[A]$ assertions and $(\tau_1^{(t,\phi)} >_p \tau_2^{(t,\phi)})[A]$ or $(\tau_1^{(t,\phi)} >_r \tau_2^{(t,\phi)})[A]$ is managed by means of rules of the following shape.

$$\frac{(\tau_1^{(t,\phi)} >_p \tau_2^{(t,\phi)})[A] \quad (\tau_1^{(t,\phi)} >_r \tau_2^{(t,\phi)})[A]}{(\tau_1^{(t,\phi)} >_a \tau_2^{(t,\phi)})[A]}\ P-R$$

The above rule can be reproduced, analogously, for the accuracy as related to recall and to precision.

Total obfuscation and *total gift* have a specific logical interpretation. This intuitions can be formalised as follow (we omitted the analogous rule for gift, for the sake of space). After the rule, we propose a formalised example.

$$\frac{\tau_1^{(t_1,\phi_1)} : A(c) \qquad (\tau_1^{(t_1,\phi_1)} \uparrow_a \tau_2^{(t_2,\phi_2)})[B] \qquad t_1 < t_2}{\sim \tau_2^{(t_2,\phi_2)} : B(c)}\ totalObf$$

Example 2. We execute Elisa (Eli) to a sample (of an individual named John – referred in the formula as J), with the explicit purpose of testing it for HIV. We execute the test on Monday, under the history of no previous test. The test results positive. Now, since Elisa has no false negatives, but has false positives, and is not particularly accurate, we execute Western-Blot (WB) on Tuesday to reveal the same property. Western-Blot is executed with the history of Elisa, that does not interfere with it. The test results negative. Now, since Western-Blot has not false negatives, we conclude that the sample is HIV-free.

$$\frac{Eli^{(Mon,\emptyset)} : HIV(J) \quad WB^{(Tue,Eli)} : \neg HIV(J) \quad Mon < Tue \quad \Box\_WB^{(Tue,Eli)} \quad \Box\_Eli^{(Mon,\emptyset)}}{\neg HIV(J)}$$

Example 3. Pediatricians often administer antibiotics (An) to children to reveal the nature of an illness. If the child (C) is resistant to the cure, we conclude that the nature is viral (Vir). This prevents further antibiogram tests (Bio), that will result always negative.

$$\frac{An^{(Mon,\emptyset)} : Vir(C) \quad B^{(Tue,An)} : \neg Bio(C) \quad An^{(Mon,\emptyset)} : Vir(C) \uparrow_a B^{(Tue,An)} : \neg Bio(C)}{\sim B^{(Tue,An)} : \neg Bio(C)}$$

Finally, we observe a strong relation between tests and *resources*. In other words, in a more refined framework, a test could reasonably *consumes* a resource in revealing a property. This reflect what effectively happens in a number of laboratory experiments. The above can be connected to a potential *logical linear flavour*.

6 Related Work

Problems of reasoning with diagnostic issues have received significant attention, in particular in the community of Artificial Intelligence in medicine. From the pioneering works of Reiter [17] and Davis [4] the discipline has been strictly connected with two mainstream concepts: case-based reasoning (see [16] for ample and recent references to this approach) and with statistical methods as applied to reasoning, inspired by the original investigation of Johnson et al. [10], and further studied (again, see [11,15] for recent studies).

Although these investigations have been deeply looking at the problems from these two different viewpoints, the value of this study is indeed in combining the two approaches.

Many studies on non-monotonic reasoning are demanded to deal with problems that are quite similar in nature to the ones we dealt with here. In particular it is important to deal with the matter of *preference change* that is crucial to the decision process in medicine. It is common to treat data learned from test, whose meaning *depends* upon further conclusions taken from the data themselves. Consider the case in which a diagnostic process reaches the conclusion that one particular patient suffers of a specific disease. We further can conclude that the patient has not that disease, exactly because another test gives out the opposite conclusion.

Recent studies by some of the authors have dealt with preference revision in non-monotonic frameworks [3,5–9,19], a topic that will be considered within the further investigations following this study, with specific attention to the reason for which we employ preferences in diagnostic reasoning, namely to capture the intuition underlying the abiity of the person doing the diagnosis itself. For instance, in the case of medical diagnosis, this includes the informal knowledge of the environmental aspects, the anamnestical analysis and also the perception of the relevance of a symptom with respect to another one, because of current general situation of the population (the so-called social context).

References

1. Caleiro, C., Viganò, L., Volpe, M.: A labeled deduction system for the logic UB. In: Proceedings of the 20th International Symposium on Temporal Representation and Reasoning, pp. 45–53 (2013)
2. Cristani, M., Burato, E., Gabrielli, N.: Ontology-driven compression of temporal series: a case study in SCADA technologies. In: Proceedings of DEXA Workshop, Turin, Italy (2008)
3. Cristani, M., Rotolo, A.: Meaning negotiation with defeasible logic. Smart Innov. Syst. Technol. **74**, 67–76 (2017)
4. Davis, R.: Diagnostic reasoning based on structure and behavior. Artif. Intell. **24**(1–3), 347–410 (1984)
5. Governatori, G., Olivieri, F., Rotolo, A., Scannapieco, S., Cristani, M.: Picking up the best goal an analytical study in defeasible logic. In: Lecture Notes in Computer Science (including subseries Lecture Notes in Artificial Intelligence and Lecture Notes in Bioinformatics), vol. 8035, pp. 99–113 (2013)
6. Governatori, G., Olivieri, F., Scannapieco, S., Cristani, M.: Superiority based revision of defeasible theories. In: Lecture Notes in Computer Science (including subseries Lecture Notes in Artificial Intelligence and Lecture Notes in Bioinformatics), LNCS, vol. 6403, pp. 104–118 (2010)
7. Governatori, G., Olivieri, F., Scannapieco, S., Cristani, M.: The hardness of revising defeasible preferences. In: Lecture Notes in Computer Science (including subseries Lecture Notes in Artificial Intelligence and Lecture Notes in Bioinformatics), LNCS, vol. 8620, pp. 168–177 (2014)
8. Governatori, G., Olivieri, F., Scannapieco, S., Rotolo, A., Cristani, M.: Strategic argumentation is NP-complete. Front. Artif. Intell. Appl. **263**, 399–404 (2014)
9. Governatori, G., Olivieri, F., Scannapieco, S., Rotolo, A., Cristani, M.: The rationale behind the concept of goal. Theory Pract. Log. Program. **16**(3), 296–324 (2016)
10. Johnson, P.E., Durán, A.S., Hassebrock, F., Moller, J., Prietula, M., Feltovich, P.J., Swanson, D.B.: Expertise and error in diagnostic reasoning. Cogn. Sci. **5**(3), 235–283 (1981)
11. Lawson, A.E., Daniel, E.S.: Inferences of clinical diagnostic reasoning and diagnostic error. J. Biomed. Inform. **44**(3), 402–412 (2011)
12. Manning, C.D., Raghavan, P., Schütze, H.: Introduction to Information Retrieval. Cambridge University Press, New York (2008)
13. Masini, A., Viganò, L., Zorzi, M.: A qualitative modal representation of quantum register transformations. In: ISMVL 2008, pp. 131–137 (2008)
14. Masini, A., Viganò, L., Zorzi, M.: Modal deduction systems for quantum state transformations. Mult. Valued Log. Soft Comput. **17**(5–6), 475–519 (2011)
15. McShane, M., Beale, S., Nirenburg, S., Jarrell, B., Fantry, G.: Inconsistency as a diagnostic tool in a society of intelligent agents. Artif. Intell. Med. **55**(3), 137–148 (2012)
16. McSherry, D.: Conversational case-based reasoning in medical decision making. Artif. Intell. Med. **52**(2), 59–66 (2011)
17. Reiter, R.: A theory of diagnosis from first principles. Artif. Intell. **32**(1), 57–95 (1987)
18. Rish, I.: An empirical study of the naive Bayes classifier. In: IJCAI Workshop on Empirical Methods in AI (2001)

19. Tomazzoli, C., Cristani, M., Karafili, E., Olivieri, F.: Non-monotonic reasoning rules for energy efficiency. J. Ambient Intell. Smart Environ. **9**(3), 345–360 (2017)
20. Viganò, L.: Labelled Non-Classical Logics. Kluwer Academic Publishers, Dordrecht (2000)
21. Viganò, L., Volpe, M., Zorzi, M.: A branching distributed temporal logic for reasoning about entanglement-free quantum state transformations. Inf. Comput. **255**, 311–333 (2017)
22. Viganò, L., Volpe, M., Zorzi, M.: Quantum state transformations and branching distributed temporal logic. In: 21st International Workshop, WoLLIC 2014, Valparaíso, Chile, 1–4 September 2014. Lecture Notes in Computer Science, vol. 8652, pp. 1–19 (2014)

Scalability of Dynamic Lighting Control Systems

Leszek Kotulski and Igor Wojnicki(✉)

Department of Applied Computer Science, AGH University of Science and Technology, Al. Mickiewicza 30, 30-059 Krakow, Poland
wojnicki@agh.edu.pl

Abstract. The lighting standards allow to dim the lighting when the road traffic decreases. A control system gathers information from sensors and generates proper dimming levels for lighting points. The *Dual Graph Grammars* has been proposed as a formal background to maintain the information structure for such a control system. It results in separation of sensors structure from lighting infrastructure. It enables taking into account complex geographical distribution of sensors and logical dependencies among them, which leads to more precise and energy efficient control. What is more important it decreases the control system's computing power requirements by reducing the problem size during run-time. The approach has been verified in practice by deployment to a control system which manages 3,768 light points. Experimental results show a reduction of the computation time by a factor of 2.8 in this case and quickly grows when number of sensors increases. It makes the control system to be scalable in IoT environments.

Keywords: Graph transformations · Dual graph grammar
Road lighting · Street lighting · Efficiency

1 Introduction

The market of modern lighting systems based on LED technology still develops. McKinsey predicts that in 2020 the LED market share will be more then 70% [1]. The lighting standards [2,3] allow to lower the lighting class when the traffic intensity is decreases. The possibility of efficiently dimming LED lamps in broad range creates technological background for implementing this aspect of the lighting norms.

A dynamic control concept has been validated by using a test bed of 3,768 light points in Kraków, Poland and resulted in more than 70% energy consumption reduction [4]. It is split into 40% being an effect of more efficient light source technology (LED instead of HPS) and 30% which comes from optimization of the lighting design by AI systems and introduction of the dynamic control system [5]. AI design prepares multiple variants of optimal lighting settings for all required

G. Jezic et al. (Eds.): KES-AMSTA-18 2018, SIST 96, pp. 156–163, 2019.
https://doi.org/10.1007/978-3-319-92031-3_15

lighting classes (in this case M2, M3, M4 and M5) and the ambient lighting levels. The control system applies the proper variant according to determined (by sensors) conditions; the theoretical background, namely the Control Availability Graph (CAG), was introduced in [6]. The dynamic control adjusts lighting levels on demand, based on the current needs [7,8]. There are two types of detectors taken into consideration: traffic intensity and ambient light. Based on data from the sensors appropriate lighting class, defined by lighting standards, is selected, which in turn leads to appropriate dimming of selected luminaires.

The CAG structure stores all information necessary to control such a system, which is:

- sensor infrastructure (represented as graph nodes),
- light point distribution (represented as graph nodes),
- lighting segments identifying streets to be illuminated (represented as nodes),
- lighting configurations (design variants, represented as graph nodes),
- light point dimming values for each of the considered variant (represented as edges).

By processing this structure appropriate actions, namely luminaire dimming, are carried out.

The main motivation for the paper is to increase scalability of the proposed graph-based control system. It has been achieved by developing a formal tool: *Dual Graph Grammar* [9] which enhances graph processing efficiency. Particularly, it addresses the problems described below.

A general efficiency problem for graph processing algorithms regards their computational complexity that depend on the whole graph size, which is polynomial. Thus processing efficiency is an important issue, especially that such a graph has to take into consideration hundred thousands of lighting points.

The second problem arises from the fact that gathering different types of information without logical separation creates problems with maintenance or modification of such. Modifications take place mainly in the context of sensor infrastructure which becomes more and more complex in time, and delivers more and more information which is a result of Smart City initiatives [10]. In Kraków, the number of ambient light sensors for outdoor lighting has increased 15 times over just one year (2016). Instead of a single sensor in an electric cabinet which supplies three circuits of 50 luminaires each, there is one for every 10 luminaires, mounted on the lamp pole. Similarly, the number of traffic intensity sensors has increased as well. For a typical intersection, the number of traffic intensity detectors (induction loops) increased by a factor of 2. For example, in case of a 4-way intersection with 10 inbound and 6 outbound lanes, it has increased from 10 to 20.

In the paper we present results of computational complexity analysis of applying the proposed new formalism: *Dual Graph Grammars*. It allows to introduce separate heterogeneous information into two graphs and creates a possibility of cooperation between them in a form of synchronization. The following section presents formal properties of the proposed formalism.

2 Dual Graph Grammar Representation

Graphs are formally generated by graph grammars, that ensures their correctness.

Definition 1. *Let us define the following. A* graph grammar Ω *is a tuple:*

$$\Omega = (\Sigma_\Omega, \Gamma_\Omega, \Delta_\Omega, \Phi_\Omega, S_\Omega, \Pi_\Omega)$$

where:

- Σ_Ω *is the set of node labels,*
- $\Delta_\Omega \subset \Sigma_\Omega$*, is the set of terminal node labels,*
- Γ_Ω *is the set of edge labels,*
- Φ_Ω *is the set of transformation rules,*
- S_Ω *is the starting graph,*
- Π_Ω *is the validation graph grammar condition, that verifies the current state of the graph.*

Let us introduce the *Dual Graph Grammar* representation, in which two graph grammars are used to generate a consistent model in such a way that they are separate graphs with common elements. The synchronization of the both graphs is provided by the validation properties associated with each of the grammars. As a result, the two graphs have separate purposes: one is used for control, and the other – to express relationships among sensors.

Definition 2. *A* Dual Graph Grammar $[\Psi, \Theta]$ *is a pair of graph grammars such that and the following condition is fulfilled:*

- $\Sigma_\Psi \cap \Sigma_\Theta \neq \varnothing$.
- *Applications of transformation rules on a graph are atomic. It means that two productions on a single sub-graph are mutually exclusive.*

Definition 3. *A* Dual Graph Grammar $[\Psi, \Theta]$ *is synchronized if both validation conditions* Π_Ψ *and* Π_Θ *are met.*

Let us explain how the proposed *Dual Graph Grammar* notation supports the problem of modeling dynamic lighting control. The basis for this concept is the notion of a *virtual detector*. It represents a data source with values calculated based on information from actual detectors. There is only one virtual detector of a given type in relation with a lighting segment. An example graph is given in Fig. 1. Nodes labeled by *dva* represent virtual ambient light detectors. Similarly, nodes labeled by *dvt* represent virtual traffic intensity detector. We assume that *dva* and *dvt* are the only node labels that belong to $\Sigma_\Psi \cap \Sigma_\Theta$. The virtual detectors are grayed.

The Ψ graph grammar will provide the logical relationships among entities; it enables control by defining a new Control Availability Graph: CAG_Ψ. Its role is to:

- maintain information about the lighting infrastructure to be controlled,
- optimize the internal structure of the lighting system (for more details see [11]),
- evaluate the current CAG_Ψ parameters to provide actual control.

The Θ grammar will define the evaluation rules for virtual detectors – a Detector Structure Graph (DSG_Θ). Its role is to:

- maintain information about the detectors' logical structure and evaluation conditions of sensory data resulting in virtual detectors' values,
- evaluate the virtual detectors' values.

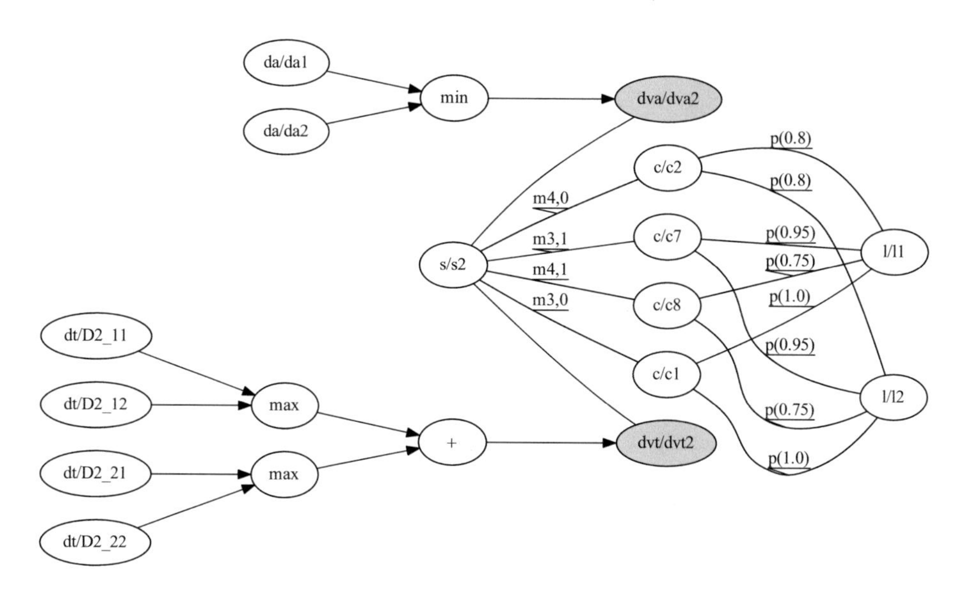

Fig. 1. A combined CAG_Ψ and DSG_Θ, ambient light intensity detectors da, traffic intensity detectors dt, arithmetic functions max, min, $+$, virtual traffic intensity detectors dvt, virtual ambient light detectors dva, road segments s, lighting configurations c, luminaires l.

Let us present important aspects of CAG_Ψ and DSG_Θ definitions.

Definition 4. *CAG_Ψ is generated by the Ψ grammar. It is defined as an attributed graph over the set of node labels Σ_Ψ and the set of edge labels Γ_Ψ such as:*

$$CAG_\Psi = (V_\Psi, E_\Psi, lab^V_\Psi, lab^E_\Psi, \Sigma_\Psi, \Gamma_\Psi, att^V_\Psi, att^E_\Psi, A^V_\Psi, A^E_\Psi)$$

where:

- $\Sigma_\Psi = \{s, l, dvt, dva, c\}$ *is a set of node labels, where:*

   - *s represents segments,*
   - *l represents luminaires,*
   - *dvt and dva represent virtual traffic and ambient light intensity detectors, respectively,*
   - *c represents a configuration, which is a set of adjustments, associated with a relevant set of luminaires at a given segment,*
- $\Gamma_\Psi = \{(m2, x), (m3, x), (m4, x), (m5, x)\}$ *is a set of edge labels, where: $m2$, $m3$, $m4$, $m5$ are lighting classes, $x \in \langle 0, 6 \rangle$ represents ambient lighting level.*

Each segment s is connected by an edge with exactly one dvt and one dva. Formally, the validation condition Π_Ψ of the grammar is as follows:

$$\begin{aligned}\forall v_1, \in V_\Psi : lab^V_\Psi(v_1) &= l, \\ &\exists! v_2, \in V_\Psi : lab^V_\Psi(v_2) = dvt, \\ &\exists! v_3, \in V_\Psi : lab^V_\Psi(v_3) = dva \\ &\exists v_4 \in V_\Psi : lab^V_\Psi(v_4) = s, \\ &\exists v_5 \in V_\Psi : lab^V_\Psi(v_5) = c, \\ &\{(v_3, v_4), (v_2, v_4), (v_4, v_5)(v_5, v_1)\} \subset E_\Psi \end{aligned} \tag{1}$$

Definition 5. *DSG_Θ is generated by the Θ grammar. It is defined as an attributed directed graph over the set of node labels Σ_Θ and the empty set of edge labels Γ_Θ, such as:*

$$DSG_\Theta = (V_\Theta, E_\Theta, lab^V_\Theta, \Sigma_\Theta, att^V_\Theta, A^V_\Theta)$$

where:

- $\Sigma_\Theta = \{dvt, dva, dt, da, +, -, *, /, min, max, val\}$ *is a set of node labels, where:*
   - *dvt and dva represent virtual traffic and ambient light intensity sensors, respectively,*
   - *dt and da represent actual traffic and ambient light intensity sensors, respectively,*
   - *$+, -, *, /, min, max$ represent arithmetic operators,*
   - *val represents constant arbitrary values.*

Formally, the validation condition Π_Θ makes sure that there are no non-terminals in the graph:

$$\forall v \in V_\Theta : lab^V_\Theta(v) \in \Delta_\Theta \tag{2}$$

While DSG$_\Theta$ is being built there may be vertices with labels being-non terminals. In such a case, the graphs are not in a synchronized state since according to Definition 3, the validation conditions are not met.

Let us present how applying the *Dual Graph Grammar* concept to the CAG representation influences processing efficiency of the control system.

3 Graph Properties

The CAG graph is interpreted by the control system in order to provide power setting for each of the luminaires for given traffic and ambient light intensities. The optimal solution is to be found in a multi-dimensional space spanned over available road segments, luminaries, sensors and configurations. Thus, formally the computational complexity is polynomial i.e. $O(n^4)$ where n is a number of nodes in the CAG graph, since a sub-graph to be identified consists of such four nodes representing the above. Real application of CAG results in a large graph size, since number of luminaires and sensors reach hundred of thousands for cities. What is more, with smart city concepts being implemented the number of sensors constantly grows.

Thus, one of the important feature of the *Dual Graph Grammar* is that it separates the sensor layer (DSG_Ψ) from the control layer (CAG_Θ) by introducing *virtual sensors*. This ensures that the growth of the sensory infrastructure will not require changes in the CAG_Θ graph, which is also important from the maintenance point of view.

It should be noted that the definition of the Ψ graph grammar which generates the DS_Ψ complies with arithmetical expression parse trees. It enables evaluation of any virtual sensor value with linear computational complexity $O(n)$.

The computational effectiveness of the control algorithm has been evaluated in the real environment of the pilot deployment in Krakow, Poland which regards 3,768 luminaries. The number of luminaires is the main parameter indicating the scale of deployment. Number of segments is derived from it (statistically there is 23 lamps per segment), as well as the number of available configurations (statistically 3 lighting classes per segment). There is constant 2 virtual sensors per segment. The number of sensors has been varying and it will grow in time. We consider a starting value of 0.7 sensors per lamp (as initially for the pilot) up to 3.5 sensors per lamp. It was verified that the real computation time comparing the dual graph grammar approach with separated of DSG_Ψ and CAG_Θ with the CAG graph has been reduced by 2.8 times. The computation time is estimated by $\#G^p$ function where $\#G$ is a number of edges in the graph. The increase in computational efficiency is calculated by the following function:

$$i.e. = \frac{(\#CAG)^p}{(\#CAG_\Psi)^p + \#CAG_\Theta}$$

If $ie = 2.8$ then p is equal to 2.7.

Let's note that the function of the type $\frac{(a+b)^p}{a^p+b}$ is horizontally asymptotic. For p greater than 2 it quickly converges for a and b greater then 10. It means that the number of luminaries alone does not influence the computational efficiency. A strong influence comes from dependencies between number of luminaries and sensors. Since number of sensors is expected to grow in future, due to IoT and Smart City deployments, the estimated computation time reduction grows much quicker. Details are given in Table 1.

Table 1. Number of sensors per lamp and computation time reduction relationship.

Number of sensors per lamp	0.7	1.4	2.1	2.8	3.5
Computation time reduction	2.83	6.56	12.43	20.85	32.17

4 Conclusions

There are two major benefits of applying the *Dual Graph Grammar* approach to modeling outdoor lighting control systems. First, it enables to clearly separate control logic (CAG_{Ψ}) from complex relationships among sensors (DSG_{Θ}). Since a formal graph grammar for synchronization can be defined, the interactions between the two graphs are straightforward. The graphs can be processed independently as long as the synchronization conditions are met. Furthermore, it reduces the risk of update issues or further development of the model. When the updates regard deploying new sensors or retiring old ones the DSG_{Θ} is modified only by the corresponding Θ grammar. It guarantees that the control logic (CAG_{Ψ}) remains intact. On the other hand any update of the control logic requirements, which would cause the control system to behave differently, will be supported by Ψ graph grammar. Such an update will take place if, for example, lighting standards are updated (it is currently an issue in Poland, since the norms are being updated from 2004 to 2014 revision).

Second, it provides significant processing power reduction as a result of graph size reduction. In other words computation time decreases, because the computational complexity of the control algorithms, which depends on the graph size, is polynomial. The improvement of the processing efficiency leads to better scalability of the control system.

An increase of number of sensors per luminaire up to 3.5 is expected. However, if the control system performance is not increased by a factor of 32.17, as indicated in Table 1, its applicability to successfully control lighting would be questionable.

Acknowledgments. Funding: this work was supported by the AGH University of Science and Technology grant number 11.11.120.859.

References

1. McKinsley and Company: Lighting the Way: Perspectives on the Global Lighting Market (2011)
2. CEN: CEN/TR 13201-1:2004, Road lighting. Selection of lighting classes. Technical report. European Commitee for Standardization, Brussels (2004)
3. CEN: CEN/TR 13201-1:2014, Road lighting – Part 1: Guidelines on selection of lighting classes. Technical report. European Commitee for Standardization, Brussels (2014)
4. Sędziwy, A., Kotulski, L.: Towards highly energy-efficient roadway lighting. Energies **9**(4), 263 (2016)

5. Wojnicki, I., Ernst, S., Kotulski, L.: Economic impact of intelligent dynamic control in urban outdoor lighting. Energies **9**(5), 314 (2016)
6. Wojnicki, I., Ernst, S., Kotulski, L., Sędziwy, A.: Advanced street lighting control. Expert Syst. Appl. **41**, 999–1005 (2013)
7. Guo, L., Eloholma, M., Halonen, L.: Intelligent road lighting control systems. Technical report. Helsinki University of Technology, Department of Electronics, Lighting Unit (2008)
8. Merlino, G., Bruneo, D., Distefano, S., Longo, F., Puliafito, A., Al-Anbuky, A.: A smart city lighting case study on an openstack-powered infrastructure. Sensors **15**(7), 16314–16335 (2015)
9. Wojnicki, I., Kotulski, L.: Improving control efficiency of dynamic street lighting by utilizing the dual graph grammar concept. Energies **11**(2), 402 (2018)
10. Neirotti, P., De Marco, A., Cagliano, A.C., Mangano, G., Scorrano, F.: Current trends in smart city initiatives: some stylised facts. Cities **38**, 25–36 (2014)
11. Wojnicki, I., Kotulski, L.: Street lighting control, energy consumption optimization. In: Rutkowski, L., Korytkowski, M., Scherer, R., Tadeusiewicz, R., Zadeh, L.A., Zurada, J.M., (eds.) Artificial Intelligence and Soft Computing - 16th International Conference, ICAISC 2017, Zakopane, Poland, 11–15 June 2017, Proceedings, Part II. Lecture Notes in Computer Science, vol. 10246, pp. 357–364. Springer (2017)

Automatic Detection of Device Types by Consumption Curve

Claudio Tomazzoli[1(✉)], Matteo Cristani[1], Simone Scannapieco[2], and Francesco Olivieri[3]

[1] Dipartimento di Informatica, Università di Verona, Strada Le Grazie 15, Verona, Italy
{claudio.tomazzoli,matteo.cristani}@univr.it

[2] R&D Department, Real T, Viale venezia 7, Verona, Italy
simone.scannapieco@realt.it

[3] Data61, 70-72 Bowen Street, Spring Hill, Brisbane, QLD 4000, Australia
francesco.olivieri@data61.csiro.au

Abstract. This work deals with the problem of automatic detection of device types given only the power consumption curve, which can be obtained by means of a cheap measurer applied to the device itself. We defined a novel method to detect these types and we describe it in details, providing ground truth evidence coming from the application of the method to real world data. We tested the method against two different set of data coming from two separate and different environments, the first located in Italy and the second in Germany, and we provide experimental results to support the method.

1 Introduction

The need for energy is constantly increasing, while the traditional sources of energy are limited. Both the civil society and the industry are becoming more eco-friendly and new sources of energy are being explored.

These new sources such as wind or solar energy need investments, and also cannot totally replace the existent sources of energy in industrial consumption. Actual transmission infrastructures can hold only a certain amount of energy unless they are upgraded with even more relevant expenses. These considerations lead to the understanding of the importance of energy saving when dealing with the problem of energy management.

Energy consumption concerns millions of houses, schools, hospitals, and business buildings, each one implementing its own policies with little or no knowledge about others and their energy behavior.

Several technologies for energy saving have been developed in the past; these implement solutions based either on specific techniques of computer technology or on the integration of computer technologies and electrical engineering solutions. In particular, there have been developed solutions which consider the *a priori* declared consumption behavior of devices connected to a system. Recently

G. Jezic et al. (Eds.): KES-AMSTA-18 2018, SIST 96, pp. 164–174, 2019.
https://doi.org/10.1007/978-3-319-92031-3_16

have been proposed a few solutions that learn the consumption behavior of these devices, by direct measures of the appliances' actual load, and allow for the setting of rules [7–9,20]. Each device is associated with an expected electrical consumption and, based on currently active ones, the command unit estimates the electrical consumption of the whole building. In this way, a rule can be set so that the command unit does not allow the activation of further electric loads if the electrical consumption of the overall installation reaches a predetermined threshold.

These rules are useful when they implement an automatic behaviour for the devices. The mapping from real word to the abstract computer model of an installation is usually human made, so that there is a considerable risk of misalignment over time due to errors, laziness and simple inaccuracy in updating the parameters. To avoid this risk we face an even greater complexity, because to do so we should be able to deduce automatically which appliances are part of a given installation.

As a consequence, the problem of devising a method to automatically detect device type is an interesting and challenging one; in this paper we deal with this problem.

The paper is organized as follows: we discuss in Sect. 2 the motivation for this work, while in Sect. 3 we describe the method employed and present its main steps. We give out experimental results that support the method correctness in Sect. 4, and revise the current relevant literature and related work in Sect. 5. Finally, we sketch some conclusions in Sect. 6.

2 Motivation

When technologies for energy saving are in place, the burden of determining the best behavior lies on the shoulder of the person in charge of energy management; however it is not trivial to define an optimal set of rules. A possible solution is the adoption of best practices, which are hard to define being all installations different.

While everyone can rightfully claim to be entitled to implement his own rules, he can surely benefit from the knowledge of best practices for his energy configuration. Every office is different from the others, but two offices equipped with ten personal computers, two printers, one vending machine and seven light sources can be considered similar, from the electrical energy consumption point of view; thus their efficiency can be compared. The behavior of the one with a cheaper electricity bill might be considered better than the other one, and its rules should be regarded as best practice.

A general best practice cannot therefore exist, but different best practices which fit for different groups of similar installations may exist. For this purpose, we have to create groups, or clusters, of similar installations, examine each group and extract the rules that are acceptable for all the plants in a particular group. Accordingly, the method relies on three main steps:

1. clustering all the plants;
2. find the better performing one on each cluster;
3. make the synthesis of the rules for the set of measures of these plants, so that these rules can be applied to other similar installations.

Clustering all the plants under observation. In real life not all installations are the same: in one home we can have one fridge, forty lamps and one oven (such as in an average flat) while in one other we can have two fridges, two ovens and twenty lamps (for instance, in a house with a garden). We have to define a procedure to determine whether a plant is similar to another one, so that they can be considered as belonging to a common group. In other words, the problem is to define a clustering for these installations. In computer science and in machine learning in particular, the term "clustering" leads effortlessly to think of multidimensional vector space in which cluster of points are defined as points whose distance is smaller than a given value. This, in turn, indicates that if we might be able to define a reasonable vector representation of an installation, this model can be effectively used to solve the problem of grouping plants by similarity.

Consider a system S with three installations p_1, p_2, p_3 where in plant p_1 there are two printers, ten computers and one coffee machine, in p_2 there are one fridge, two televisions, one washing machine and one oven, while in p_3 we have five computers, one coffee machine, one fridge and one printer. Each device can be considered as belonging to a *class* and the value of the ith component of the vector representing a plant will be number of appliances of the corresponding class. In this case, the classes will be $C =$ {Printer, Computer, Fridge, Coffee machine, Television, Washing machine, Oven} and the vector representation of the installations will be $p_1 = (2, 10, 0, 1, 0, 0, 0)$ $p_2 = (0, 0, 1, 0, 2, 1, 1)$ $p_3 = (1, 5, 1, 1, 0, 0, 0)$

Once we have the feature representation of the systems, we can apply the well known k-means clustering algorithm [18] to find the group of similar systems.

Finding the better performing plant. We are now able to find the best performing plant, one per cluster. It is a straightforward task given only the energy bill over a few months, but it can also be performed in detail using the energy data collected over time from the appliances in the plant. The desired best performing plant is simply the installation whose consumption is minimal with respect to all other plants in the same cluster over a given interval of time.

Rules synthesis from consumption data. Data collected from devices are relative to the power consumption of the electric load itself and to the time at which the data has been collected. Without lack of generality we consider these to be a dataset $D = d_1, ..., d_k$ in which every element has the form $d_i = \{plant, device, powervalue, timestamp\}$ so that each record bears information identifying the device (within an installation) and its consumption at a given time. Given that a configuration of a plant at a given time is the set of the statuses of all belonging devices at that time, we extract from that dataset all the configurations of a plant in a given interval of time (say, a week). The times at which a change occurs in the configuration of the plant are defined "relevant moments".

In a given installation, we shall have several records with the tuple of the form [device, power, timestamp] where we can assume timestamps are almost synchronized. We shall than record, for each relevant moment, which device was on and which was off and we can verify if there are times when some device is active while another one is inactive. This leads to think to association rules and therefore we might try and mine rules from data using the knowledge borrowed from the study of "association rules mining".

What we achieve is a set of powerful rules which do not only rely on time constraints but also on contemporary associations of active devices, so that we can infer rules such as "when the oven is on, the washing machine shall be off".

Automatically detecting device type. Environments are not immutable and energy load change over time sometimes in an unpredictable way; a trivial example being an office in which vending machines and printers may be added or removed without previous notice to the energy manager. To ensure that the method will work the composition of the installation at any given time shall be given for granted. The constant update of the connected device database is a time consuming task and the resulting process is likely to be prone to a number of errors. This lead to the risk of invalidating the whole method based on best practices, so a solution to the problem of "having the correct representation of which are the actual loads in a plant" is in order. A possible way of solving the problem is using machine learning to automatically detect and recognize a device based on measurements.

3 The Framework

Consider the power consumption of a device measured at several moments. This can be viewed as a function of power over time. We give an alphabetical representation of the power in a given interval of time representing only the angles between one measure and the following one. We use the angle because we want to be insensitive to scaling; a big fridge and a small fridge are similar appliances, with similar power figures, only the bigger one is scaled because it consumes much more power; the same applies to lights and all other electric loads. We establish a mapping function from two consecutive measure to a letter of the alphabet (as in Fig. 2) so that the "trace" of a given device (like a printer) in a given interval of time shall be in the form of a sequence of characters, like for example OOOOVOOOOOPOOOOOOOOOOVOOOOOPO.

We are mainly interested in changes, so dropping to a single "O" all the sequences of multiple "O"s we have a string in the contracted form which will looks like "OVOPOVOPO".

After the transformation, a sequence of energy measures is a sequence of strings like the one above, or in other terms the measures over a given time frame are transformed in sequence of words, something like a little text footprint

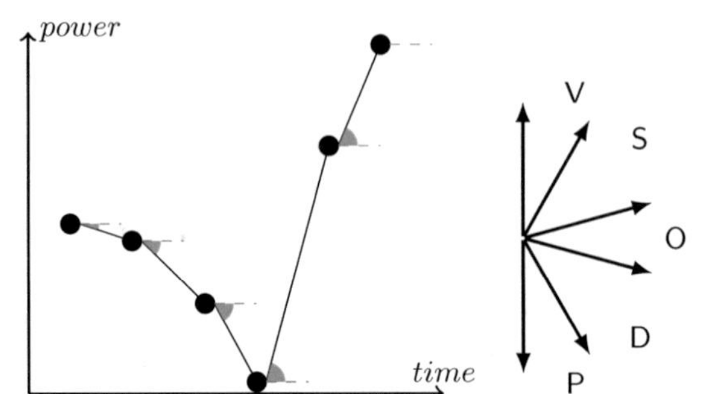

Fig. 1. Points are power values of a device at given times

Fig. 2. Alphabetical transformation given the angle

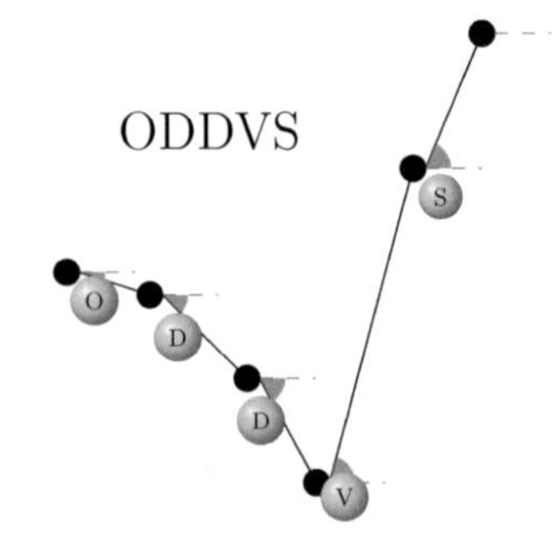

Fig. 3. Alphabetical representation of power over time

of the device under inspection. Consider, for instance, the function of power expressed in Fig. 1. This function is expressed in string form by means of the above mentioned transformation as in Fig. 3.

Here we can apply algorithms from the well known field of text mining to extract which devices are similar; devices with many words in common are more than similar than ones with few. The bag-of-words model is a simplifying representation used in natural language processing and information retrieval. In this model, a text is represented as an unordered collection of words, disregarding grammar and even word order. Using this model we shall have bag of energy words for each load, an example of which is given in Table 1.

Table 1. Example bag of energy words

Word	Test devices			
	Coffee machine	Printer	Boiler	Acquarium
OVOPO	1	35	0	0
OVPDO	1	1	0	0
OSDO	35	1	0	0
OVPO	92	36	0	0
OVDPO	0	1	0	0
OSSPO	1	2	0	0
ODPO	1	0	0	0
OVSO	2	2	0	0
OPO	0	0	77	2
OVO	0	0	77	2

The Naive Bayes text classifier is a simple probabilistic classifier which is based on Bayes theorem with strong and naive independence assumptions. It is one of the most basic text classification techniques and is based on the assumption that the position of the words in a text is irrelevant, which is exactly the case in word. We use a Naive Bayes text classifier to classify electric loads: given a set of known devices, which can grow over time, we can classify unknown ones by comparing them with known (labeled) ones, thus automatically detecting device type.

4 Experimental Results

We tested the hypothesis, that a device type can be inferred looking only at the consumption curve, using electrical consumption data taken from the real word.

Our test dataset included data, taken every half minute, from ten different devices over a period of over one year. We used time frames of one hour, one day, one week and extracted the text footprint of energy consumption of each device. We computed the trace of each device for every single day, so that one feature vector is the result of only one day of observation. Our dictionary of different "words" counted more than 68.000 items the vast majority of which occurred only once or twice; to avoid dimensionality problems we considered only those word which appeared at least three times in all the textual traces. This reduced the problem's size at least tenfold, so that the resulting vector space ended up having vectors with 3706 components. We than extracted the bag of word for each footprint, a sample of which is given in Table 1.

We executed the test using the Bayesian Naive classifier and a 10-fold cross-validation, obtaining the following results, better detailed in terms of accuracy by class in Table 2, while in Table 3 we provide the confusion matrix. Generally speaking we have 842 (the 75.5%) of correctly classified instances and 274 (the 24.5%) of incorrectly classified ones.

Table 2. Detailed accuracy by class

	Precision	Recall	F-measure	ROC area
Printer	0.974	0.516	0.675	0.881
Aquarium	0.524	0.253	0.341	0.92
Coffee machine	0.959	0.954	0.957	0.996
Fridge	0.773	0.98	0.864	0.993
Boiler	0.604	0.898	0.722	0.947

As can be seen, different devices often show textual footprints that can be taken apart one from the other. We have been able, at present time, to recognize printers, coffee machine, boilers but we have not been able to distinguish between different resistive loads with simple behavior like lamps or aquariums.

Table 3. Confusion matrix

	Classified as				
	a	b	c	d	e
a = Printer	113	12	7	48	39
b = Aquarium	1	43	0	15	111
c = Coffee machine	0	0	209	10	0
d = Fridge	2	1	2	248	0
e = Boiler	0	26	0	0	229

The CoSSMic[1] project is a public European project whose aim is to allow household and neighborhood optimisation and power sales to the network, in addition to a higher degree of predictability of power deliveries for the large power companies. Among the deliverables of the project there are data regarding measures of electrical consumptions in several small businesses and private households relevant for household- or low-voltage-level power system modeling. The starting point for the time series, as well as data quality, varies between households, with gaps spanning from a few minutes to entire hours. Measurements were initially conducted in 3-min intervals, later in 1-min intervals. Data for both measurement resolutions are published separately in large CSV files whose dataset contains load information according to the following naming conventions:

- heat_pump: Heat pump energy consumption
- dishwasher: Dishwasher energy consumption
- washing_machine: Washing machine energy consumption
- refrigerator: Refrigerator energy consumption
- freezer: Freezer energy consumption

This data package contains the measurements of the German trial site in Konstanz going from October 2015 up to February 2017.

We executed the test using the Bayesian Naive classifier and a 10-fold cross-validation, obtaining the following results, better detailed in terms of accuracy by class in Table 4, while in Table 5 we provide the confusion matrix. Generally speaking we have 478 (the 69.38%) of correctly classified instances and 211 (the 30.62%) of incorrectly classified ones.

5 Related Work

To the best of our knowledge, the present work is one of the first attempts to define a method for automatic detection of device types given only the power consumption curve. A few recent investigations [16,24] showed how the problem might be dealt with smart electricity meters, and devise a general framework for

[1] http://cossmic.eu/.

Table 4. Detailed accuracy by class

	Precision	Recall	F-measure	ROC area
washing_machine	0.453	0.368	0.406	0.843
dishwasher	0.42	0.638	0.507	0.875
freezer	0.834	0.824	0.829	0.943
heat_pump	0.958	0.729	0.828	0.884
refrigerator	0.882	0.926	0.903	0.951

Table 5. Confusion matrix

	Classified as				
	a	b	c	d	e
a = washing_machine	113	12	7	48	39
b = dishwasher	53	68	17	1	5
c = freezer	41	74	1	10	0
d = heat_pump	16	1	126	2	8
e = refrigerator	3	1	3	2	112

smart homes based on intelligent agents. However, the effort we put on is only based on the consumption curve and it is therefore applicable to every situation in which consumption data can be read by a local detector, such as a smart plug.

Different efforts have been spent for decreasing energy consumption [6,19]. The first one aims at conceptual models of human behaviour in energy saving contexts e.g., exploiting practical reasoning and psychological theories for a responsible use of energy consuming elements [25]. Energy saving is envisioned as an optimization problem where the aim is to discover programming models and algorithms that efficiently lower energy usage and ensure a proper task scheduling in domestic environments [1], or large-scale buildings [27]. In [18] the relation of the buildings energy consumption information which is directly correlated with the production of CO_2 emissions have been considered. The analyses of energy consumption information can lead to the gathering of business-relevant data. For instance, in [18] authors show the linear correlation between the incoming of a person/family with the consumption of electricity through a statistic approach. Another way for decreasing energy consumptions is to teach users to be eco-friendly; in [26] authors describe a sociological study where tests and analyses are performed in a way to decrease the consumption of energy and to get the dynamic energy consumption indicators. Conflicting rules and rules with exceptions are common in natural language specification employed to describe the behaviour of devices operating in a real-world context. In [10,21] authors provide a general method for rule processing using defeasible rules as applied to human interfaces to smart environments, in the specific case of energy saving. Information and communication technology industry is one of the industries that is putting several distinct efforts for decreasing its impact on the global environment [3]. There is a need to reduce the substantial electricity consumption of networks themselves [15]. A lot of researches on sustainable information and communications technology have been focused on new technologies, for example optical IP networks [2,23], data center architecture and design [11].

6 Conclusions and Further Work

We recognized the importance of a correct representation of the power plant; since the device database update process is prone to errors, we defined a measurement based algorithm which use machine learning to automatically detect and recognize electric loads. In this paper we go further than has been done in [22] where authors highlight the problem and envision a draft solution for clustering different installations which can be defined as having similar energy needs. We tested it against two different set of data coming from the real world, from two environments the first located in Italy and the second in Germany. We plan to test the method and algorithms against a larger dataset and over a longer time frame and to include the method in a more general method whose aim is to help making correct automatic energy consumption behaviours.

A second line of extension is based upon the usage of the method of machine learning devised here along with techniques of automated deduction based on non-monotonic frameworks, as already explored by authors in [21]. A promising opportunity is the investigation of Hopf et al. [16] where smart meters are used to provide information about devices. Other recent attempts have been made to include agents into methods for energy curve interpretation [4,5]. The usage of agent-based defeasible reasoning has already been applied by some of the authors in other investigations regarding business process compliance [12–14,17] whilst the combined use of machine learning techniques and nonmonotonic reasoning has yet not been investigated and it is one of the main effort we are taking further.

References

1. Anvari-Moghaddam, A., Monsef, H., Rahimi-Kian, A.: Optimal smart home energy management considering energy saving and a comfortable lifestyle. IEEE Trans. Smart Grid **6**(1), 324–332 (2015)
2. Baliga, J., Ayre, R., Hinton, K., Sorin, W.V., Tucker, R.S.: Energy consumption in optical IP networks. J. Lightwave Technol. **27**(13), 2391–2403 (2009)
3. Baliga, J., Hinton, K., Ayre, R., Tucker, R.S.: Carbon footprint of the internet (2009)
4. Bicego, M., Farinelli, A., Grosso, E., Paolini, D., Ramchurn, S.D.: On the distinctiveness of the electricity load profile. Pattern Recognit. **74**, 317–325 (2018)
5. Bicego, M., Recchia, F., Farinelli, A., Ramchurn, S.D., Grosso, E.: Behavioural biometrics using electricity load profiles, pp. 1764–1769 (2014)
6. Breheny, M.: The compact city and transport energy consumption. Trans. Inst. Br. Geogr. **20**(1), 81–101 (1995)

7. Cristani, M., Karafili, E., Tomazzoli, C.: An ambient intelligence technology for energy saving. In: IEA-AIE 2014 Proceedings. Springer (2014)
8. Cristani, M., Karafili, E., Tomazzoli, C.: Energy saving by ambient intelligence techniques. In: 2014 17th International Conference on Network-Based Information Systems (NBiS), pp. 157–164. IEEE (2014)
9. Cristani, M., Karafili, E., Tomazzoli, C.: Improving energy saving techniques by ambient intelligence scheduling. In: Proceedings of the 2015 IEEE 29th International Conference on Advanced Information Networking and Applications (AINA 2015), vol. 1, pp. 324–331, Los Alamitos, California, Conference Publishing Services (CPS) – IEEE Computer Society (2015)
10. Cristani, M., Tomazzoli, C., Olivieri, F., Karafili E.: Defeasible reasoning about electric consumptions. In: Proceedings of the 30th IEEE International Conference on Advanced Information Networking and Applications (AINA-2016), pp. 885–892 (2016)
11. Fan, X., Weber, W-.D., Barroso, L.A.: Power provisioning for a warehouse-sized computer. In: Proceedings of the 34th Annual International Symposium on Computer Architecture, ISCA 2007, pp. 13–23. ACM, New York (2007)
12. Governatori, G., Olivieri, F., Rotolo, A., Scannapieco, S., Cristani, M.: Picking up the best goal an analytical study in defeasible logic. LNCS (LNAI and LNBI), vol. 8035, pp. 99–113 (2013)
13. Governatori, G., Olivieri, F., Scannapieco, S., Cristani, M.: Designing for compliance: norms and goals. LNCS (LNAI and LNBI), vol. 7018, pp. 282–297 (2011)
14. Governatori, G., Olivieri, F., Scannapieco, S., Rotolo, A., Cristani, M.: The rationale behind the concept of goal. Theor. Pract. Logic Program. **16**(3), 296–324 (2016)
15. The Climate Group: Smart 2020: Enabling the low carbon economy in the information age (2008)
16. Hopf, K., Sodenkamp, M., Kozlovkiy, I., Staake, T.: Feature extraction and filtering for household classification based on smart electricity meter data. Comput. Sci. Res. Dev. **31**(3), 141–148 (2016)
17. Olivieri, F., Governatori, G., Scannapieco, S., Cristani, M.: Compliant business process design by declarative specifications. LNCS (LNAI and LNBI), vol. 8291, pp. 213–228 (2013)
18. Prez-Lombard, L., Ortiz, J., Pout, C.: A review on buildings energy consumption information. Energy Build. **40**(3), 394–398 (2008)
19. Santamouris, M., Papanikolaou, N., Livada, I., Koronakis, I., Georgakis, C., Argiriou, A., Assimakopoulos, D.N.: On the impact of urban climate on the energy consumption of buildings. Sol. Energy **70**(3), 201–216 (2001). Urban Environment
20. Scannapieco, S., Tomazzoli, C.: Ubiquitous and pervasive computing for real-time energy management and saving. In: Barolli, L., Enokido, T. (eds.) Innovative Mobile and Internet Services in Ubiquitous Computing, pp. 3–15. Springer International Publishing, Cham (2018)
21. Tomazzoli, C., Cristani, M., Karafili, E., Olivieri, F.: Non-monotonic reasoning rules for energy efficiency. J. Ambient Intell. Smart Environ. **9**, 345–360 (2017)
22. Tomazzoli, C., Cristani, M., Olivieri, F.: Automatic synthesis of best practices for energy consumptions. In: Proceedings of the Tenth International Conference on Innovative Mobile and Internet Services in Ubiquitous Computing, pp. 1–8. IEEE CPS (2016)
23. Tucker, R.S., Parthiban, R., Baliga, J., Hinton, K., Ayre, R.W.A., Sorin, W.V.: Evolution of WDM optical IP networks: a cost and energy perspective. J. Lightwave Technol. **27**(3), 243–252 (2009)

24. Valogianni, K., Ketter, W., Collins, J., Zhdanov, D.: Enabling sustainable smart homes: an intelligent agent approach (2014)
25. Vastamäki, R., Sinkkonen, I., Leinonen, C.: A behavioural model of temperature controller usage and energy saving. Pers. Ubiquitous Comput. **9**(4), 250–259 (2005)
26. Wood, G., Newborough, M.: Dynamic energy-consumption indicators for domestic appliances: environment, behaviour and design. Energy Build. **35**(8), 821–841 (2003)
27. Xu, Z., Jia, Q.-S., Guan, X., Xie, X.: A new method to solve large-scale building energy management for energy saving. In: 2014 IEEE International Conference on Automation Science and Engineering, CASE 2014, New Taipei, Taiwan, 18–22 August 2014, pp. 940–945. IEEE (2014)

Business Process Management

Advantages of Application of Process Mining and Agent-Based Systems in Business Domain

Michal Halaška(✉) and Roman Šperka

School of Business Administration in Karvina, Department of Business Economics and Management, Silesian University in Opava, Univerzitní nám. 1934/3, 733 40 Karviná, Czech Republic
{halaska, sperka}@opf.slu.cz

Abstract. Businesses and business decisions are getting driven by the information gained from the data more and more nowadays. The number of businesses supporting and managing their processes through the use of information systems and new technologies is growing every day. Even though, there is still a lot of rigidity in the implementation of new technologies. There is a great potential for the use of two of so far not so common disciplines in a business domain, which complement each other. That are process mining and multi-agent systems. Thus, in this paper, we are going to demonstrate the possible utilization of both process mining and multi-agent approaches in business domain. To demonstrate it, we use multi-agent simulator of trading company called MAREA. We analyzed implemented company model with the use of process mining. Process mining was used in two different ways. Firstly, to validate the workflow of the process model. Secondly, to analyze bottlenecks in company's business processes and the impact of marketing campaigns on these business processes.

Keywords: Process mining · MAREA · Multi-agent systems
Business · Model · Business process

1 Introduction

Companies all over the world are adopting new technologies. Some of them slower, some of them faster, but the reality is that modern companies will depend on technologies more and more. It is complicated to imagine businesses that are situated in modern economies or try to penetrate on markets that do not use the Internet, and other technologies, or even some sort of information system. Trends like digitalization, Internet of Things, orientation on data and many others are continuously changing the companies and their approaches towards their businesses. Companies are being challenged on the global markets thanks to globalization and the rise of e-commerce and other factors. Global competition pushes companies to continuously innovate, to continuously deliver something new to the customers, to raise the value of products for the customers, to be more efficient, to produce products with lower costs and so on and so forth. One of the things – from many – to achieve all of this, is to support and control their business processes with information systems completed with new technologies. And those information systems are not used only by big companies nowadays, but also

G. Jezic et al. (Eds.): KES-AMSTA-18 2018, SIST 96, pp. 177–186, 2019.
https://doi.org/10.1007/978-3-319-92031-3_17

by medium and small sized businesses. Kollár [1] argue that the implementation of innovation, information and communication technologies is crucial for companies. Similarly to them, we also believe that now is the time for companies, which want to succeed against the global competition, to adopt some new techniques that can help them to make the right decisions, to improve, to be successful and more competitive. And those techniques are from the area of process mining and the area of multi-agent systems (MAS) or agent based modelling and simulation (ABMS) respectively. Especially, if we consider the raise of process-centric orientation of the companies and the integral role of technologies in designing, modelling, optimizing and managing business processes within and across companies. And why exactly those disciplines? As we show and discuss later, both disciplines have a lot to offer to the businesses. And the MAS allow companies to apply intended changes without any risks. The reason we present both fields in this paper is because they complement each other very well. In the first and second section, we introduce both approaches with respect to business domain and state advantages and disadvantages of both approaches. In the third section, we introduce multi-agent simulator of a trading company MAREA. Finally, we present practical use of both approaches in business domain. In conclusion, we summarize and discuss our findings.

2 Process Mining

One of the major concepts that stays in the background of this paper is the term business process. We consider business process to be a set of activities that take one or several inputs and transform those inputs into the output that is of the value to the customer. Process mining is a group of techniques, which focus on extracting hidden patterns and useful information from data sources about processes. It is a subset of Business Process Management (BPM) and data science, thus process mining connects two particularly trendy areas in business practice, data-centric and process-centric approach.

As we said earlier, process mining is partly data science, thus it is highly dependent on the quality of data. This type of data is produced by information systems called process-aware information system (PAIS) like, e.g., ERP systems, CRM systems, etc. [2]. PAIS are information systems that support or control the whole end-to-end process. That means they cover processes from the very start to the very end of the process, not just particular activities that are happening in the process. The data for the process mining analysis have to have particular form known as log or event log [3]. Adjustment of the data usually takes a lot of effort. The first problematic part that one needs to address is during the extraction of the data from different data sources like, e.g., databases, data warehouses, etc. One has to select the suitable data from big amount of data and compose it in the appropriate way to get logs that truly represent analyzed process. It is not necessary to have PAIS, but otherwise the extraction and composition of event logs is even more difficult. The whole process of extraction, selection and composition of event logs is described in detail in work of Aalst [4] along with twelve points guideline for creation of event log.

The main advantage of process mining is that the analysis is conducted on the real company data. With the use of classical tools of BPM for modelling the business process, there is always a chance of a possible bias or ignorance from the side of management. Another advantage is that thanks to the support of many process mining tools one does not have to have such a high level of knowledge in neither process mining nor BPM. The pseudo-advantage is that discovery and conformance checking is possible without knowing basically anything about the underlying process. But of course, one always needs to confront the process mining analysis and the results with the reality to be able to evaluate the process mining analysis and conduct the enhancement of the process if needed.

The weakness of the process mining is that the process mining analysis can only be as good as provided data. Another disadvantage is the possible complications with extraction of data from data sources. The disadvantage might also be that the process mining does not cover the whole BPM life-cycle.

3 Agent-Based Systems and Their Use in Business Domain

The first of all, we address the difference between MAS and ABMS approaches. MAS are systems formed by multiple interacting software autonomous (intelligent) agents within an environment. On the other hand, ABMS are formed by multiple interacting software agents within an environment. There exist great penetration between MAS and ABMS. If we consider some general definitions of both approaches above, as you can see, the only difference between them is that ABMS does not require software agents to be intelligent. For the purpose of this paper, difference between both approaches is irrelevant and thus if we refer to the one of the approaches it is applicable for both of them. Also in the rest, we refer to both of them as agent-based systems (ABS).

ABS are appropriate tool for modelling and simulation of complex systems, such as, e.g., business company. ABS could be used to understand dynamics of highly complex systems [5]. ABS enable us to examine how micro-level processes affect macro-level outcomes and vice versa [6, 7]. There are four main business areas, where ABS were successfully used:

- supply chains (see, e.g., Sun et al. [8]),
- marketing management (see, e.g., North et al. [9]),
- automated trading (see, e.g., Li and Shi [10]),
- managerial science (see, e.g., Wall [11]).

Business processes can be also considered as highly complex systems. Thus, it seems advisable to use ABS for modelling and simulation purposes of business processes.

Twomey and Cadman [12] summarized the main strengths of ABS approach as lower degree of limitations like, e.g., linearity or stationarity. Another strength is that ABS provides more realistic models through the use of interacting entities, which can have high degree of heterogeneity, and the fact that ABS provides us a possibility of quite natural representation between model and simulated system, and maintenance and refinement. ABS can be useful for all approaches of process mining analysis, not just for enhancement. Another advantage is that once the model is implemented, it can be reused how many times is needed.

ABS has also some weaknesses. Similarly to process mining, ABS needs adequate and quality data. Besides that ABS has demand for programming skills and technological skills. One of the biggest weaknesses of ABS in general is validation and verification of such systems, and with more complex models both verification and validation is getting harder. But the advantage to use both approaches – process mining and ABS – together is that process mining can help with validation and verification of such models.

4 Research Methodology

In the text to follow, we describe the methodology used in this research and MAREA framework as the source of the data for process mining analysis. After that, we demonstrate the use and the advantages of both process mining and ABS for business domain.

4.1 MAREA

In this paper, we use multi-agent modelling and simulation software called MAREA (Multi-Agent REA). The framework models ongoing processes of a trading company. It is based on REA ontology. It consists of two parts: simulation editor and ERP editor. Simulation editor is used to create simulation models, while ERP editor is used for storing and manipulating with data (Cash level, Turnover, Gross profit, Profit, Supplies).

We model trading company that is selling goods, concretely LAN cables. The whole model consists of following agents: sales representative agents, purchase representative agents, customer agents, supplier agents, accountant agent (takes care of bookkeeping of the company), manager agent (manages the sales representative agents and calculates KPIs) and disturbance agent (responsible for historical trend analysis of sold amount). All agents are designed according to multi-agent approach and the interaction between agents is based on the FIPA contract-net protocol [13].

The model of the company in MAREA is dependent on all above mentioned agents; nevertheless, it is based on negotiation between customer agents and sales representative agents. The behavior of the customer is determined by decision function below [14]. Decision function for i-th customer determines the quantity that i-th customer accepts. If $x_i <$ *quantity* demanded by customer, the customer realizes that according to his preferences and budget, offered quantity is not enough, he rejects sales quote.

$$x_i^m = \alpha_i \frac{m_i}{p_x} \tag{1}$$

x_i^m – quantity offered by m-th sales representative to i-th customer,
α_i – preference of i-th customer (randomized),
m_i – budget of i-th customer (randomized),
p_x – price of the product x.

4.2 Analysis of MAS Simulations of the Company Using Process Mining

In the previous sections we theoretically introduced both AMS and process mining approach, stated their strengths and weaknesses and presented the reasons why they should be used in a business domain. Now, we present the use of both approaches together. Besides multi-agent simulator MAREA, we use process mining tool Disco to analyze overall processes involved and mainly selling process, where the customer – sales representative agent negotiation is involved. In our simulations, we use three factors to differentiate between scenarios: (1) number of customers – either 100 or 500; (2) number of sales representatives – either 1 or 2; and (3) involvement of marketing campaign – either yes or no. Thus, in our set up we experiment with 8 different simulation scenarios and in the analysis we work with 10 simulation runs for each scenario. The marketing campaign is in the form of advertisement (e.g., PPC (Pay Per Click), mailing campaign, etc.), mainly because it is easier to work with them in terms of costs and their effects on customers. The effect of advertisement on the customers in our simulation model is that they are willing to pay slightly higher prices. The marketing campaign is carried out regularly during the simulation run.

Validation of the Workflow of the Company's Model
First of all, we can use process mining for validation of workflow between agents in the company to check if it is according to our desire. Figures 1 and 2[1] show the overall workflow of the company's simulation. As one can see, we can separate the overall company workflow into three processes. Figure 1 shows on the right the activities involving the managing of the company like, e.g., decisions about education of sales and purchase representatives, determining the prices of goods, etc. The managing process involves activities of management agent and accounting agent. The process on the left in Fig. 1 is the supply process. The core activity in the supply process is the negotiation between purchase representatives and supply agent (or in other words vendors). And finally, Fig. 2 shows the sales process. In the sales process the sales representative agents and customer agents are involved. After we validated the workflow of the processes we can analyze the company processes.

Analysis of Company's Business Processes Using Process Mining and ABS
In the text to follow, we analyze sales process, where we found two problematic parts in the company's sales process. They both occur over all simulations.

Sales Request Revoked Bottleneck
The first bottleneck is the number of sales requests revoked (Fig. 2 green oval on the right), which means that a lot of customers that visit the store or their web pages buy nothing from the company. As one can see in Table 1[2], it is especially problematic in the scenarios with 500 customers, where it can drop up to 35%, while in case of 100

[1] Figure 2 shows only part of the sales process with bottlenecks. However, "Sales quote acceptance" activity is followed by the series of sequential activities {"Material request", "Productio request", "Sales order", "Bonus payment", "Production ready", "Stock level"}.

[2] The *w* in the scenario name means without marketing campaign, *m* means with marketing campaign, *1 s* means one sales representative and *2 s* means two representatives, and *100c* means 100 customers and *500c* means 500 customers in particular scenario.

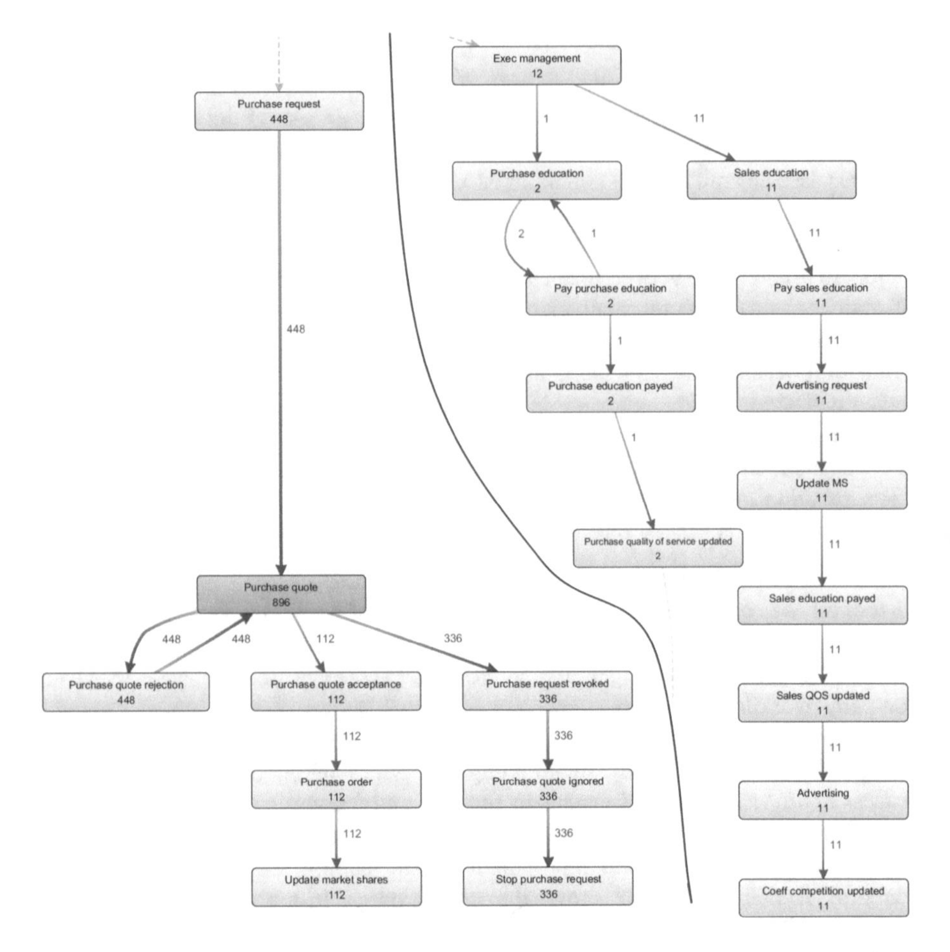

Fig. 1. Supply (on the left) and managing processes (on the right)

customers the number of active customers does not drop below 85%. The reason for this drastic difference is that one sales representative is able to handle 100 customers, but has a hard time with 500 customers, as we can see from Table 1. And even if we add sales representative (scenario m1s500c and m2s500c) we jump only to 40% of active customers. The reason for this is that the first sales representative, throughout all scenarios with two sales representatives, handles 4 times and sometimes up to 8 times more sales requests than the second sales representative. This is fine in scenarios with 100 customers, but causes problems in scenarios with 500 customers.

This behavior has its consequences. If we compare KPIs like Cash level or Profit of appropriate scenarios according to involvement of marketing campaign (w1s100c and m1s100c, w1s500c and m1s500c, etc.), we can see that the marketing campaign is more effective in scenarios with 100 customers as company achieves higher profits and levels of Cash level due to use of marketing campaign. On the contrary, marketing campaign is less effective in the scenarios with 500 customers, where company

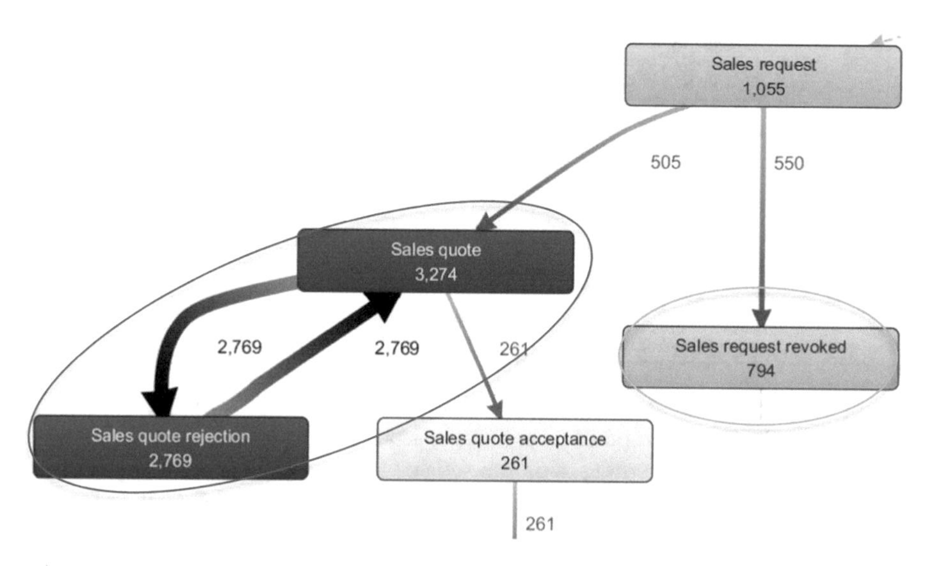

Fig. 2. Problematic part of the sales process

achieves higher profits and cash levels without marketing campaign. This is due to two reasons: (1) marketing campaign for 500 customers is more costly than for 100 customers and (2) the primary reason is that company is wasting money on customers, which cannot be handled and is not able to offer them a quote.

Impact of Marketing Campaigns on Sales Process and Sales Quote Bottleneck
So far, we were analyzing the part of the process in the green oval in Fig. 2. Now we analyze the second bottleneck in the red oval (left oval) in Fig. 2. Here is a bottleneck that is significantly lowering the performance of the company. This problem again occurs throughout all scenarios and simulation runs. The problem is that the deals between customers and sales representatives are being edited too many times due to, e.g., price negotiation, missing information, etc.

Another interesting thing that would be very hard to discover without process mining is the indirect impact of marketing activities in the managing process (Fig. 1) of the sales process, concretely on the bottleneck in the red oval from Fig. 2.

If we compare averages of values of number of "Sales quote rejection" to number of "Sales quote" ratio of all simulation runs for every scenario (see Table 2 and Fig. 2 left oval), we can see the 12 to 15% drop in values of averages of ratios with respect to scenarios where the marketing campaign was involved. The comparison regarding involvement of marketing campaigns is done using relevant scenarios, thus we compare values of scenarios rows of Table 2 (w1s100c and m1s100c, w1s500c and m1s500c, etc.).

This means that the marketing activities are unintentionally making the sales process more efficient. If we look at Table 3 we can see that according to ANOVA the improvement is statistically significant. One can also see that number of sales representatives does not have impact on the ratio of edited quotes. But marketing campaign

Table 1. Average KPIs for scenarios and its simulation runs

Scenario	Cash level	Profit	Number of active customers
w1s100c	29837,38	24837,38	85,10
w1s500c	43571,30	38571,30	237,50
w2s100c	34466,15	29466,15	90,40
w2s500c	44211,93	39211,93	262,80
m1s100c	39298,45	34298,45	86,30
m1s500c	36300,11	31300,11	177,70
m2s100c	37068,13	32068,13	89,70
m2c500c	37293,15	32293,15	206,10

is not the only factor. Number of customers has also impact on the ratio of edited quotes. Nevertheless, according to omega squared measuring the size of the effect or in other words the association, the effect of marketing campaign is distinctly higher than that of number of customers. In the last section, we comment on some flaws and weaknesses, and discuss the results and their implication.

Table 2. Average ratios of number of "Sales quote rejection" to number of "Sales quote"

Without marketing campaign	Average of ratios	With marketing campaign	Average of ratios
w1s100c	0,8527	m1s100c	0,7024
w1s500c	0,7831	m1s500c	0,6435
w2s100c	0,8400	m2s100c	0,7010
w2s500c	0,7536	m2s500c	0,6269

Table 3. P-values of ANOVA test and omega squared values

Factor	P-value	ω^2
Marketing campaign	4,58.10^-24	0,7270
Number of sales representatives	0,4434	–
Number of customers	4,42.10^-05	0,1814

5 Discussion and Results

Firstly, we presented the reasons, why we think it is necessary for companies, their management and their employees to adopt new technologies. The main reason is to hold on with global markets and to be able to face global competition, and how technologies like information systems, internet, etc., are nowadays more of necessity then advantage. Today advantages are developed by the use of data and process centric approaches by the management of the companies. We think that if the companies want to push their competitive advantage even further, it is time to start looking for new techniques like, e.g., process mining and ABS. Further, we briefly introduced both approaches with their strengths and weaknesses.

After that, we presented research methodology and used software tools MAREA and Disco. In our case study, we applied process mining to analyze business processes of modelled company. Firstly, we validated if the company's model workflow truly meets our requirements. Secondly, we discovered bottlenecks in sales process and explained its causes, namely the very high number of unrealized sales and the high number of edited sales quotes. We were also able to reveal the impact of hiring new sales representative on the sales process without actually having to hire someone. We were also able to reveal the impact of marketing campaigns without actually having to spend money on any marketing campaign. We demonstrated that with the suitable model of the company's processes and proper use of process mining, company's management would be able to make more informed decisions regarding trying to solve inefficiency in the sales process by hiring new employees or using marketing campaigns more efficiently based on, e.g., number of expected customers.

Acknowledgement. The work was supported by the SGS project of Silesian University in Opava, Czechia.

References

1. Kollár, I., Král, P., Laco, P. 2016. A few notes on deployment of supervised corporate financial distress prediction models in small enterprises. In: Conference Proceedings of 19th International Scientific Conference "Applications of Mathematics and Statistics in Economics", pp. 204–213. Univerzita Mateja Bela: Banská Bystrica (2015)
2. van der Aalst, W.M.P., Dumas, M., Ouyang, C., Rozinat, A., Verbeek, E.: Conformance checking of service behavior. ACM Trans. Internet Technol. **8**(13), 1–30 (2008)
3. Medeiros, A.K.A. de, Aalst, W.M.P. van der, Weijters, A.J.M.M.: Workflow mining: current status and future directions. In: On the Move to Meaningful Internet Systems 2003: CoopIS, DOA, and ODBASE. LNCS, vol. 2888, pp. 389–406. Springer, New York (2003)
4. van der Aalst, W.M.P.: Extracting Event Data from Databases to Unleash Process Mining. In: Brocke, J. vom, Schmiedel, T. (eds.) BPM - Driving Innovation in a Digital World, Management for Professionals, pp. 105–128. Springer International Publishing (2015)
5. Runje, B., Krstić Vukelja, E., Stepanić, J.: Agent-based simulation of measuring the quality of services. Tech. Gaz. **22**(6), 1561–1566 (2015)
6. Siebers, P., Aickelin, U., Celia, H., Clegg, C.: A multi-agent simulation of retail management practices. http://dx.doi.org/10.2139/ssrn.2831284. Accessed 14 May 2018
7. Terano, T.: Beyond the KISS principle for agent-based social simulation. J. Soc. Inf. **1**(1), 175–187 (2008)
8. Sun, J., Tang, J., Fu, W., Wu, B.: Hybrid modeling and empirical analysis of automobile supply chain network. Phys. A **473**, 377–389 (2017)
9. North, M.J., Macal, C.M., Aubin, J.S., Thimmapuram, P., Bragen, M., Hahn, J., Karr, J., Brigham, N., Lacy, M.E., Hampton, D.: Multiscale agent-based consumer market modeling. Complexity **15**(5), 37–47 (2010)
10. Li, G., Shi, J.: Agent-based modeling for trading wind power with uncertainty in the day-ahead wholesale electricity markets of single-sided auctions. Appl. Energy **99**, 13–22 (2012)

11. Wall, F.: Agent-based modeling in managerial science: an illustrative survey and study. RMS **10**(1), 135–193 (2014)
12. Twomey, P., Cadman, R.: Agent-based modelling of customer behaviour in the telecoms and media markets. Info **4**(1), 56–63 (2002)
13. Sandita, A.V., Popirlan, C.I.: Developing a multi-agent system in JADE for information management in educational competence domains. Procedia Econ. Financ. **23**, 478–486 (2015)
14. Vymětal, D., Ježek, F.: Demand function and its role in a business simulator. Munich Personal RePEc Archive, 54716 (2014)

Modelling the Validation Process of Enterprise Software Systems

Robert Bucki[1] and Petr Suchánek[2(✉)]

[1] Institute of Management and Information Technology, Bielsko-Biala, Poland
rbucki@wsi.net.pl
[2] School of Business Administration in Karviná, Silesian University in Opava, Karviná, Czech Republic
suchanek@opf.slu.cz

Abstract. The paper highlights the problem of the criteria-based assessment of the software system taking into account their sustainability, maintainability, and usability. The extended iteration model of creating software as well as its subsequent implementation is emphasised. The mathematical model for validating the software by highly qualified experts is introduced. It is followed by presenting the general pseudocode on the basis of which it is possible to program the software responsible for choosing the appropriate experts who are able to validate the software types.

Keywords: Software development · Software quality
Criteria-based assessment · Software expert · Software validation
Validation process · Mathematical model · Extended iterative model
Model for validating

1 Introduction

One of the major areas of modern computer science since its inception has been software development. Software development methods are constantly evolving in line with new possibilities and requirements for information and communication technologies (ICT) as well as development in areas such as project management, automation, artificial intelligence, mathematical modeling, logic, etc. Over the years, software engineering has helped programmers and researchers resolve various issues, develop products within a certain timeframe, and create specific universal or dedicated solutions. Improving the effectiveness of software development can be achieved through the use of experience and proven methods that were used in the past to create concrete products successfully implemented in practice. These methods need to be further developed and adapted to current possibilities and needs. Therefore, it is important to identify individual steps and processes during product software creation in order to visualize and subsequently evaluate their individual sub-relationships and dependencies on the basis of which new and more efficient procedures and methods can be developed much easier. For this purpose, a large number of special software tools are available these days [1–3], however, they are often very complex and expensive [4]. The project manager is responsible for a complex IT project which is the starting point

G. Jezic et al. (Eds.): KES-AMSTA-18 2018, SIST 96, pp. 187–196, 2019.
https://doi.org/10.1007/978-3-319-92031-3_18

for creating a new software product. He plans the project course, defines objectives and assumptions, manages the development and monitors and evaluates the whole life cycle of the development and the resulting software product in collaboration with the project team. The comprehensive IT project consists of requirements analysis and feasibility study, detailed specification, proper project definition (components, finance, personality, time schedule, etc.), programming, testing (verification) and validation. The starting point for the software creation is relevant data about the target group of future users or specific users (the set of companies or a specific company) [5]. The first case is more complex because it requires increased complexity, universally applicable functionality as the software must accommodate a very large number of target groups. It is also necessary to put big emphasis on customization options which is characterised by mass features in this case [6]. An important aspect of IT projects is an appropriate methodology while the choice of methodology is always heuristic [7]. It is possible to find a wide range of methods and specific recommendations in scientific resources [8, 9], however, there are recognized and proven methodologies such as RUP, XP, ITIL, CMMI, FDD, SCRUM, MSA, MDD, MDA, RAD, PRINCE2 and others. The human factor is a key component of all IT software development projects. It is important not only for software development itself but also for subsequent phases, which are mainly testing, implementation, validation and, last but not least, the operation itself. The importance of the human factor in these contexts is mentioned, for example, in [10]. Generally speaking, the software IT project is an important strategic choice which is selected on the basis of the so-called software development paradigm which contains basic assumptions, approaches, and methodology commonly accepted by members of a developing team. It is also a cognitive framework shared by all involved team members. It is equipped with its own set of tools accompanied by clearly defined methods and procedures which enable the team to define software development life cycle. All necessary engineering concepts concerning the software development have to be applied. Moreover, software can be created for a target audience or for a specific customer. The usefulness and purposefulness of the software product must be verified, which is the basic purpose and task of validation throughout the software development process. The software development paradigms known as software engineering paradigms are used by developers to select the most adequate strategy to develop the ordered software. It is important to apply all engineering concepts pertaining to the software development. Implementing the criteria-based assessment model, where assessment is based upon pre-established parameter-based objectives achievement, is determined by the success in meeting stated goals. Otherwise, a software system which is subject to validation requires thorough amendments, which consume a lot of time and increase the cost of compliance. Generally, validation of the software can be considered as the result of a set of verifiers for each area (Fig. 1).

The key issue is to demonstrate a clear link between existing processes and software modules to ensure their support [11]. Today, validation can be routinely performed through a variety of standard procedures and recommendations [12], e.g. using automation [13] both in the development of software by a single developer or within a geographically deployed development team creating one product [14]. The methodology of software development is increasingly based on mathematical models allowing the effective quantification of important parameters. This also applies to validation

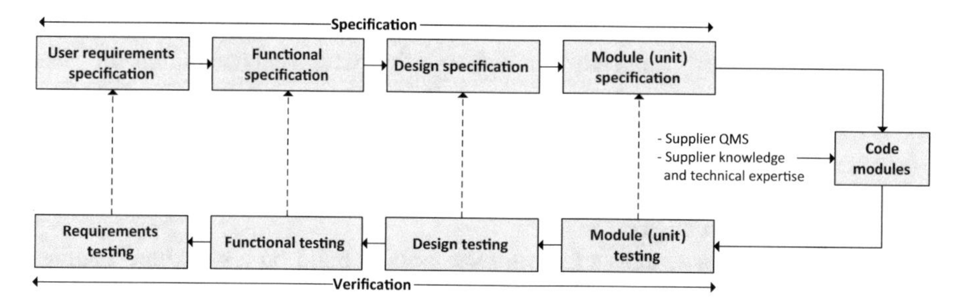

Fig. 1. General scheme of software validation Source: [18]

which is the main content of this paper whose main goal is to present a specific mathematical model of software product validation process. The first part lists the default parameterisation which is further applied to the iterative model described schematically and subsequently mathematically. An important part is the expression of the general pseudocode process which can be used as a basis for programming the final functionality applicable for a variety of purposes as one of the modules of complex validation tools.

2 Criteria-Based Assessment

A complex criteria-based assessment of the ready software takes into account such aspects as its sustainability, maintainability, and usability which require quality measurement resulting in software improvement in its poor areas or not proper performance when high-level decisions are unavoidable [15]. The more characteristics are satisfied, the more sustainable the software is. There is a need to match all qualities with the appropriate weight.

The sustainability and maintainability assessment criteria are classified as follows:

(i) identity project/software identity (indicator: clearness and uniqueness);
(ii) copyright (indicator: easiness to see the ownership of the project/software);
(iii) licensing (indicator: adoption of appropriate license);
(iv) governance (indicators: easiness to understand the project run and the development of the software management);
(v) community (indicators: evidence of both current and future community);
(vi) accessibility (indicators: evidence of both current and future ability to download);
(vii) testability (indicator: easiness to test correctness of the source code);
(viii) portability (indicator: usability on multiple platforms);
(ix) supportability (indicator: evidence of both current and future developer support);
(x) analysability (indicator: easiness to understand at the source level);
(xi) changeability (indicator: easiness to modify and contribute changes to developers);

(xii) evolvability (indicator: evidence of both current and future development);
(xiii) interoperability (indicator: interoperability with other required or elated software).
(xiv) The usability assessment criteria are grouped as follows:
(xv) understandability (indicator: easiness of understanding software);
(xvi) documentation (indicator: comprehensiveness, appropriateness, well-structured user documentation);
(xvii) buildability (indicator: straightforwardness to build on a supported system);
(xviii) installability (indicator: straightforwardness to install on a supported system);
(xix) learnability (indicator: easiness to learn how to use the software functions).

3 Extended Iterative Model

The iterative model, best thought of as a cyclical process, is a particular implementation of a life cycle software development life cycle (SDLC) focusing on an initial, simplified implementation, which then progressively gains more complexity and a broader feature set until the final system is complete. The key advantages of the iterative model include: inherent versioning, rapid turnaround, and easy adaptability.

The main disadvantages of the iterative model include: costly late-stage issues, increased pressure on user engagement, and feature creep.

The iterative method is the concept of incremental development describing the alterations made throughout the design and implementation stage of each new iteration [16, 17]. The initial planning phase is preceded by a thorough analysis of the real environment. Each stage can be repeated over and over letting the completion of the cycle incrementally if all detected errors are removed. This allows each next iteration to be better than the previous one. Iteration stages as well as the implementation stages are presented in detail in Fig. 2. The extended iteration model consists of the software creation stage and its implementation stage.

4 Problem Description

A need to compare and subsequently evaluate software systems arises more than often, which results in the idea to define a set of software containing systems of the same described type. It is always problematic to decide whether software systems are similar in nature. Nevertheless, it is necessary to recognise and choose the most adequate experts who are in the possession of professional knowledge letting them evaluate software systems of a certain type. The following model proposes a solution to this kind of problem.

Let us assume that the m-th software system, $m = 1, \ldots, M$ is to be validated and subsequently compared with another available one. However, the m-th software system must be compared only with the software system within the same n-th type of software systems, $n = 1, \ldots, N$.

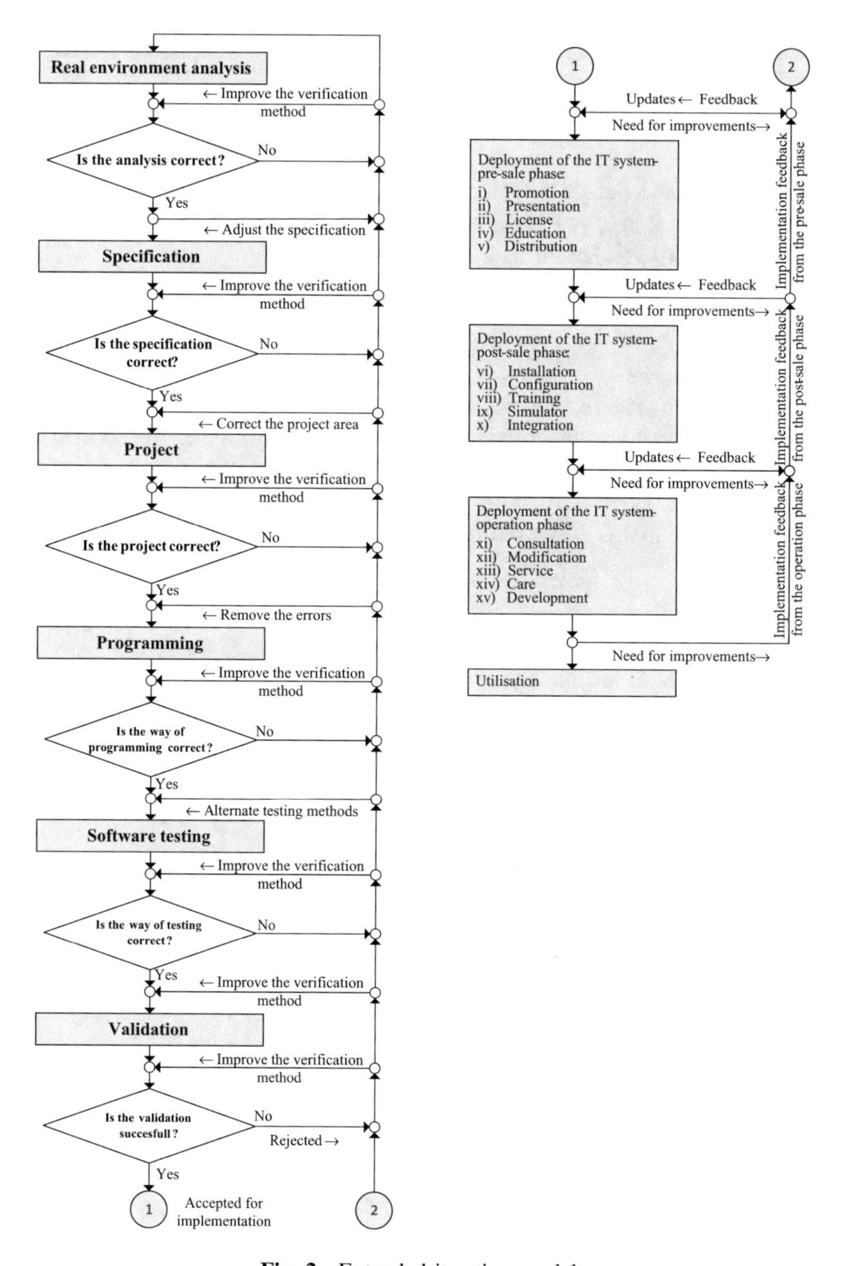

Fig. 2. Extended iteration model

Let Π be the matrix of software systems which are subject to the validating process (1):

$$\Pi = \left[\pi_{m,n}\right],\ m = 1,\ldots,M,\ n = 1,\ldots,N \tag{1}$$

where: $\pi_{m,n}$ - the m-th software system of the n-th type of software systems.

At the same time: $\pi_{m,n} = 1$ if the m-th software system of the n-th type of software systems is subject to validation; $\pi_{m,n} = 0$ otherwise.

Let Q be the vector of criteria for validating the software system quality (2):

$$Q = [q_\alpha],\ \alpha = 1,\ldots,A \tag{2}$$

where: q_α - the α-th criterion for validating the software system quality.

There is a need to introduce the assignment matrix of criteria to the n-th type of software systems meaning that it is possible to evaluate the m-th software system only on condition that it is placed within the n-th set of software systems, $n = 1,\ldots,N$. Then the discussed matrix takes the following form (3):

$$\Gamma = \left[\gamma_{\alpha,n}\right],\ \alpha = 1,\ldots,A,\ n = 1,\ldots,N \tag{3}$$

At the same time: $\gamma_{\alpha,n} = 1$ if the n-th type software systems is subject to the validation procedure according to the α-th criterion; $\gamma_{\alpha,n} = 0$ otherwise.

Let P^n be the matrix of parameters for validating the quality of software system within the n-th type of software systems (4):

$$P^n = \left[p^n_{\alpha,i}\right],\ \alpha = 1,\ldots,A,\ i = 1,\ldots,I,\ n = 1,\ldots,N \tag{4}$$

where: $p^n_{\alpha,i}$ - the i-th parameter within the α-th criterion for validating the quality of the n-th type software systems.

At the same time: $p^n_{\alpha,i} = 1$ if the i-th parameter within the α-th criterion for validating the quality of the n-th type of software systems is implemented; $p^n_{\alpha,i} = 0$ otherwise.

Let K be the maximal number of stages required to find the appropriate number of experts to evaluate the software systems. The time span between stages k and $k+1$ equals a defined period of time needed to modify the set of experts. The stage $k+1$ begins while the next course of drawing experts is initiated. For practical reasons, it seems impossible to wait too long for establishing a sufficient number of experts so the discussed time span should be defined in days maximally. Prolonging this time makes it impossible to either start the implementation process or return to the corrective procedures within the software creating phase.

Let us introduce the vector of all available expert staff able to evaluate software systems (5):

$$B^k = \left[b^k_j\right],\ j = 1,\ldots,J,\ k = 1,\ldots,K \tag{5}$$

where: b^k_j - the j-th expert available to evaluate software systems at the k-th stage.

Moreover, the j-th expert must achieve a certain level of qualifications to be able to evaluate the n-th type of software systems, i.e. not every expert can validate the n-th type of software system after entering the k-th stage. Nevertheless, the level of qualifications of the j-th expert either improves or decreases throughout the pass of time which may result in either qualifying or removing the j-th expert from the set of experts being able to validate the n-th type of software system. The j-th expert may also resign from collaborating within the team of experts, which means they are not taken into account. It is assumed that the resignation may come into being before resuming the subsequent search for experts at the stage $k+1$. This results in the need to propose the set of experts being able to validate the n-th type of software system at the k-th stage. Additionally, some experts are qualified enough to validate more types of software systems.

Let E^k be the matrix for evaluating the n-th type of software systems by the available staff (6):

$$E^k = \left[e^k_{j,n}\right], j = 1, \ldots, J,\ k = 1, \ldots, K,\ n = 1, \ldots, N \tag{6}$$

where: $e^k_{j,n}$- the adjustment of the j-th staff member to evaluate the n-th type of software systems at the k-th stage.

At the same time: $e^k_{j,n} = 1$ if the j-th staff member is qualified enough to evaluate the n-th type of software systems at the k-th stage; $e^k_{j,n} = 0$ otherwise.

Let us assume that the m-th software system of the n-th type set of software systems requires m_η experts, where $1 \leq \eta \leq N$.

The matrix $\Omega(\eta)$ determines the number of experts needed to validate the given software system (7):

$$\Omega(\eta) = \left[\omega(\eta)^k_n\right],\ n = 1, \ldots, N,\ k = 1, \ldots, K \tag{7}$$

where: $\omega(\eta)^k_n$ - the number of experts required to validate the n-th type of the software systems at the k-th stage.

There is a need to determine the following:

- the minimal number of experts $\eta_{n(\min)}$ required to validate the n-th type of software systems;
- the maximal number of experts $\eta_{n(\max)}$ required to validate the n-th type of software systems.

The required number of experts who are able to validate the n-th type of software systems must meet the following condition:

$$\eta_{n(\min)} \leq \omega(\eta)^k_n \leq \eta_{n(\max)} \tag{8}$$

If there are not enough qualified experts to validate the n-th type of software system, it is necessary to modify vector B^k as long as $k = K$. Then in case of not determining the required number of experts an alternative approach to the problem has to be put forward. Let us propose the general pseudocode for choosing qualified experts able to validate the n-th type software system:

(i) Introduce: $M, N, Q, K, P^n, B^k, E^k$.
(ii) $k = 1$
(iii) $m = 1$
(iv) $n = 1$. Go to (vi).
(v) $n = n + 1$
(vi) Does the m-th software system belong to the n-th set of software systems? If *Yes*, go to (viii). Otherwise, go to (xi).
(vii) Add the m-th software system to the n-th set of software systems.
(viii) Modify $\Pi = \left[\pi_{m,n}\right]$
(ix) $n = N$? If *Yes*, go to (x). Otherwise, go to (v).
(x) $m = M$? If *Yes*, go to (xii). Otherwise, go to (xi).
(xi) $m = m + 1$. Go to (iv).
(xii) $j = 1$. Go to (xiv).
(xiii) $j = j + 1$
(xiv) $n = 1$. Go to (xvi).
(xv) $n = n + 1$
(xvi) Is the j-th expert able to evaluate the n-th type of software systems? If *Yes*, go to (xvii). Otherwise, go to (xxiv).
(xvii) Add the j-th expert to the set of experts able to evaluate the n-th type of software systems.
(xviii) Modify $E^k = \left[e^k_{j,n}\right]$
(xix) $n = N$? If *Yes*, go to (xx). Otherwise, go to (xv).
(xx) $j = J$? If *Yes*, go to (xxi). Otherwise, go to (xiii).
(xxi) $\underset{n}{\exists} \eta_{n(\min)} \leq \omega(\eta)^k_n \leq \eta_{n(\max)}$? If *Yes*, go to (xxvi). Otherwise, go to (xxii).
(xxii) If $\omega(\eta)^k_n = 0$, go to (xxiv). If $\omega(\eta)^k_n > \eta_{n(\max)}$, go to (xxvi).
(xxiii) $k = K$? If *Yes*, go to (xxiv). Otherwise, continue to (xxv).
(xxiv) Determine the expert staff in an alternative way and modify $E^k = \left[e^k_{j,n}\right]$. Go to (xxvii).
(xxv) $k = k + 1$. Go to (iii).
(xxvi) Determine $\omega(\eta)^k_n = \eta_n$.
(xxvii) Validate the m-th software system in accordance with $p^n_{\alpha,i}$.
(xxviii) Report.

5 Conclusions

The paper focuses on presenting the extended iteration model which is used while creating software systems in contemporary times. This sophisticated model also comprises the phase of analysing the real environment which precedes the specification and the project. The phases of programming and subsequent testing are enlarged by the presence of the validation phase. Finally, the implementation stage, divided into pre-sale phase, post-sale phase, and the operation phase, is described in detail. The validation phase is supported by the mathematical model which is followed by the general pseudocode allowing building a software system responsible for choosing qualified experts who are able to validate the defined type of software. It is based on the assumption that there are at least two software systems which are subject to the validation process. The given criteria are evaluated on condition they are described by certain defined parameters. Another issue worth emphasising is the need to focus on the staff of experts who are skilled enough to evaluate the defined type of software. Moreover, the qualifications of the considered experts change within the pass of time which needs to be included in the validation process of the software systems. The proposed model requires its programmed software version enabling us to choose the best available reviewers to verify software products of various types. This model could also be implemented for other purposes e.g. determining reviewers of scientific papers which are put forward to be printed in conference proceedings or choosing project experts who have to decide whether the project is worth considering or not. It is expected that the quality of the review process depends on the current experts' knowledge which has to be confirmed by the preceding verification process.

Acknowledgement. This paper was supported by the project SGS at the Silesian University in Opava, School of Business Administration in Karvina.

References

1. Sathiyanarayanan, M., Alsaffar, M.: Small multiples Euler-time diagrams for software engineering. Innov. Syst. Softw. Eng. **13**(4), 299–307 (2017). https://doi.org/10.1007/s11334-017-0298-x
2. Gravino, C., Host, M.: Guest editorial of special section on software engineering and advanced applications in information technology for software-intensive systems. Inf. Softw. Technol. **93**, 246–247 (2018). https://doi.org/10.1016/j.infsof.2017.10.011
3. Baginski, J., Bialas, A.: Validation of the software supporting information security and business continuity management processes. Complex Syst. Dependability **170**, 1–17 (2012)
4. Buller, K.: A pragmatic method for assessing systems architectures during the architecture generation process with a focus on repurposing business software to systems engineering. Complex Adapt. Syst. **61**, 153–159 (2015). https://doi.org/10.1016/j.procs.2015.09.178
5. Menard, P.A., Ratte, S.: Concept extraction from business documents for software engineering projects. Autom. Softw. Eng. **23**(4), 649–686 (2016). https://doi.org/10.1007/s10515-015-0184-4
6. Modrak, V., Bodnar, S., Soltysova, Z.: Application of axiomatic design-based complexity measure in mass customization. In: 26th Cirp Design Conference, pp. 607–612 (2016). https://doi.org/10.1016/j.procir.2016.04.196
7. Sharma, B., Nag, R., Makkad, M.: Process performance models in software engineering: a mathematical solution approach to problem using industry data. Wirel. Pers. Commun. **97** (4), 5367–5384 (2017). https://doi.org/10.1007/s11277-017-4783-1

8. Harrison, R., Bener, A.B., Mericli, C., Turhan, B.: Guest editorial: special issue on realising artificial intelligence synergies in software engineering. Autom. Softw. Eng. **24**(4), 789–790 (2017). https://doi.org/10.1007/s10515-017-0224-3
9. Kuhrmann, M., Fernandez, D.M., Daneva, M.: On the pragmatic design of literature studies in software engineering: an experience-based guideline. Empir. Softw. Eng. **22**(6), 2852–2891 (2017). https://doi.org/10.1007/s10664-016-9492-y
10. Gren, L., Goldman, A.: Useful statistical methods for human factors research in software engineering: a discussion on validation with quantitative data. In: 9th IEEE/ACM International Workshop on Cooperative and Human Aspects of Software Engineering, pp. 121–124 (2016). https://doi.org/10.1145/2897586.2897588
11. Aversano, L., Grasso, C., Tortorella, M.: Validation of a measurement framework of business process and software system alignment. In: Proceedings of the 12th International Conference on Enterprise Information Systems, Information Systems Analysis and Specification, vol. 3, pp. 470–474 (2010)
12. Kerry, E.: Software engineering practices to improve code quality and prepare for validation. In: Proceedings of the 5th Frontiers in Biomedical Devices Conference and Exposition, pp. 85–86 (2010)
13. Zawistowski, P.: The method of measurement system software automatic validation using business rules management system. In: Photonics Applications in Astronomy, Communications, Industry, and High-Energy Physics Experiments (2015). https://doi.org/10.1117/12.2205929
14. Ali, N., Lai, R.: A method of software requirements specification and validation for global software development. Requir. Eng. **22**(2), 191–214 (2017). https://doi.org/10.1007/s00766-015-0240-4
15. ISO/IEC 25010:2011.: International Organization for Standardization, 20 December 2017. https://www.iso.org/standard/35733.html
16. Ghahrai, A.: Iterative Model, 18 December 2017. https://www.testingexcellence.com/iterative-model/
17. Powell-Morse, A.: Iterative Model: What Is It and When Should You Use It? 18 December 2017. https://airbrake.io/blog/sdlc/iterative-model
18. https://www.tutorialspoint.com/software_testing_dictionary/validation_testing.htm

Design and Implementation of Intelligent Agents and Multi-Agent Systems I

Multi-agent System for Forecasting Based on Modified Algorithms of Swarm Intelligence and Immune Network Modeling

Galina A. Samigulina(✉) and Zhazira A. Massimkanova

Institute of Information and Computational Technologies, Almaty, Kazakhstan
galinasamigulina@mail.ru, masimkanovazh@gmail.com

Abstract. The use of modern achievements of artificial intelligence in the creation of innovative information forecasting technologies is an urgent task. The article is devoted to the development of multi-agent system for forecasting based on modified algorithms of swarm intelligence and artificial immune systems approach. The construction of an optimal immune network model is one of the most important tasks at solving the problem of image recognition and prediction based on artificial immune systems. The problem of preliminary processing and selection of informative descriptors is solved on the basis of swarm intelligence algorithms. Selection of informative descriptors is carried out based on a multi-algorithmic approach, which allows to choose the algorithm of swarm intelligence, in which the generalization error will be minimal after immune network modeling. An algorithm of functioning of the multi-agent system for forecasting has been developed and a description of the main agents has been given. The modeling results have been presented and a comparative analysis for various algorithms of swarm intelligence has been performed.

Keywords: Multi-agent system · Swarm intelligence
Immune network modeling

1 Introduction

Nowadays the development of innovative information technologies based on bioinspired intellectual approaches and the development of a multi-agent paradigm for processing and analyzing large data, for optimizing and solving prediction problems is an urgent task. The most widely used are intelligent technologies with the use of neural networks, evolutionary algorithms, artificial immune systems (AIS), swarm intelligence (SI) algorithms, etc.

Particular attention all over the world is paid to AIS based on the principles of information processing by protein molecules. This approach uses the principles of theoretical immunology, which are used to solve various applied problems [1]. There are many publications on this topic [2]. The work [3] presents the AIS approach for predicting time series. The problem of filling missing data is being solved and the noisy parameters values sensitivity analysis is being carried out. A new model based on AIS for predicting the electrical load of the power system is considered in work [4]. The proposed model has been tested on real data and the comparative analysis has been

G. Jezic et al. (Eds.): KES-AMSTA-18 2018, SIST 96, pp. 199–208, 2019.
https://doi.org/10.1007/978-3-319-92031-3_19

carried out with other models based on intellectual methods. The obtained results have confirmed the effectiveness of the proposed model of AIS.

The SI algorithms such as ant and bee colonies algorithms, particle swarm algorithms, the algorithm of bacteria moving and many others are actively developing. These approaches are successfully used at solving optimization problems and are characterized by the ability to solve the problem of complex search for optimal solutions quickly and accurately. The SI algorithms differ from classical methods by greater reliability in the search for a global maximum, by simplicity of software implementation and by greater accuracy [5].

The basis of SI methods is a multi-agent system (MAS), which consists of a set of interacting agents. Each agent performs certain tasks. Different agents cooperate with each other when collecting and processing information and perform the optimal solution. The undoubted advantages of MAS are ability to perform quick approximate result, stability to changes of input parameters and scalability.

There are many MAS based on SI algorithms. This approach is especially widely used for constructing an optimal set of descriptors and for reducing low information data, which significantly improves the forecasting quality and saves computing resources. For example, a new approach for features selection based on the integration of genetic algorithm and particle swarm algorithm is proposed in work [6]. The article [7] describes ant colony optimization for the identification of informative features. The proposed algorithm increases the accuracy of the solution and eliminates premature convergence.

The SI algorithms for solving applied problems of bioinformatics, such as data clustering, prediction of protein structure and molecular docking are considered in work [8]. The article [9] is devoted to the construction of the procedure of adaptive selection of parameters entering the input of ant algorithm for predicting the structure of the protein. In the article [10] a combined approach based on the ant algorithm and on the adapted Bayesian classifier is proposed in order to improve the efficiency of the obtained model. The medical data about heart diseases was used as an example. The research [11] uses ant colony optimization and back propagation neural network classification method for data preprocessing of 6-channel electroencephalogram. By comparing other feature selection, algorithms or classifiers it is important to underline the validity and versatility of the proposed combination. The article [12] proposes a hybrid MAS based on particle swarm optimization for diagnosis and monitoring neurological diseases. The research [13] uses ant colony optimization and particle swarm optimization for distributed MAS. The MAS is composed of relatively simple agents with highly decentralized and self-organized behaviors. The results demonstrate that the proposed hybrid swarm intelligence model is feasible, efficient, and robust to coordinate a distributed MAS. The article [14] is devoted to an artificial bee colony (ABC) algorithm in MAS for the information extraction from hyper-spectral images. The results indicate that the proposed agent-based ABC algorithm can effectively solve the problem of information extraction.

In work [15] PLANTS (protein-ligand ant system) system was developed based on Max-Min ant system algorithm for the detection of biologically active ligands in protein-ligand docking problem. The article [16] is devoted to the joint use of the Ant Colony Optimization-Multiple linear regression to reduce the descriptor space at the construction of quantitative structure-property relationship.

All the above meant confirms the relevance of this research area. Modern methods of artificial intelligence are widely used for computer modeling of new drug compounds with pre-defined properties. Forecasting quantitative structure-activity relationship (QSAR) of drugs and identifying the links between the compounds structure and their activity is an actual problem. For example, the work [17] is devoted to the application SI algorithms for solving optimization problems in QSAR. One of the important step of prediction process of QSAR is the allocation of informative descriptors to reduce the size of the descriptor space. The creation of new medicinal compounds is associated with the use of different approaches for the analysis of large data sets, therefore there is a need to develop an effective specialized multi-agent information system for conducting scientific researches that implements modified algorithms of swarm intelligence and AIS approach.

The structure of the article is as follows: Sect. 2 describes problem statement of the research. Section 3 is devoted to the description of immune network technology and SI algorithms for solving the problem of informative descriptors allocation and prediction of QSAR chemical compounds. Section 4 is devoted to the development of the multi-agent information system for forecasting and for the description of agent functions. Section 5 presents the results of modeling on the example of sulfonamides database with different duration of action. Section 6 presents the conclusion. At the end of the article, there are given the references.

2 Problem Statement

The problem statement is formulated as follows: it is necessary to develop a multi-agent information system for conducting scientific researches in order to create the optimal immune network model based on swarm intelligence algorithms and on ontological approach, as well as for solving the problem of image recognition and prediction using AIS algorithms.

Let us introduce the following definition: under the optimal structure of the immune network there is meant a network constructed on the basis of the weight coefficients of the selected informative descriptors most fully characterizing the state of the system depending on the underlying factors affecting the functioning of this system. The criterion is the maximum storage of information at a minimum number of descriptors [18]. The aim of the construction of an optimal immune network model is to remove the low informative descriptors that have significant errors, which reduce the quality of the forecasting and the training time of AIS on the basis of the decrease of the descriptor space dimension.

3 Solution Methods

The immune network technology for prediction of the structure-property/activity relationship of chemical compounds, which consists of the stages of preliminary data processing, immune network training, image recognition, energy error estimation and chemical compounds pharmacological properties prognosis was developed in work [19]. Intellectual technology based on immune network modeling allows to analyze hidden interactions between descriptors. At the stage of preliminary data processing there is performed normalization, completeness and reliability of the descriptors check. The selection of informative descriptors is performed on the basis of algorithms of swarm intelligence in accordance with the concept of multi-algorithmic approach, in which several algorithms are used. An algorithm with the minimum value of generalization error is selected after the immune network modeling, on the basis of the comparison of the prediction results. The application of SI algorithms makes it possible to shorten the time for training the immune network by the creation of an optimal immune network model and to exclude low information descriptors. There are many modifications of classical SI algorithms. The ant colony algorithm has several modifications [20], for example, AntSrank, Max-min ant system, Elitist ant system, etc. The modified particle swarm algorithms [21]: CoPSO, Fully informed PSO, Inertia Weighted PSO, etc. are also developed. They differ from each other by the speed of convergence, by the presence of additional parameters, etc. The development of OWL (Web Ontology Language) models for the systematization of the used algorithms [22] is also actual.

4 Multi-agent Information System for Forecasting

The following agents were created during the development of the multi-agent system for forecasting (Table 1): database agent, manager agent, ontological agent, ant colony agent, particle swarm agent, image recognition agent, AIS forecasting agent and assistant agent.

Below given an algorithm of the information system functioning:

Step 1. Connecting the database. The system supports databases in the .csv format.
Step 2. Selection of SI algorithm based on modified algorithms of ant colonies and particles swarm. There are no universal methods of the optimal set of descriptors creation, therefore the system provides the possibility of selecting the SI algorithm for descriptors preprocessing.
Step 3. Entering the coefficients (the number of populations, the number of iterations, weight and speed).
Step 4. Multidimensional data processing and informative descriptors selection.
Step 5. Creation of the optimal immune network model based on the selected set of descriptors. Redundancy and low informative descriptors reduce the forecasting quality, therefore selecting new chemical compounds for drug candidates it is important to take into account the informative nature of the descriptor.
Step 6. Immune network training.

Step 7. Image recognition based on the singular value decomposition.
Step 8. Estimation of AIS energy errors.
Step 9. Forecasting the structure-property/activity relationship of chemical compounds. Selection of candidates for new chemical compounds with given properties for further researches.
Step 10. Output of the result.

Table 1. Agents description.

Agent	Description of the agents function
Database agent	– Creation of a descriptors database (DB) – Work with DB
Manager agent	– Implementing the relationship between agents – Information transmitting organization – Coordination of the work of agents
Ontological agent	– Formation of OWL model of ant colony algorithms – Formation of OWL model of particle swarm algorithms – Creation of the OWL model of AIS – Structuring of input and output data
Assistant agent	– Forming of tips at entering of algorithm parameters – Support at operating in a software environment
Ant colony agent	– Solution of the problem of informative descriptors selection based on ant colony algorithms 1. Implementation of the classical BasicACO algorithm: – agent population creation; – arbitrary sharing of agents between nodes; – fitness-functions calculating and determination of the pheromone amount; – migration of agents; – evaporation of pheromone; – updating the local and global amount of pheromone; – checking the completion condition; – determination of the edge weight and the preservation of the best position of the agent; 2. Implementation of the modified AntSrank algorithm 3. Implementation of the modified Max-min ant system algorithm 4. Implementation of the modified Elitist ant system algorithm – reduction of low informative descriptors
Particle swarm agent	– The solution of the problem of informative descriptors selection on the basis of particle swarm algorithms 1. Implementation of the classical BasicPSO algorithm: – agent population creation; – generation of initial positions and velocities of agents; – fitness-functions calculation; – determination of the best position of the agent; – migration of agents; – updating agents' positions and speeds;

(*continued*)

Table 1. (*continued*)

Agent	Description of the agents function
	– updating the agent's best position; – checking the completion condition; – maintaining the best position of the agent 2. Implementation of the modified CoPSO algorithm 3. Implementation of the modified Fully informed PSO algorithm 4. Implementation of the modified Inertia Weighted PSO algorithm – reduction of low informative descriptors
Image recognition agent	– Implementation of AIS technology and the creation of an optimal immune network model – creation of matrixes of standards and matrixes of images formed from time series (descriptors) – training of AIS with the teacher – singular value decomposition (SVD) – determination of binding energies between formal peptides – solution of the problem of image recognition based on the determination of the minimum value of the binding energy and forecasting
AIS forecasting agent	– Estimation of energy errors of AIS – Averaging of the potentials by the homologies – Calculation of the native (functional) structure energy by average homologies – Calculation of the Z-factor (the average amount of standard deviations between the native structure energy and the energy of the randomly chosen stacking of the chain). – Determination of prediction risk factors.

5 Modeling Results

Intellectual technology of immune network modeling includes work with chemical compounds descriptors databases. Let consider medicinal compounds - sulfonamides, which refer to drugs of wide antibacterial action. As an example, there was used a fragment of the Sul_1 database on the basis of the Mol-Instincts and PubChem [23] resource consisting of chemical compounds of the sulfanilamide group and of more than 800 descriptors. Sulfanilamides are classified according to the duration of pharmacological activity: short, medium and long-acting. Table 2 shows a fragment of the sulfonamide database. There are considered the descriptors such as: CoMMA displacement, sum of geometrical distances between NN, HASA-2/sqrt (TMSA), Squared Moriguchi octanol-water partition coefficient, max bond order (H), etc.

At modeling using the classical ant algorithm the size of the population is 100, the number of iterations is 50, c1 (the amount of pheromone) = 1, c2 (pheromone evaporation) = 2, report frequency = 50. As a result, 39 informative descriptors were selected from 820 ones. Figure 1 shows the visualization of the selected informative descriptors.

Table 2. A fragment of the sulfonamide database.

Class	CoMMA displacement	Sum of geometrical distances between NN	HASA-2/sqrt (TMSA)	Squared Moriguchi octanol-water partition coefficient	…	Max bond order (H)
short_acting	0,40	19,32	0,69	0,80	…	0,91
short_acting	0,49	12,81	0,82	3,10	…	0,93
medium_acting	0,33	0,00	0,70	0,49	…	0,93
…	…	…	…	…	…	…
long_acting	0,92	7,02	0,48	4,06	…	0,94

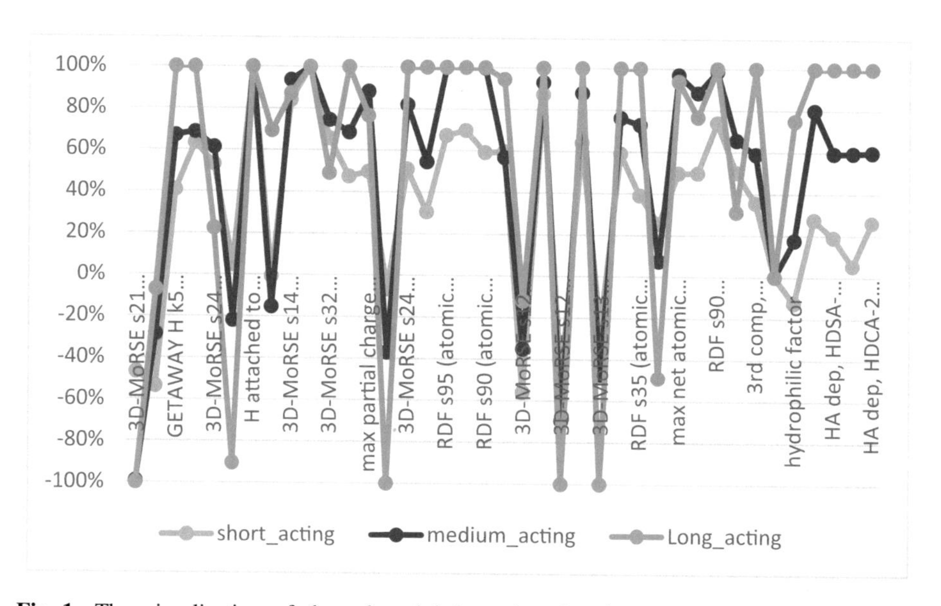

Fig. 1. The visualization of the selected informative descriptors based on the ant colony algorithm with a population size of 100

If at modeling the population size is 200 (Fig. 2) then, as a result, 16 informative descriptors are selected from 820 ones.

Thus, as a result of the modeling, the weight of each feature is calculated. The more weight, the more informative it is. Therefore, with a large number of populations and iterations the number of selected informative features decreases.

Table 3 shows the results comparison of the informative descriptors selection of sulfonamides based on the algorithms of ant colonies and particles swarm at the population size of 100 and 200.

The effectiveness of SI algorithms depends on a number of parameters, which include: the population size, the number of iterations, etc. The larger the swarm size is, the greater the varieties of potential solutions are.

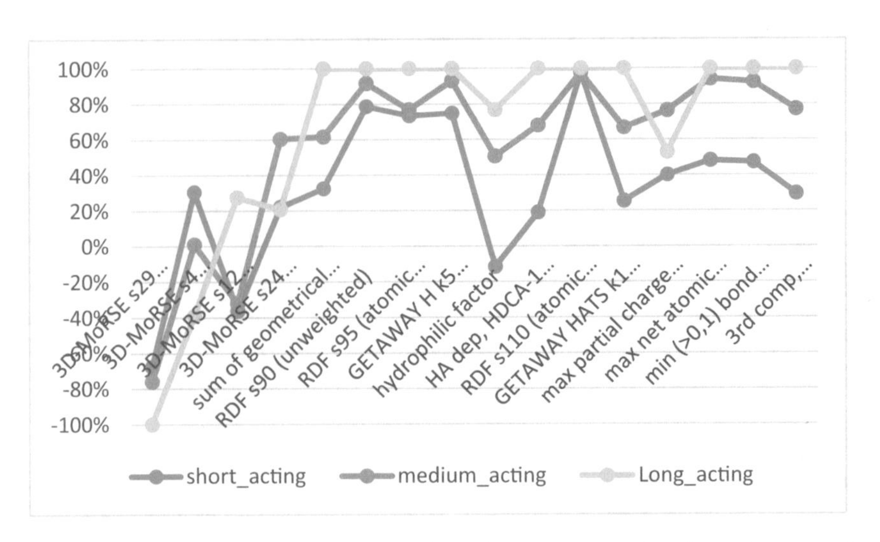

Fig. 2. The visualization of the selected informative descriptors based on the ant colony algorithm with a population size of 200

Table 3. Comparative analysis of the sulfonamides modeling results

Coefficients	Ant colony algorithm		Particles swarm algorithm	
Population size	100	200	100	200
Number of iterations	50		50	
Number of pheromone	1		1	
Evaporation of pheromone	2		2	
Report frequency	50		50	
Sulfonamides modeling results				
General number of descriptors	820	820	820	820
Number of informative descriptors	39	16	95	38

At a large amount of populations and iterations the number of selected informative descriptors decreases. Agents explore the search space more detailed, therefore the algorithms show the best results. Similarly, the information system works with the particle swarm algorithm.

After the selection of informative descriptors the optimal immune network model creation and AIS training are carried out.

Next, an algorithm of image recognition based on the singular value decomposition is implemented and the estimation of the AIS energy errors is performed. After the immune network modeling based on a comparison of the prediction results an SI algorithm with a smallest generalization error is chosen. In conclusion, the selection of candidates of medicinal compounds of the sulfanilamide group with predetermined pharmacological properties is carried out for further researches. The use of this technology can significantly reduce the time and financial costs in the developing of new drugs.

6 Conclusion

The performed modeling shows that the preliminary data processing based on the informative features selection using SI algorithms, as well as the created optimal immune network models significantly improve the forecasting quality at solving the problem of image recognition based on AIS. In the multi-agent information system for forecasting it is possible to use other methods of artificial intelligence [24], which allows to select quickly the best algorithm for preliminary data processing for a specific set of descriptors based on the minimum value of the AIS generalization error. Moreover, the developed multi-agent system can be used to solve the problem of forecasting in other areas, for example in industry, economics and education.

The research is carried out according to the grant of the Science Committee of the Ministry of Education and Science of the Republic of Kazakhstan under the grant No. AP05130018 on the theme: Development of cognitive Smart - technology for intelligent complex objects control systems based on artificial intelligence approaches (2018–2020).

References

1. Timmis, J., Neal, M., Hunt, J.: An artificial immune system for data analysis. BioSystem **55** (1), 143–150 (2000)
2. Dasgupta, D.: Recent advances in artificial immune systems: models and applications. Appl. Soft Comput. J. **11**, 1574–1587 (2011)
3. Dudek, G.: Artificial immune system for forecasting time series with multiple seasonal cycles. In: International Conference on Computational Collective Intelligence, pp. 468–477. Springer, Berlin (2011)
4. Dudek, G.: Artificial immune system with local feature selection for short-term load forecasting. IEEE Trans. Evolutionary Comput. **21**, 116–130 (2017)
5. Hinchey, M.G., Sterritt, R., Rouff, C.: Computer society «From Ants to People: an Instinct to Swarm». Swarms Swarm Intell. **40**, 111–113 (2007)
6. Ghamisi, P., Benedikstsson, J.A.: Feature selection based on hybridization of genetic algorithm and particle swarm optimization. Geosci. Remote Sens. Lett. **12**, 309–313 (2014)
7. Liu, Y., Wang, G., Chen, H., Zhao, Z., Zhu, X., Liu, Z.: An adaptive fuzzy ant colony optimization for feature selection. J. Comput. Inf. Syst. **7**, 1206–1213 (2011)
8. Agrawal, S., Silakari, S.: A review on application of Particle Swarm Optimization in Bioinformatics. Curr. Bioinform. **10**, 401–413 (2015)
9. Niu, D., Wang, Y., Wu, D.D.: Power load forecasting using support vector machine and ant colony optimization. Expert Syst. Appl. **37**, 2531–2539 (2010)
10. Bouktif, S., Hanna, E.M., Zaki, N., Khousa, E.A.: Ant colony optimization algorithm for interpretable bayesian classifiers combination: application to medical predictions. PLOS ONE **9**(2). https://doi.org/10.1371/journal.pone.0086456. Accessed 12 Jan 2018
11. Erguzel, T.T., Ozekes, S., Gultekin, S., Tarhan, N.: Ant colony optimization based feature selection method for QEEG data classification. Psychiatry Investig. **11**(3), 243–250 (2014)
12. Kaur, A., Sikander, S.C.: A hybrid multi-agent based particle swarm optimization for telemedicine system for neurological disease. In: Recent Advances and Innovations in Engineering. IEEE, India (2016). https://doi.org/10.1109/icraie.2016.7939527. Accessed 07 Jan 2018

13. Meng, Y., Kazeem, O., Muller, J.C.: A swarm intelligence based coordination algorithm for distributed multi-agent systems. In: Integration of Knowledge Intensive Multi-Agent Systems, IEEE, USA (2007). https://doi.org/10.1109/kimas.2007.369825. Accessed 12 Jan 2018
14. Yang, L., Sun, X., Zhang B., Chi, T.: An multi-agent combined artificial bee colony algorithm to hyper-spectral image end member extraction. In: Hyperspectral Image and Signal Processing: Evolution in Remote Sensing. IEEE, Japan (2015). https://doi.org/10.1109/whispers.2015.8075439. Accessed 12 Jan 2018
15. Korb, O., Stützle, T., Exner, T.E.: PLANTS: application of ant colony optimization to structure-based drug design. In: International Workshop on Ant Colony Optimization and Swarm Intelligence ANTS 2006, pp. 247–258 (2006)
16. Atabati, M., Zarei, K., Borhani, A.: Ant colony optimization as a descriptor selection in QSPR modeling: Estimation of the k-max of anthraquinones-based dyes. J. Saudi Chem. Soc. **20**, 547–551 (2016)
17. Khajeh, A., Modarress, H., Zeinoddini-Meymand, H.: Application of modified particle swarm optimization as an efficient variable selection strategy in QSAR/QSPR studies. J. Chemom. **26**, 598–603 (2012)
18. Samigulina, G.A.: Immune Network Modeling Technology for Complex Objects Intellectual Control and Forecasting System: Monograph. Science Book Publishing House, USA (2015)
19. Samigulina, G.A., Samigulina, Z.I.: Drag design of sulfanilamide based on immune network modeling and ontological approach. In: Proceedings of the 10th IEEE International Conferences on Application of Information and Communication Technologies AICT 2016, Azerbaijan (2016). www.aict.info/2016. Accessed 2017/11/17
20. Sorin, C.N., Constantin, O., Claudiu, V.K., Carabulea, I.: Elitist ant system for route allocation problem. In: Proceedings of the 8th WSEAS International Conference on Applied informatics and communications, Greece, pp. 62–67 (2008)
21. Li, S., Hsu, C., Wong, C., Yu, C.: Hardware/software co-design for particle swarm optimization algorithm. Inf. Sci. **181**, 4582–4596 (2011)
22. Samigulina, G.A., Massimkanova, Z.A.: Ontological models of swarm intelligence algorithms for immune network modeling of drugs. Bull. Al-Farabi KazNU **1**(93), 92–104 (2017)
23. https://www.molinstincts.com. Accessed 21 Apr 2017
24. Samigulina, G.A., Samigulina, Z.I.: Immune Network Technology on the basis of random forest algorithm for computer-aided drug design. In: Bionformatics and Biomedical Engineering, Spain, pp. 50–61 (2017)

Multi-Agent System Model for Diagnosis of Personality Types

Margarita Ramírez Ramírez(✉), Hilda Beatriz Ramírez Moreno, Esperanza Manrique Rojas, Carlos Hurtado, and Sergio Octavio Vázquez Núñez

Universidad Autónoma de Baja California, Baja California, Mexico
{maguiram, ramirezmb, emanrique, sergio.vazquez}@uabc.edu.mx, Isc.carlos.hurtado@gmail.com

Abstract. In this paper we make the proposal of a multi-agent system model that makes the diagnosis of personality types presented by individuals based on established questionnaires.

This proposal uses technology of intelligent agents to make the diagnosis or identification of personality types through an approach that is different from traditional approaches, which is based on the analysis of the knowledge base and the information captured from the user with concrete actions that the agents resolve through their communication capabilities adapting to the needs of the environment.

To make a diagnosis, the agents use reasoning rules stored in a knowledge base and process information received by the agents of their environment.

Keywords: Multi-Agent systems · Information technologies · Diagnosis

1 Introduction

On the subject of health, be it physical or mental, the adequate diagnosis of diseases is essential for achieving efficient results. Similarly, it's very important to have the correct identification of the personality traits that individuals present, in order to achieve the full development of their abilities. The area of information technologies can be an important support in carrying out these activities. In this area, the changes that have occurred are many; to name a few, there are new ways of storing information from medical records and files that allow the retrieval of information at all times, performing surgeries, diagnosing in the field of psychology, and operations guided by technological devices [1].

There are different technologies that have become support for experts in the health area, a diagnostic support tool for medical personnel or for people who lack the appropriate infrastructure to carry out the observation of a user; an example is the use of multi-agent systems: these systems favor a correct diagnosis since the agents enable breaking down the conventional processes into tasks that are solved by using the communication capabilities of the agents. It's possible to integrate an indefinite number of variables when making the differential diagnosis and to adapt to the needs that a changing environment requires.

G. Jezic et al. (Eds.): KES-AMSTA-18 2018, SIST 96, pp. 209–214, 2019.
https://doi.org/10.1007/978-3-319-92031-3_20

The modeling of this type of systems depends to a great extent on the specific needs that are intended to be attended, in most cases the problem is decomposed in a similar way and focuses on decision-making mechanisms that the diagnostic agent adopts to integrate a result. The other agents that make up the system work to provide data obtained from the environment to this agent. This article addresses the design of a multi-agent system that allows differential diagnosis or identification of personality type, makes use of decision trees to achieve an identification, which will be based on questionnaires applied by experts to users, employees or patients, in order to facilitate the identification of the personality type of the person to whom the study is performed.

2 Agents

The term agent is a concept that combines several disciplines ranging from artificial intelligence, software engineering, databases, distributed systems, to fields of knowledge such as psychology, sociology, medicine, economic theories, etc.

An agent is a computer system located in some environment, within which it acts independently and in a flexible manner to meet its objectives. In addition to the interaction with the environment, an agent is characterized by the following properties [2]:

- Autonomy. An agent has the ability to act without direct human or other agents' intervention.
- Sociability. It's possible to interact with other agents using a common language that serves for communication.
- Reactivity. The agent is in an environment, the space from which he perceives stimuli and before which he reacts in a preset time.
- Initiative. An agent is capable of undertaking and taking the initiative to act guided by the defined goals.

Agents have the characteristics that they are capable of taking initiative, they are capable of sharing and communicating knowledge, they maintain social character, they are also capable of cooperating and negotiating as well as of committing to common goals.

An intelligent agent develops in an environment from which he receives information via sensors and takes concrete and rational actions through activators or actuators (Fig. 1).

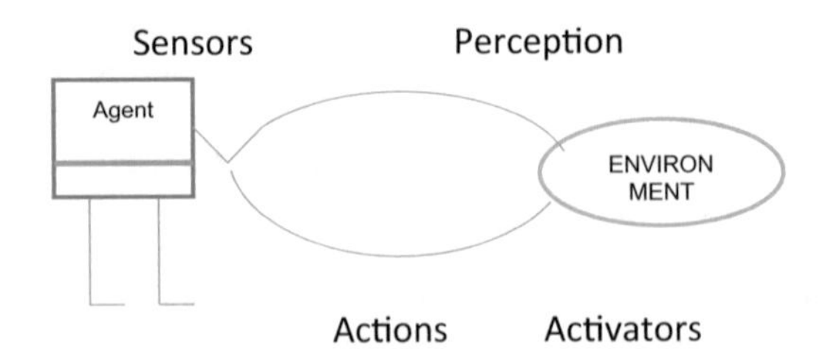

Fig. 1. Intelligent agent

3 Multi-Agent Systems (MAS)

A Multi-Agent System (MAS) is a set of autonomous agents that are generally heterogeneous and independent, that work together, integrating capabilities and resources to achieve the expected functionality. These systems have the ability to interact in a common environment and are able to coordinate, share knowledge and negotiate to achieve the desired goal and solve specific problems.

This network of agents goes beyond the capabilities or individual knowledge of the agents [3]. It is possible to identify a Multi-Agent System as an organized society composed of semi-autonomous agents that interact with each other, either to collaborate in the solution of a set of problems or in the achievement of a series of individual or collective goals [4], These systems, composed of multiple computational elements called agents that interact with each other, are responsible for the coordination of the intelligent behavior of a group of autonomous agents, who have the ability to coordinate their knowledge, goals, skills, decision making and plans [5].

In the previous concepts we find similarities, coincidences and it is possible to consider that these systems are part of a new technological trend. Their capabilities to solve problems that require coordination and communication exceed the object-oriented model in many aspects, allowing the construction of dynamic systems capable of adapting to the changes suffered by their environment [6].

A MAS has several features, including: modularity, supports the bases of Software Engineering and allows the integration of systems. It's possible to solve some problems through the interaction of multiple agents, the use of distributed resources, as well as the distribution of knowledge such as concurrent engineering, manufacturing, health care, among others.

Another important feature of these systems is sharing and communicating knowledge, which is achieved through a common knowledge representation language. This representation of common knowledge is known as ontology. A specific definition for the term ontology is presented by Jimenez, who mentions that an ontology is "a specification of objects, concepts and relationships of a certain area of interest" [7].

The agents that make up a multi-agent system are autonomous; however, they work together. Each agent has incomplete information or capabilities, once these capacities are integrated, the goals are achieved. Agents can dynamically decide which tasks they should perform and which agents will perform each task. In a multi-agent system, interactions must be focused as the basis for understanding the behavior of a system and its evolution.

A multi-agent system monitors users' activities; an agent can read, listen to conversations, recognize patterns generated in conversations, and they can deduce information and goals from past experiences and processed data.

The agent seeks the achievement of its objectives, makes decisions and can decompose objectives into sub-objectives, in addition to performing different tasks. To meet the objectives, it needs collaboration with other agents and capacities such as negotiation, delegation and coordination; agents need location services, agents communicate with the user through interfaces.

The basic characteristics of MAS are the management of distributed information, communication and coordination between autonomous entities; a good option to consider is the design of a patient monitoring system [8].

4 Design of the Multi-Agent System Model

An intelligent multi-agent system model is made up of agents that carry out specific activities such as perceiving, evaluating and diagnosing situations based on the information obtained from the database repositories. In Fig. 2, the elements that make up the model of the multi-agent system for the diagnosis of personality types are shown.

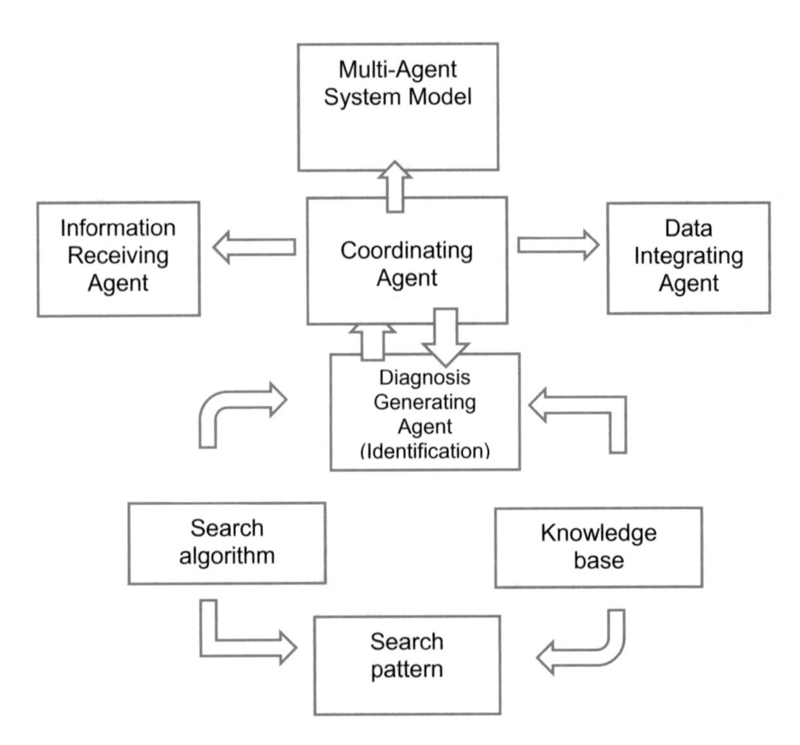

Fig. 2. Model of the Multi-Agent diagnostic system

The multi-agent system model for diagnosing personality types consists of a system that can perform the diagnosis of an identified personality, in particular the definitions and personality disorders. The system is composed of a set of subsystems of intelligent agents, which interact and deal with the specific things to perceive, analyze, evaluate and present the preliminary diagnosis. This model has an agent responsible for receiving the information through questions posed to patients through questionnaires, an agent that integrates the information received, a coordinating agent, responsible for having control between the communications generated between the agents, and an agent responsible for generating the diagnosis or identification of the personality type that the user presents once it has analyzed the information received by the data agent.

5 Description of Agents

Below is a description of each of the agents integrating the proposed model.

a. *Information Receiving* agent. This agent captures information obtained through questionnaires applied to users (patients). There are defined questionnaires that, once applied, make it possible to obtain a diagnosis or identification of the personality type of the person who has answered the test.
b. *Data Integrating* agent. This agent is responsible for keeping track of all users to whom the tests are applied and distributes the information obtained from the Information Receiving agent.
c. *Diagnosis Generating* agent. This agent is able to generate a diagnosis or identification of the personality of the individual once it performs the analysis of the data obtained from the Information Receiving agent and stores it by means of the Data Integrating agent and sets up a comparison with the established knowledge base, in which the characteristics and types of response are described and based on them the different types of personality are determined.
d. *Coordinating* agent. This agent coordinates the agents involved in the system and is responsible for determining personality type identification presented based on answers given to the questions made.

6 Conclusions

A multi-agent system model is extremely useful in areas where it's necessary to diagnose or identify a classification based on an information analysis and an established knowledge base. Such systems allows the different processes to be broken down into specific tasks, which are carried out using the capabilities of communication agents, agents of interaction with the environment and other agents, as well as the ability to consider a large number of variables to integrate a differential diagnosis and adapt to situations in a changing environment.

This paper deals with the design of a model of a diagnosis or identification of personality type based on the use of decision trees considering information obtained in standard questionnaires applied to determine the predominant personality type in a person, in such a way that it's possible to place an individual in an appropriate role or identify a person's reactions based on their personal characteristics.

Intelligent agents use reasoning algorithms with software patterns, which access a knowledge base and determine personality types.

The integration of different agents and appropriate technologies offer benefits that can be exploited in different areas and make appropriate decisions.

References

1. Ramírez, M.R., Moreno, H.B.R., Millán, N.C.O., Núñez, S.O.V., Garza, A.A.: Big Data and health "clinical records" Smart Innovation, Systems and Technologies. Chen, Y.-W. et al. (eds.), vol. 71, pp. 12–18, Springer, Cham (2018)
2. Wooldridge, M., Jennings, N.R.: Intelligent agents: theory and practice. Knowl. Eng. Rev. **10** (2), 115–152 (1995)
3. Sycara, K.P., Roth, S.F., Sadeh, N., Fox, M.S.: Resource Allocation in Distributed Factory Scheduling. In: Zweben, M., Fox, M.S. (eds.) Intelligent Scheduling, pp. 29–40. Morgan Kaufman Publishers, San Francisco (1991). Accessed 06 Feb 2018. http://www.cs.cmu.edu/afs/cs/user/katia/www/katia-home.html
4. Jiménez, J.: Un Modelo de Planificación Instruccional usando Razonamiento Basado en Casos en Sistemas Multi-Agente para entornos integrados de Sistemas Tutoriales Inteligentes y Ambientes Colaborativos de Aprendizaje.Tesis Doctoral. Universidad Nacional de Colombia (2006)
5. UNAL: Universidad Nacional de Colombia. Inteligencia Artificial Distribuida, Disponible en. Consultado Abril de 2012. http://www.virtual.unal.edu.co/cursos/ingenieria/2001394/docs_curso/contenido.html
6. Reyes, V., Rivera, G.M.I.: Prediagnóstico de enfermedades neurológicas a través de un sistema multiagente Ciencias Computacionales
7. Jiménez, A.: Ontologías para comunicación entre agentes Disponible en. Consultado el 26 de marzo de 2011. http://alfonsojimenez.com/computers/ontologiaspara-comunicacion-entre-agentes/
8. Cruz, J. et al.: Sistema Multiagentes para el Monitoreo Inteligente. In: Müller-Karger C., Wong S., La Cruz A. (eds.) IV Latin American Congress on Biomedical Engineering 2007, Bioengineering Solutions for Latin America Health. IFMBE Proceedings, vol 18. Springer, Heidelberg (2007)
9. Liu, G., Lee, K.Y., Jordan, H.F.: TDM and TWDM de Brujin networks and suffflenets for optical communications. IEEE Trans. Comput. **46**, 695–701 (1997)

Towards a Multi-Agent System for an Informative Healthcare Mobile Application

Carlos Hurtado[1](✉), Margarita Ramirez Ramirez[2], Arnulfo Alanis[1], Sergio Octavio Vazquez[2], Beatriz Ramirez[2], and Esperanza Manrique[2]

[1] Departamento de Sistemas y Computacion, Instituto Tecnologico de Tijuana, Calzada Tecnologico S/N, Tomas Aquino, 22414 Tijuana, BC, Mexico
{carlos.hurtado,alanis}@tectijuana.edu.mx

[2] Facultad de Contaduria y Administracion, Universidad Autonoma de Baja California, Mesa de Otay, 14418 Tijuana, Baja California, Mexico
{maguiram,sergio.vazquez,ramirezmb,emanrique}@uabc.edu.mx

Abstract. Since Multi-Agent Systems (MAS) are composed of sets of acting autonomous and pro-active agents, they provide suitable concepts to the construction of self-organizing processes, promising robust and adaptive system behaviors being recommender agents one of the most used for its characteristic. A recommender agent in a mobile app should be aware of user interaction with the interface and make accurate recommendations of its preferences. In this paper, we propose a prototype of a multi-agent recommender system for a mobile healthcare app based in two agents the information retrieval agent, to get user interaction and the recommender agent to make recommendations with the information retrieved about healthy food diet.

Keywords: Multi-agent system · Recommendation agent · e-Health JADE-LEAP Android

1 Introduction

The use of information technologies has grown day by day, they have incorporated and changed the way we communicate, live and do things, one of the technologies that has helped to empower and grow these technologies is the use of smartphones, due to the features they have: always connected, easy to use, allow to see accurate information in real time, but above all for their portability.

Due to these properties presented by smartphones today we find apps for virtually anything, from entertainment, shopping, productivity and health, talking about health issues apps have played a key role as they help people to perform various activities, for example, to have a record of their physical activity, helps them to have control of their weight, pressure, sugar among other things.

Poor nutrition and low physical activity are one of the main causes of overweight in Mexico, which has led to diseases such as diabetes and high blood pressure, which are

G. Jezic et al. (Eds.): KES-AMSTA-18 2018, SIST 96, pp. 215–219, 2019.
https://doi.org/10.1007/978-3-319-92031-3_21

the most common in the population and in which the federal government invests the most according to the Mexican Social Security Institute (IMSS).

Within health mobile applications, there are various categories such as informative, monitoring, for diagnosis and treatment among others, so an application that shows information to people about food consumption, their kilo calorie amount, as well as a calculator of the body mass index and ideal weight was developed to treat to problems mentioned above.

In the first phase, the application only displays information to users but we believe that it is better that it can adapt to the user's preferences and in turn make food recommendations that the patient should consume according to the day of the week, in this way you can take a personalized diet depending on each person likes, thus using intelligent agents that recognize the user's behavior and make recommendations based on that information is proposed.

This paper is organized as follows. Section 2 presents the definition of intelligent agents and multi-agent systems. Section 3 explains the structure of Android operating system and the tools needed to develop multi-agent systems (JADE and JADE-LEAP). Section 4 shows and describes the multi-agent recommender system and the functions of each agent and Sect. 5 explains the conclusion and future works.

2 Intelligent Agents and Multi-Agent Systems

According to Wooldridge [1] agents are defined as computer systems that are located in some environment and are able to take actions autonomously to meet their design objectives. Intelligent agents [1, 2] are defined as agents who can react to changes in their environment, have social communication skills and the ability to use computational intelligence to achieve their goals while being proactive. Agents are active, task-oriented, modeled to perform specific tasks, are able to take autonomous actions and make decisions. The agent modeling paradigm can be seen as a stronger encapsulation of localized computational units that perform specific tasks.

By combining multiple agents in one system to solve a problem, the resulting system is a multi-agent system (MAS) [1]. These systems are composed of agents that solve problems that are simpler than a system in general, they can communicate with each other and assist each one in reaching bigger and more complex goals. These problems that software developers previously thought were very complex [3] now can be solved. Multi-agent systems have been used to predict the stock market [4], industrial automation [5], online learning systems [6] and health systems.

2.1 Using MAS to Support e-Health Services

The use of multi-agent systems in e-Health can help reduce medical visits and waiting lists drastically. Agents are also good tools to help patients follow preventive strategies and support day-to-day self-care. In fact, the proactive nature of agents helps create a bond of trust between patients and the care system, making agents constantly send valuable information to patients, without the need for explicit demands.

With respect to the services provided quality, the most important contribution of multi-agent systems lies in providing correct information to highly specialized health professionals. The proactive nature of the agents and their semantic interoperability supports this need with the possibility of feeding users with information acquired from various sources and adapted to the patient. Thanks to information from health records, accessibility increases rapidly, unifies the information at each stage of the medical process and improves the continuity of care.

2.2 Recommendation Agents

The recommendation agents (RA) are software agents that obtain the interest or preferences of users or products, implicitly or explicitly, and make recommendations. The RAs have been used in various areas, such as education, e-commerce and health. The RAs are characterized as decision-making systems (DSS), as in the DSS in the RAs systems, a person provides an input used as a criterion to search information in a database, generates notifications and recommendations to users.

3 The Android Platform

Android is an open operating system designed to be used in mobile devices such as smartphones, tablets and others. Currently the company that develops Android is Google, Inc., this system includes the Linux kernel and its drivers, an abstraction layer to access the sensors, a virtual machine called Dalvik Virtual Machine, applications and libraries.

3.1 JADE and JADE-LEAP

JADE (Java Agent DEVELOPMENT Framework) is a software framework fully implemented in the Java language that simplifies the implementation of multi-agent systems [7]. The JADE-LEAP add-on was used to solve the integration of these agents in Android platform is a major player in the operating system store for mobile and embedded devices such as smartphones, tablets and others.

From the perspective of development of applications and users, JADE-LEAP is the same as JADE in terms of APIs, and also in terms of run-time administration. Developers can implement their JADE agents in JADE-LEAP and vice versa without changing code. There is no need for a guide or documentation of the JADE-LEAP API, since the one used in JADE applies to JADE-LEAP. However, you must take into account that the JADE and JADE-LEAP containers cannot be combined on one platform.

As the Fig. 1 shows, we can easily integrate agent technology in Android with JADE-LEAP.

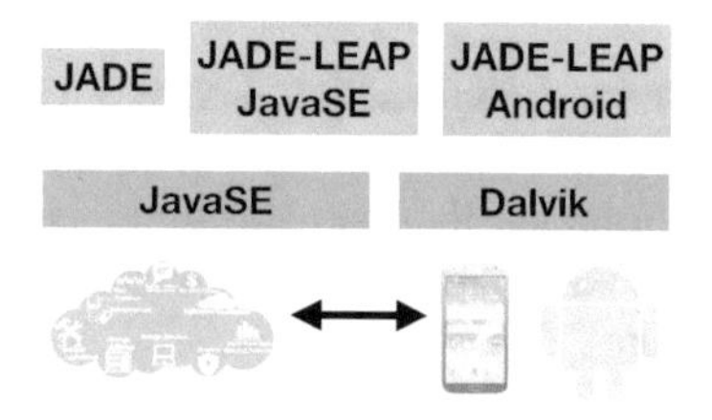

Fig. 1. Android and JADE-LEAP integration.

4 The Multi-Agent Recommender System

Figure 2 shows the multi-agent recommender system, in which the user first login in the application interface, this data is stored in the user information database, later the recommendation agent sends information to the interaction screen, this display foods to be consumed on a specific day based on their preferences, after user interaction with the application, the information retrieval agent sends it to the database so the next time the user opens the application, the recommendation agent reads the information from the interaction database and makes the recommendations based on these previous data.

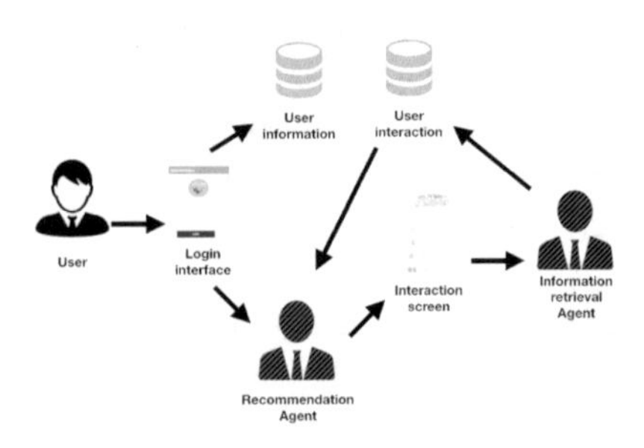

Fig. 2. Multi-agent recommender system scheme.

Figure 3 describes the system model, which focuses on the information retrieval agent and the recommendation agent which work as follows:

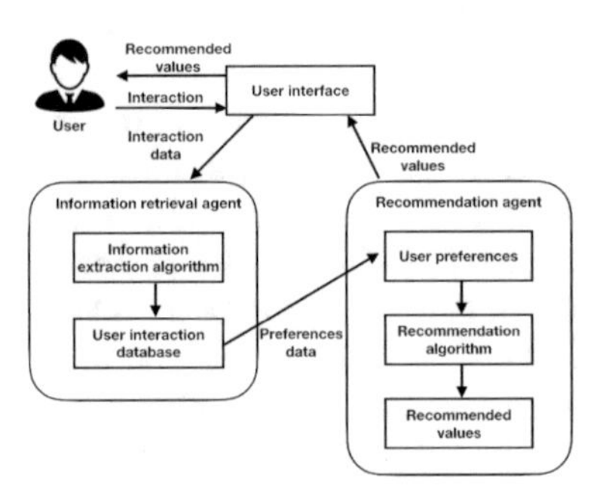

Fig. 3. System model.

- Information retrieval: obtains the interaction data of the user, which goes through an extraction algorithm to be later stored in the interaction database.
- Recommendation agent: obtains the user interaction data from the database and based on these values, an algorithm is used to recommending later recommending values to the user interface.

5 Conclusion and Future Work

In this article, a multi-agent system to recommend nutritional information is presented, after analysis we conclude that we can incorporate multi-agent recommender systems to the Android app we already develop through programming in JADE-LEAP which is compatible with java the programming language of Android.

With the implementation of MAS the mobile application will obtain user interaction information and based on that, will make recommendations so the food consumed will be adjusted to your preferences and needs; the multi-agent system approach for solving the problem is successful in reducing the information overload while recommending relevant information about nutrition to the users.

As future work, we intend to implement the MAS to the health application that we have developed to implement it in the Android play store and analyze its results.

References

1. Wooldridge, M.: An Introduction to MultiAgent Systems. Wiley, Chichester (2004)
2. Rudowsky, I.: Intelligent agents. Commun. Assoc. Inf. Syst. **14**, 275–290 (2004)
3. Marwala, T., Hurwitz, E.: Multi-Agent Modeling using intelligent agents in a game of Lerpa (2007). eprint arXiv:0706.0280
4. Mariano, P., Pereira C., Correira, L., Ribeiro, R., Abramov, V., Szirbik, N., Goossenaerts, J., Marwala, T. and De Wilde P. Simulation of a trading multi-agent system IEEE International Conference on Systems, Man, and Cybernetics, vol. 5, pp. 3378–3384 (2001)
5. Wagner, T.: An Agent-Oriented Approach to Industrial Automation Systems. Agent Technologies, Infrastructures, Tools, and Applications for E-Services, pp. 314–328, (2002)
6. Garro, A., Palopoli, L.: An XML Multi-agent System for E-learning and Skill Management. Agent Technologies, Infrastructures, Tools, and Applications for E-Services, pp. 283–294 (2002)
7. El Kamoun, N., Bousmah, M., Aqqal, A.: Virtual Environment Online for the Project-Based Learning Session, Cyber Journals: Multidisciplinary Journals in Science and Technology, Journal of Selected Areas in Software Engineering (JSSE), January Edition (2011)
8. Marquez, B., Y., Castañon-Puga, M., Castro, J.R., Suarez, E.D., Magdaleno-Palencia, J.S.: Fuzzy models applied to complex social systems: modeling poverty using distributed agencies. International Journal on New Computer Architectures and Their Applications, The society of Digital Information and Wireless Communications, 292–303 (2011)

Proposal of a Bootcamp's User Activity Dashboard Based on MAS

Lenin G. Lemus-Zúñiga[1(✉)], Valeria Alexandra Haro Valle[2], José-V. Benlloch-Dualde[2], Edgar Lorenzo-Sáez[1], Miguel A. Mateo Pla[1], and Jorge Maldonado-Mahauad[3,4]

[1] Instituto de Tecnologías de la Información y Comunicaciones (ITACA), Universitat Politècnica de València, Camí de Vera S/N, 46022 Valencia, Spain
{lemus,edlosae,mimateo}@upv.es

[2] Escuela Técnica Superior de Ing. Informática de la Universitàt Politècnica de València, Camí de Vera S/N, 46022 Valencia, Spain
vahaval@inf.upv.es, jbenlloc@upv.es

[3] Departamento de Ciencias de la Computación, Pontificia Universidad Católica de Chile, Santiago, Chile
jjmaldonado@uc.cl

[4] Departamento de Ciencias de la Computación, Universidad de Cuenca, Cuenca, Ecuador

Abstract. In modern work environment, "Technologies of Computation and Networks" (CNT) and "Information and Communication Technology" (ICT) have changed the way we access and produce information thereby creating the need for professionals with ICT skills. However, there is critical mass of people aged between 25 and 29 who are unemployed and only use boot camps to acquire these skills. This mode of study often requires external help to allow students reach their goal. To overcome this limitation, this article shows a dashboard with different metrics to monitor user's activity and the student can take decisions during the course. It also creates a self-regulating culture based on data to allow success in the course and facilitate student's incorporation into the job market.

Keywords: Technological bootcamps · Dashboard · Unemployment

1 Introduction

"Computer and Networking Technologies" (CNT) and "Information and Communication Technologies" (ICT) have been evolving at high speed since 1970. Consequenly, its impact has changed the way in which we access and use information. As a result, lifestyle has changed and new opportunities arrive. Social networks and e-commerce are examples of new kind of business.

In order to take advantages of these business opportunity, business needs to assure that their employees have acquired the necessary skills related to CNT and ICT.

G. Jezic et al. (Eds.): KES-AMSTA-18 2018, SIST 96, pp. 220–230, 2019.
https://doi.org/10.1007/978-3-319-92031-3_22

This need of high skilled employees impacts directly in the curricula of the Universities. However, for a professional and young people who has not obtained a bachelor of science degree (BSc), it is impossible to be aware of the new technologies, they faced the next reality: Information technology (IT) employers search for employees with the appropriate set of skills for creating applications based on CNT and ICT. As an example, in the USA, schools focused primarily on technology have emerged as the new trade schools, defining a trade school, as a technical school or a vocational school to teach a specific skill set such as electronics repair, or plumbing to people.

Such necessity has generated/motivated the creation of bootcamps, a new business opportunity where the main objective is to train unemployed young professionals into the development of applications, ranging from coding to virtual reality games.

Bootcamps are designed as a short-term course (they only last a few weeks). However, there are cases where they can be very intense, specially the days where the participants face immersion training, that means, spending the whole day doing activities related to a concrete topic.

Our research group, is interested in applying "learning analytics" to assess students gains in bootcamps. To achieve this goal we have started proposing the design of a dashboard and defining the initial metrics. This dashboard will help us to understand how students works.

This paper is organized as follows: Sect. 2 presents management tool's design related work. In Sect. 3 the dashboard design is presented. In Sect. 4 the materials, methods and procedures are presented. Section 5 is dedicated to present results. And finally, Sect. 6 presents conclusions and future work.

2 Related Work

In this section, the main concepts related to management tool design are presented. Firstly, it is important to understand the hierarchical structure of planning for the management of an organization (see Fig. 1).

Linked with those management levels, there are two main types of tools: the Balanced Scorecard (BSC) and the Dashboard (DB). These tools are described and compared in the subsequent sections.

2.1 Hierarchical Structure of the Planning of an Organization

The hierachichal structure has been obtained from the literature [4]. A summary of such hierarchy is presented below.

- **Strategic Planning:** It is where the Strategic Plan of the organization is developed in a long-term time horizon, defining the mission, vision, values and strategic challenges that will be addressed.
- **Tactical Planning:** It is where the Tactical Map is developed or the different Programs that will specify the strategic objectives that will seek to address the strategic challenges of the Strategic Plan. It will be developed in a medium / long-term time horizon.

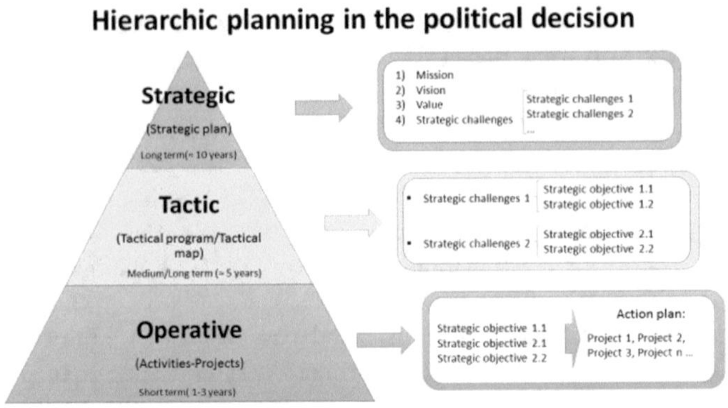

Fig. 1. Hierarchical structure of the planning of an organization.

- **Operational Planning:** It is composed of the different projects / actions included in the Action Plan to achieve the strategic objectives of the tactical map or specific program. It will be developed in a short-term time horizon.

2.2 Management Tool Types

The two main types of management tools that currently exist are:

- **Balanced Scorecard (BSC):** Management and planning technique created in 1992, by Robert S. Kaplan and David P. Norton that translates the strategy into objectives, measured through Key Performance Indicators (KPIs) and linked to the action plan that allow to align the behavior of the members of the organization with the strategy of the company.
- **Dashboard (DB):** Graphic representation tool to consolidate numbers and metrics through graphs and diagrams. It is used to show current status of the organization in an operational level.

2.3 Most Relevant Differences Between BSC and DB

Frequently, the term dashboard (DB) is used as a synonym of balanced scorecard (BSC). Both two are performance management tools and help an organization to achieve its goal, but each one has a different function.

In [5] the most important differences among them are explained, below there is a summary of such differences. First, the DB serves the organization to monitor performance at operational level while the BSC has the purpose of managing it at strategic level. The most important difference is that the dashboard measures metrics while the BSC measures KPIs (metrics linked to strategic objectives). Therefore, the function of the DB is aimed at measure the real time state of the organization to have an accurate and up-to-date knowledge of what is happening, in order to allow it to meet operational objectives (short term). On the other hand, the function of the BSC is aimed at show the evolution (usually monthly) of performance, to follow the strategy defined by the organization through the relationship between the KPIs and the strategic objectives (medium/long term). An example within the same system could be the instrument panel of the car that it indicates the state of the vehicle (speed, revolutions, temperature, etc.) as a dashboard, and the GPS that shows currently state (current location), the objective (the destination), and the process to achieve the objective (the roadmap), functioning therefore as BSC.

It is also important to explain the differences between the way each one is used. In the case of the BSC, it is marked strategic objectives based on the tactical map or program of the organization. According to this, it is elaborated the action plan to define the projects or actions that seek to achieve those objectives, and it is evaluate the achievement of the objectives with the monitoring of KPIs. In the case of the DB, it is generated events where the metrics are processed, calculated and displayed in chart or diagram form in the front-end tool in order to monitoring the most relevant metrics of the operational planning in real time.

Finally, The design process is also different. On the one hand, the BSC has a Top-Down design process, setting the strategic objectives of the Tactical Map of the organization, and then assigning their KPIs together with the definition of an action plan to achieve the objectives. On the other, the design of the DB consists in the selection of the metrics to be followed up in operational planning on the one hand and in the design of the graphs and diagrams (front-end) that will be used for the graphic representation of their values on the other hand.

3 Design of the Dashboard Tool

The basic steps to design a DB were introduced in previous section. Following, those steps will be detailed.

3.1 Process of Obtaining the Metrics. From Observation to Metrics

The first step to design the DB tool is the selection of the metrics to follow (see Fig. 2). This process has four different phases that are described below:

- **Observation:** It is the first phase of the data collection. It consists in determine what things are intended to be measured in order to obtain the value of the parameter wanted.
- **Quantification:** The second phase is based on quantifying the variables of the measure, making a mapping of the observation to numbers.
- **Measure:** The third phase is the result of quantification and it could be defined as the quantity or degree of something.
- **Metric:** The fourth phase is the calculation applied to the measure that it is necessary to obtain the wanted parameter. Some sources define it as the derivative of the measure [3].

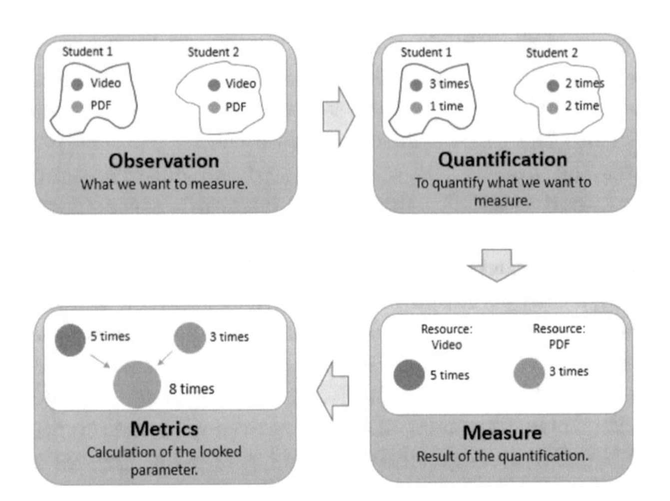

Fig. 2. Metrics' obtention process.

3.2 Design of the Front-End

The second step to design the DB tool is the design of the front-end (see Fig. 3) for the graphic representation of the metrics.

The process is described below:

1. **Grades:** This graph help the student to understand their grades using a different color, where the red means risk and the student should focus their time and resources to improve it, the yellow means precaution and the green means the students get good grades.

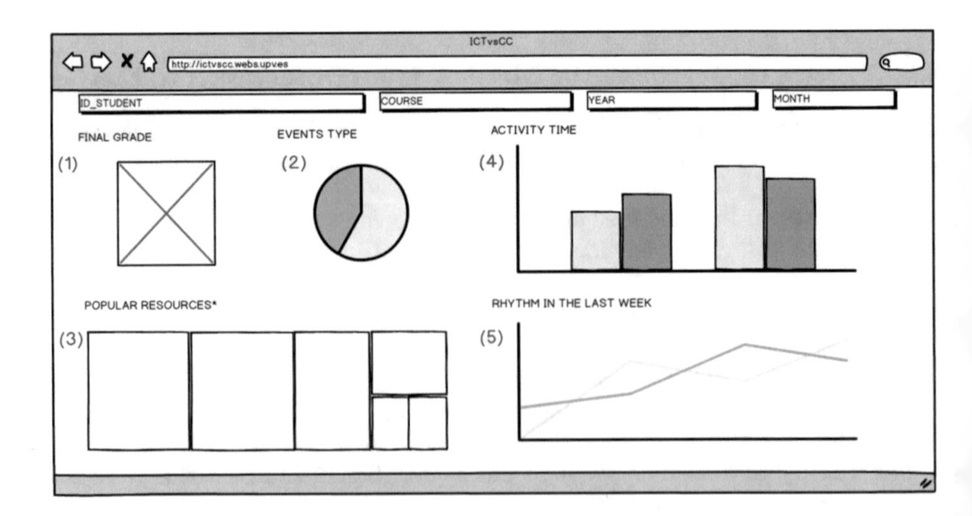

Fig. 3. Mock-up of the front-end.

2. **Events Type:** The LMS have different types the events available where the student can interact, this pie graph represents the average the time of the student in each event done by them, events like read a resource or view a video.
3. **Popular Resources:** Treemaps are an alternative way of visualizing where displaying quantities for each resource via area size, the student can view the popular resources and the teacher can identify the less popular resources to improve the content and attract the attention of the student.
4. **Activity Time:** The bar graph shows the time of student activity in the LMS distribute by months.
5. **Rhythm in the Last Week:** This line graph shows the student activity in the LMS in the last weeks compare with the average of the classmates' activity to help the student to understand if their compromise with the subject is equal, less or better than their classmates.

4 Materials, Methods and Procedures

In the next subsections it is going to be described the used materials, the method for obtaining the data provening from the LMS of the UPV, the method used, the characteristics of the course under study and the procedure done to obtain data.

4.1 Materials

The materials consist on desktop computers used by teachers, laptop computers used by students and software. The details of each item is described below:

4.1.1 Computers

Teachers participating in the experience used destktop computers with standard components, while students participating in the experience used their own laptop computer. Table 1 shows the main characteristics of the used computers.

Table 1. Main characteristics of computers used by teachers and students.

	Teacher	Student
Computer type	Desktop computer	Laptop computer
Operating system	Windows 10 enterprise	Windows 10 home
Operating system type	64 bits	64 bits
Processor	Intel core i7-4790 CPU@4 GHz	Intel Core i7-7700HQ CPU@2.80 GHz
RAM memory	16.0 GB	8.0 GB

4.2 Learning Management Systems Used at the UPV

The UPV uses a LMS based on SAKAI [XX]. In words of their designers "Sakai represents a fundamentally different approach to the learning management system. Unlike other "open" systems available today, the direction and feature set of Sakai originates from within higher education to address the dynamic needs of a global academic community. The Sakai open-source community values the participation of its contributors highly, with educators and developers from various institutions working together to turn great ideas into reality for the entire community of Sakai adopters".

4.3 Course Under Analysis

The data is being obtained from fresh students of the "Grado en Ingeniera de Computadores" taught at the UPV during the course 2017–2018. The course TCO is taught during the first semester of the degree and has 400 enrolled students.

4.4 Procedure Used to Obtain Data

The main data source is the database of the LMS SAKAI [2]. Through a special interface teachers of the course TCO can access to the data activity of the students. Teachers can retrieve predefined data or ask for specific data. Figure 4, shows the predefined reports that a teacher can obtain and Fig. 5, shows the form that teachers has to fill in order to obtain specific reports.

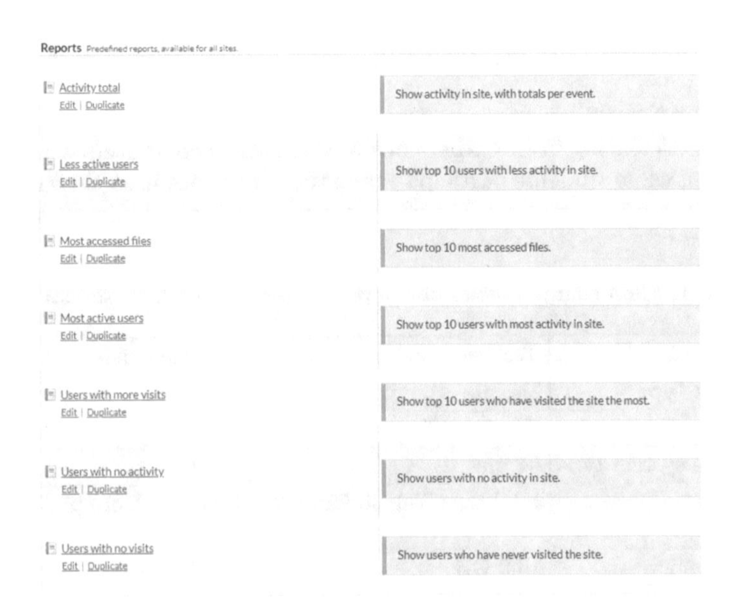

Fig. 4. Predefined reports that teachers can obtain from the LMS.

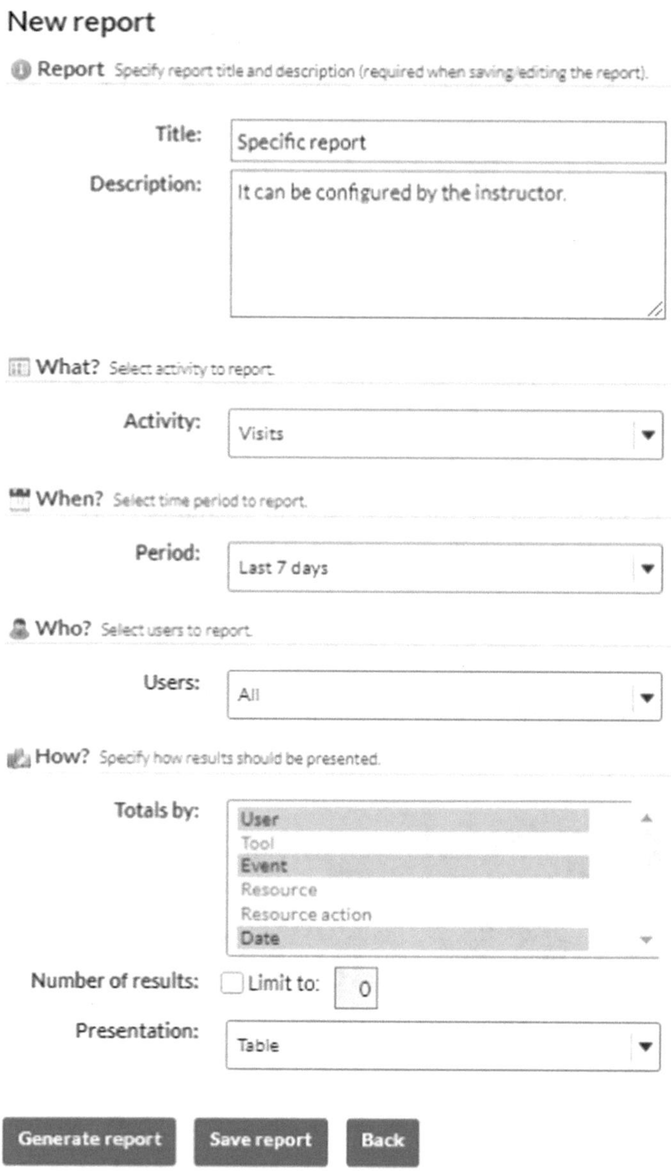

New report

Report Specify report title and description (required when saving/editing the report).

Title: Specific report

Description: It can be configured by the instructor.

What? Select activity to report.

Activity: Visits

When? Select time period to report.

Period: Last 7 days

Who? Select users to report.

Users: All

How? Specify how results should be presented.

Totals by: User
Tool
Event
Resource
Resource action
Date

Number of results: Limit to: 0

Presentation: Table

Generate report Save report Back

Fig. 5. Form used by teachers to retrieve data from the LMS.

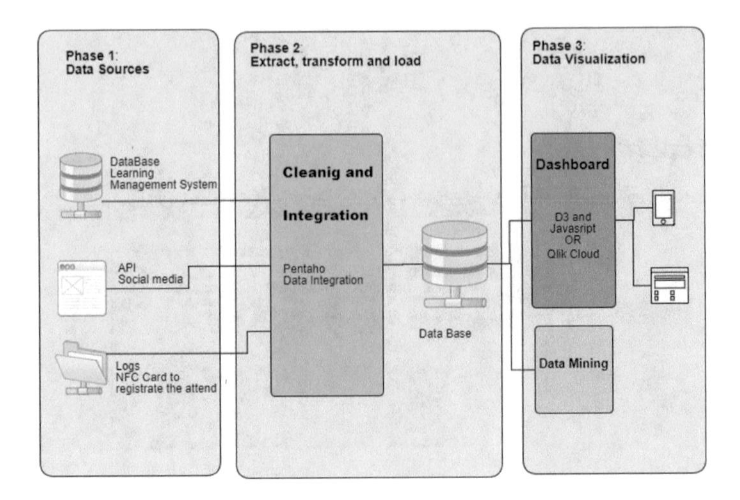

Fig. 6. Data Flow.

Once the teachers have obtained data, related with student's activity, they proceed to clean it and anonymize it and store it in a database. Using Tableau [XX] we have proceeded to plot data. Figure 6, shows how data is gathered, processed, stored and visualized.

5 Results

Using the user's activity data of TCO fresh students of the degree "Grado en Ingenieria Informatica" the aspect of the dashboard is shown in Fig. 7.

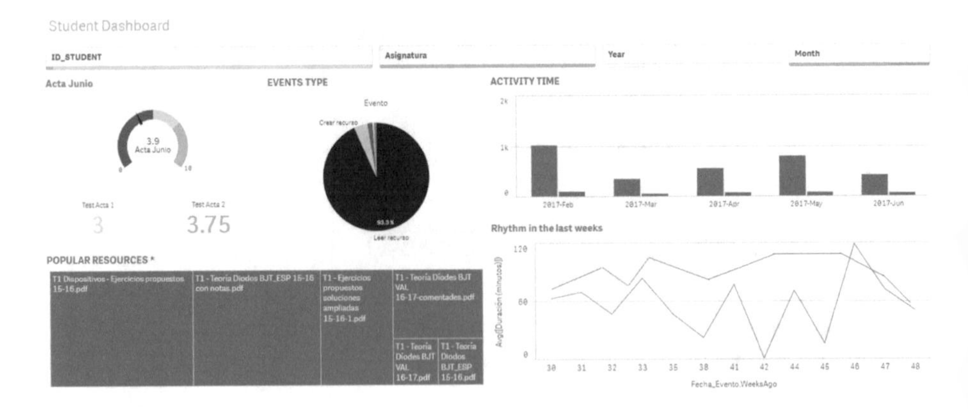

Fig. 7. Architecture of the data processing system.

The system will be designed as a multi-agent system (MAS). In a MAS we can distinguish between the services provided by the system and the actions

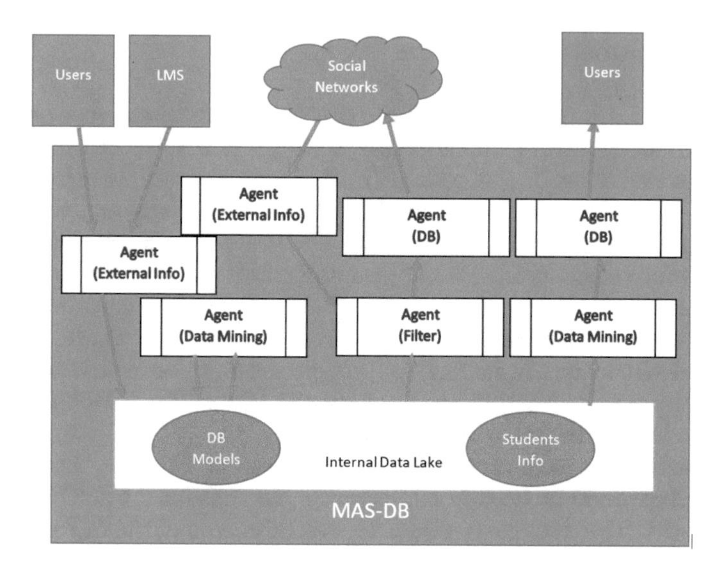

Fig. 8. Multi-agent system elements.

performed by the agents [9]. In our case, the system will be responsible for providing data storage services in addition to the services of a MAS.

The agents of the system may have different roles, although it is usual for each agent to deal with a single role at a given time. At this moment, four types of agents have been defined (see Fig. 8):

- External Info agents: They obtain information external to the MAS and include it in the Internal Data Lake. The external information will basically come from the LV of the UPV, but there are also agents that can obtain information directly from users or search social networks (Twitter, Telegram, etc.)
- Data Mining agents: agents that uses internal information to produce NEW information. The new information can be stored inside the MAS or sent to other agent to further processing.
- Filter agent: Get a set of information and removes part of it, sending the remaining to other agent.
- Dashboard Agent: an agent that acts as a dashboard i.e. shows consolidated information to users. The medium could be a web interface, a GUI or a message to a social network.

Although the structure of the design is based on MAS, the final implementation is still not defined and may not be based on an agent system [11], such as Jade [1].

6 Conclusions and Future Work

The objective of this project has been to design an operational management tool (dashboard) so that students can check their performance. DB serves its user to know the real state of the system (in this case its performance and that of the rest of students), and based on this, to make decisions and to modify the operation in the short term (for example, the use of resources at their disposal such as PDFs, videos, etc.). But it does not allow the teacher to manage the system based on strategic objectives in the medium and long term. For this purpose, a tactical management tool (Balanced Scorecard) should be designed to provide feedback to the tactical map or program, in order to link the action plan with the strategic objectives through the KPIs. The development of this tool is considered interesting as future work.

In this article we have presented the bootcamp phenomena. The design of a dashborad that could be used in a bootcamp. The dashboard has been designed using real data. The data corresponds to fresh students of the degree "Grado en Ingeniería Informática".

Besides an architecture to process the data stored in the LMS used by the UPV has been shown.

And finally we have proposed the architecture of a Multi-Agent system to automate the gathering, cleaning, anonymazyng and storing data of the LMS. Such system is under development.

References

1. Jade Site: Java Agent DEvelopment Framework
2. Sakai. Accessed 02 Feb 2018. https://www.sakaiproject.org/
3. Aleksey, S.: The difference between quantification, measure, metric, and kpi
4. Aleksey, S.: Get the big picture about balanced scorecard, its connections and roles
5. Aleksey, S.: What's the difference between a dashboard and a balanced scorecard?
6. Charleer, S., Moere, A.V., Klerkx, J., Verbert, K., De Laet, T.: Learning Analytics Dashboards to Support Adviser-Student Dialogue. IEEE Trans. Learn. Technol., 1–1 (2017)
7. Ferguson, Rebecca: Learning analytics: drivers, developments and challenges. Int. J. Technol. Enhanc. Learn. **4**(5/6), 304 (2012)
8. Few, S.: Information dashboard design. The Effective Visual Communication of Data. O'Reilly, Sebastopol (2006)
9. Pearce, J.: Multi-Agent Architecture
10. Phillip, L., George, S.: Penetrating the Fog: Analytics in Learning and Education–EDUCAUSE
11. Singh, M.P., Chopra, A.K.: Programming multiagent systems without programming agents. In: Lecture Notes in Computer Science (Lecture Notes in Artificial Intelligence and Lecture Notes in Bioinformatics, vol. 5919 LNAI, pp. 1–14 (2010)

Toward to an Electric Monitoring Platform Based on Agents

Jorge E. Luzuriaga[1(✉)], Guillermo Cortina Rodríguez[1], Karolína Janošová[2], Monika Borova[2], Miguel Ángel Mateo Pla[1], and Lenin-G. Lemus-Zúñiga[1]

[1] Institute of Information and Communication Technologies (ITACA), Universitat Politècnica de València, Valencia, Spain
jorluqui@upvnet.upv.es

[2] VŜB - Technical University of Ostrava, Ostrava, Czech Republic
https://ictvscc.webs.upv.es/en/

Abstract. Smart grids are electricity supply-networks which detect and react to local changes in usage based on feedback from digital communications technologies. Their expansion are likely to have a positive impact on our economy, social development, and especially, on environmental measures taken to counteract climate change. In addition, the application of business strategies in the electricity-production sector would facilitate the pre-diagnosis of energy consumption fluctuations. Energy-saving audits are currently carried out by specialised professionals who rely on initiatives which are costly, both in terms of time and effort.

This paper presents a platform that guides users through the process of performing online energy consumption audits in near real-time. It describes how, based on the user input, the platform auto-completes the remaining necessary data in a simple and intuitive way. We then propose the integration of two types of agents into this platform: One to ease the transformation of source information and the other to act as a warning system to detect unusual energy consumption levels. We tested our methodology by calculating and validating energy consumption from different perspectives in two different public infrastructure electrical grids belonging to the city of Llíria in Valencia (Spain).

Keywords: Agents · Smart grids · Energy-saving
Electric audit · Energy consumption

1 Introduction

Smart grids, electricity supply-networks that detect and react to local changes in usage based on feedback from digital communications technologies, are a new tool that can help to achieve societal energy-consumption targets and thus, promote competitiveness. In addition, they produce savings compared to current electrical systems, and hence, they can contribute, to some extent, to mitigating climate change [1,14].

G. Jezic et al. (Eds.): KES-AMSTA-18 2018, SIST 96, pp. 231–240, 2019.
https://doi.org/10.1007/978-3-319-92031-3_23

However, some issues regarding smart grids must still be solved, including gaining a better understanding of their potential benefits and of their potential impact on the economy and public policies. Making smart grids feasible will require implementing several changes, especially in terms of energy marketplace organisation, and will also involve deregulating other parts of the energy market, including the electricity and gas industries.

In this paper, we present a platform that guides users, from start to finish, through the process of performing audits online, in near real-time. Based on the user input, the platform auto-completes any necessary data in a simple and intuitive way so that no information is omitted. The platform was designed to allow users to perform the whole audit process, from data collection to its evaluation, according to three specific steps: (a) data acquisition; (b) information management; and (c) generation of the final report.

An important secondary part of this project corresponds to the development of agent-based technologies for the platform. These employ innovative techniques which are based on a range of experiments and whose use has been verified for energy management and electrical grids.

The remainder of this paper is structured as follows: a review of the literature with different approaches to ensure energy efficiency is described in Sect. 2. The current architecture of the electric monitoring platform in addition to our proposal toward to convert it into an architecture based on agents is outlined in Sect. 3. In Sect. 4, we present the preliminary results of real data corresponding to the year 2017 of the electrical consumption of two specific public energy contracts. In Sect. 5, we wrap up with our conclusions and future work.

2 Related Work

The following literature review summarises the state of the art in the development of agent-based technologies which can be used in electrical grids and energy management. The work included ranges from articles that outline the issues identified within the energy marketplace to publications that characterise the individual techniques, designs, and implementations of the algorithms relevant in this area.

Authors in [14] understanding the complexity of energy management solutions proposed a reflexive theoretical model that can be adjusted from time to time to predict actions and decisions.

With the aim of providing recommendations to electricity users to improve the energy efficiency and to reduce their energy consumption. We can find solutions, like in [4] where current lighting systems were substituted by the energy efficient light emitting diode (LEDs) or appliances such as the sleeping mode cycle of laboratories, air conditioners, and office computers were increased. Moreover, fast tools to personal computers designed by authors in [5] and even applications for smartphones developed by authors in [13]. However, these solutions present certain limitations due to they just work with determined parameters or inputs.

Several works that try to help in the automation of the execution, processing, and analysis of buildings' energy audit data are founded in the literature, e.g., authors in [3] offer some of the previously mentioned options in an online management system. In addition to that, authors in [12] proposed a cloud computing solution that uses specialized audit agents to allow audits processing and reporting on the fly.

For a better understanding of electrical energy usage habits for the wider sections of our society. Authors in [6] provide a depth analysis of different energy-intensive industrial sector groups, providing methods and options to eliminate and improve energy efficient management. Moreover, authors in [7] outlines that the main part of the energy waste is largely caused by domestic users. To achieve high efficiency of energy they recommend to install new effective and efficient devices and equipment.

In the terms of real energy audits, authors in [8] evaluated the situation related to the energy consumption in historical buildings of the University of Pavia, considering different aspects related to standards compliance. Similarly, an energy audit of historic buildings in Italy focused on winter season is presented in [9].

Since the current literature on energy agents and the lack of a practical approach to ensure energy efficiency, this paper focuses on solving these deficiencies.

3 Materials and Methods

Our platform was developed to set up and monitor energy consumption following a classical client-server application model with three-tier architecture. The functions like presentation, application processing, and management of the data are physically separated in the mentioned tiers. As can be seen in Fig. 1, the architecture is composed by 3 different components: sources, database and user interface.

On one hand, in the presentation tier (front-end) the information is provided to the users, in the way that users can visualize and understand better, in less time, a large volume of data. The information can be accessed through a web interface which gets data from the logic tier by using RESTful Web Services [2], then it is displayed via charts or reports. On the other hand, the back-end is deployed by the logic and data tier. Both interact to provide the information to the presentation tier. Here is where data is recovered, stored and maintained, also requests from the front-end are processed. Data can be received from two different types of sources, like: measurement values in consumption (kilowatts per hour) and electricity bills. By using Extract-Transform-Load (ETL) techniques Services [10], the platform can collect data, process and analyse it, so wrong data can be avoided. Also, data is stored in a database where in conjunction with different methods were created to access, monitor and control the data.

Our platform allows data from multiple sources and can set up and monitor remote energy consumption (remote reading electricity meters or Internet of Thing (IoT) devices, like the device shown in Fig. 2, with information transfer

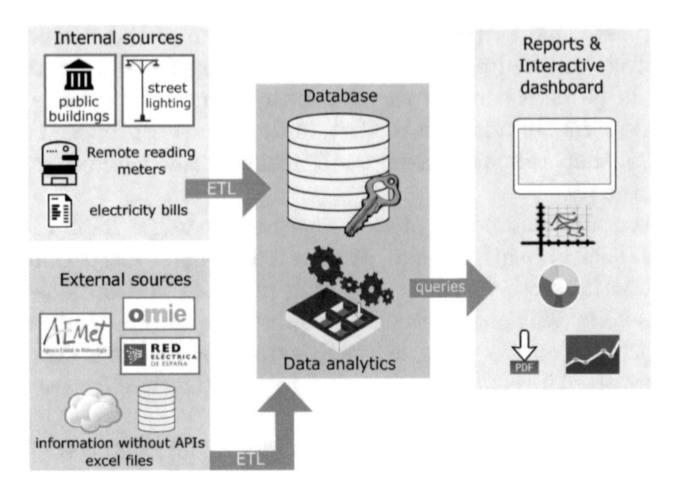

Fig. 1. Connected measurement components showing the flow of information.

ability by means of the Internet. Each meter it is related with a contract which identify it. The platform is also able to centralize information from electricity bills and consumption. Using business intelligence techniques, the information is summarized. Furthermore, this model ensures the integrity and heterogeneity of data and it must be said that as it is a three-tier architecture it is possible to scale if the demand increase.

Fig. 2. Example of remote reading electricity meters.

The benefits of our platform model architecture are: (a) maintainability, due to each tier is independent of the other tiers, so updates or changes can be carried out without affecting the application. (b) scalability, due to tiers are based on the deployment of layers thus scaling out an application is reasonably straightforward. (c) flexibility, due to each tier can be managed or scaled independently.

(d) availability, due to the modular architecture enables systems the use easily scalable components which increase availability.

An agent is a computer program that acts as a user or other program, agents can activate themselves perceiving their context without interaction of users. The idea it is to have a network were agents systematize data.

The aim of the paper is to convert the platform described before into an agent architecture with the least number of agents. We planned that the architecture does not require a multi-agent system due to the inclusion of them will complicate the functionality of our platform.

Thus, our architecture will just include two agents. The first one will oversee the data collection and the second one will take care of monitoring of the information.

3.1 Information Agent

This agent will take care about downloading information from both sources receipts and remote reading electricity meters, in an interval near real time. This agent must be able to scan from a web application database and it also will be able to insert reliable data into the database.

3.2 Monitoring Agent

This agent will be in charge of monitoring the values of energy consumption. Through data analytics and machine learning techniques the agent will have the ability to extract value from such massive data. The agent establishes automatically thresholds by using average, minimum and maximum based in previous consumption stored in the database. In case of the consumption exceed above the threshold, the agent will send an alert to the user to inform that the "day before" he has exceeded in his consumption. Also, the agent can notify the user when the consumption is going to cross the threshold based on historic data stored in the database.

4 Preliminary Results

The proposed agent-based system consists of the two types of agents instantiated in each energy contract. First, an alarm agent who set a threshold of the consumption in kWh to warns the exceedance from a certain limit value. The second agent who works with information sources, standardizing the incoming data.

The following results show the electrical consumption from different perspectives corresponding to the year 2017 of two specific public energy contracts. The numerical data were collected from real data of two buildings located in the city of Lliría. Specifically, they represent particularly (a) the citizen's attention office and (b) a group of street lighting lamps close to a high school. The electric power consumption of the municipal outdoor lighting in Spain for a municipality of size between 10,000 to 40,000 inhabitants it is 163 W per point of light [11].

For instance, Fig. 3 shows the different consumption in the same period for the year 2017, taking into account the average outdoor temperature. According to this graph, the higher consumption is during the winter months (from December to February) and the lower consumption during the months with the longest lighting days (in the summer solstice - the longest day in June in Spain has 15 h of sun).

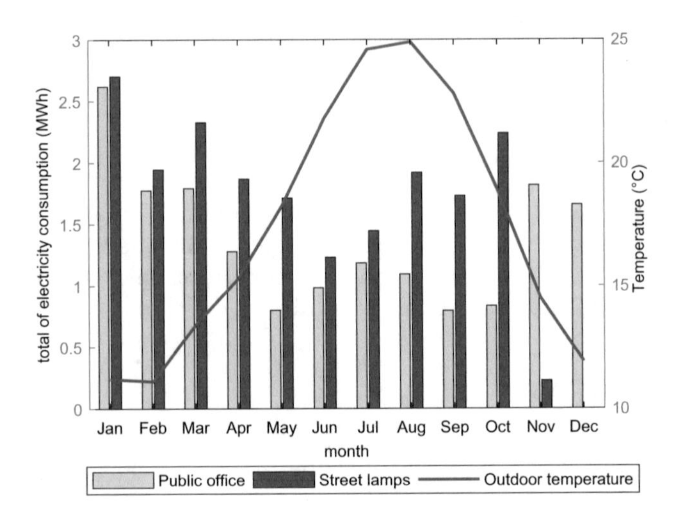

Fig. 3. Annual electricity consumption for both contracts taking into account the outdoor temperature.

To see the monthly consumption values for the buildings during the year of study, in Fig. 4 we can see the consumption for the public office at left while at right of the figure the consumption for the street lamps is showed.

As is shown in the graphs there are some red markers outside of the boxes that represent the presence of anomalous readings in a specific month. Statistically, these points are insignificant but for the machine learning algorithm used by the agent are used to identify errors in the remote meter readings.

In the public building, there is not the presence of anomalous values in the readings and they are more regular and predictable. Opposite situation is found with the street lamps measures, where for instance the month of November is incomplete, in addition to December where the system has not any reading. Thus the graph has a different course in these months.

After that, we focus on the street lamp's consumption, due that it presents several problems in the meter readings. In Fig. 5 we selected four months to show in which day the problems with the readings are presented.

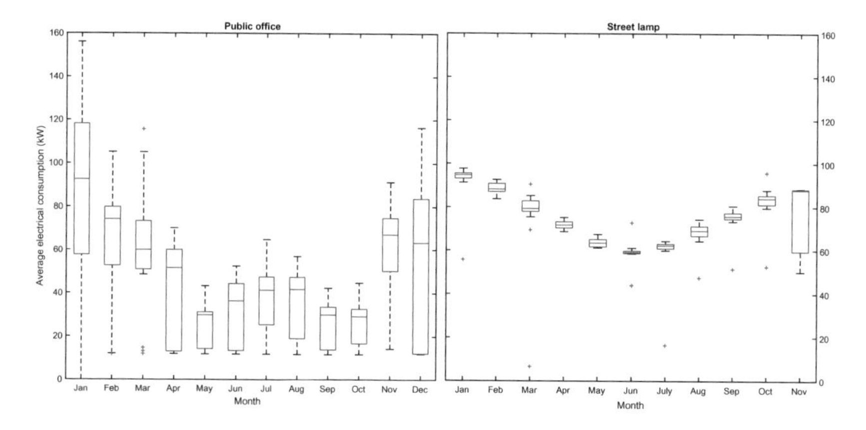

Fig. 4. Monthly energy consumption in the public office (left) and street lamps (right).

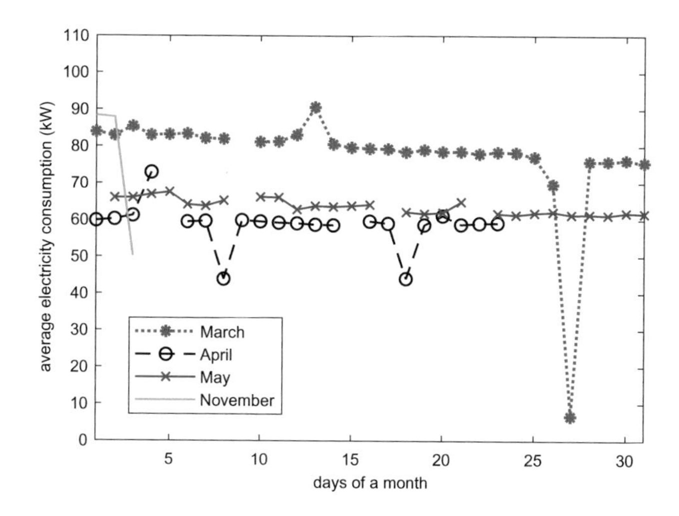

Fig. 5. Energy consumption for different months.

We choose from the previous plot three of the months that have more problems with the collected data (March, April, November) in comparison with one month that has most of the meter readings (May).

Using our energy audit system, we can just alert to the responsible person about the detection of incomplete data readings but is out of our scope found the reason to know why the remote reading fail with this specific meter.

Figure 6 shows the statistical behaviour observed for each day of the week with the average consumptions of the whole year 2017. According to the trend shown in the graph, if the consumption distribution is compared, every day have very close values among them. Figure 6 has also red markers that represent problems in the remote readings, specifically, at left of the this figure, the problems are observed in days wednesday, thursday and friday. Which is proof that in the

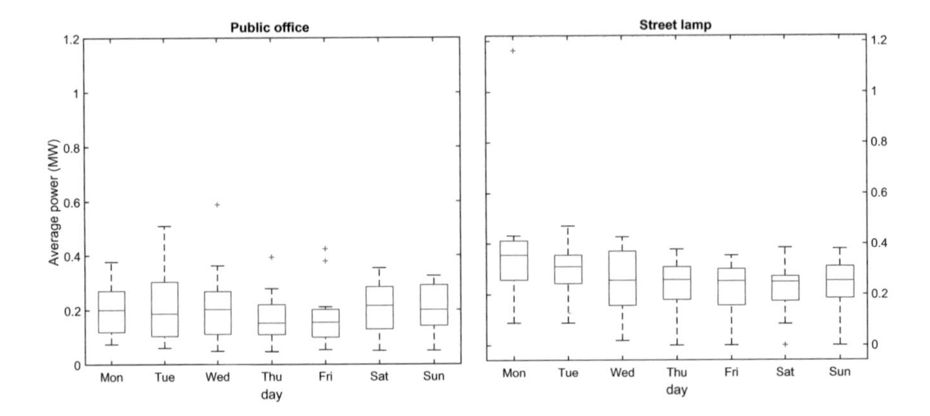

Fig. 6. Statistical behaviour observed for each weekly day for the consumption of the whole year by the public office (left) and the street lamps (right).

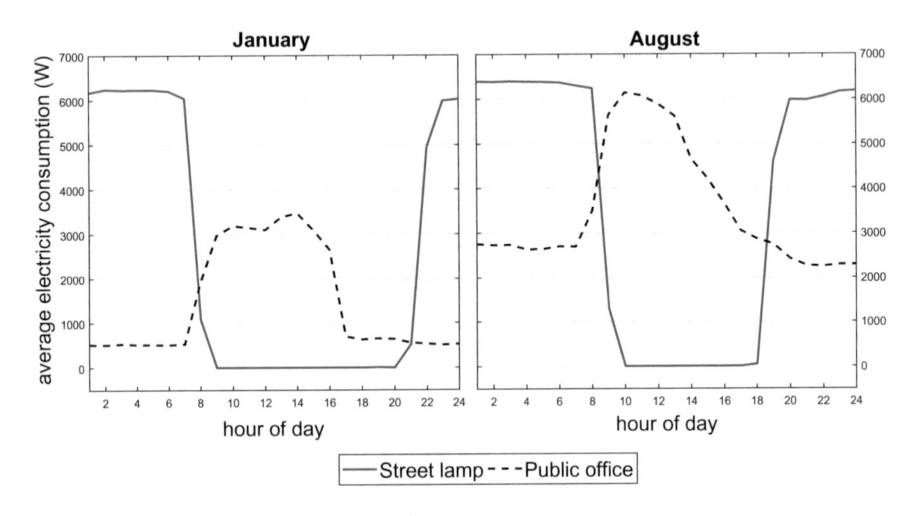

Fig. 7. The average energy consumption in January (left) and August (right) 2017 for both electrical contracts distributed along the hours of a day.

public office more issues are caused by the human factor, if it is compared with street lamps where just one point is out of the boxes.

Figure 7, presents how energy consumption changes in average during the hours of a day for a whole month, to the public office and also to the street lamps respectively.

Consequently, we choose two representative months along a year in Valencia-Spain. Firstly, August for the warmer months in summer as well as December as a representative month for the winter period. In general, in the case of the street lamps, the consumption decrease to the zero during the daylight according to the length of the day (sunshine and sunset) where the street lamps were switched off/on.

In the case of the public office, in January (left of Fig. 7) the consumption gradually increased from 07:00 to noon and after that a slightly decreasing up to 20:00, where there is no more consumption until next day. While in August (right of Fig. 7) the consumption gradually increased from 06:00 to 09:00 and the consumption remains up to closing time (16:00).

5 Conclusions

With the objective of improving energy efficiency, we developed a platform that can be used to easily set up and remotely monitor energy consumption. This platform manages data relating to energy consumption and electricity bills and uses this to provide users with knowledge that can help them to make more informed decisions.

To automate the platform as much as possible, we will soon be implementing an architecture based on two agents, one to facilitate data management and the other an alert monitoring system designed to automatically notify users, in near real-time, about any anomalous energy consumption rates. This agent architecture have the advantage that it enables our platform to carry out several different tasks without requiring additional human interventions.

Finally, as a future work, we are implementing an energy forecasting module following the guidelines above. We are convinced that this system will make effective predictions based on historical information about energy demand collected by our current platform and electricity bills.

References

1. Clastres, C.: Smart grids: another step towards competition, energy security and climate change objectives. Energy Policy **39**, 5399–5408 (2011)
2. Fu, C., Belqasmi, F., Glitho, R.: RESTful web services for bridging presence service across technologies and domains: an early feasibility prototype. IEEE Commun. Mag. **48**, 92–100 (2010)
3. Ganapathy, A., Soman, G., Manoj, G.V.M., Lekshamana, R.: Online energy audit and renewable energy management system (2017). https://www.scopus.com/inward/record.uri?eid=2-s2.0-85016213718&doi=10.1109%2FICCUBEA.2016.7860035&partnerID=40&md5=50dcd67a18905385e57728d52fd69bff
4. Getu, B.N., Attia, H.A.: Electricity audit and reduction of consumption: campus case study. Int. J. Appl. Eng. Res. **11**(6), 4423–4427 (2016)
5. Hasanah, R.N., Suyono, H.: A building audit software to support energy management and conservation. In: Proceedings of the Conference on Applied Electromagnetic Technology (AEMT), pp. 11–15 (2012)
6. ICF Consulting: Study on Energy Efficiency and Energy Saving Potential in Industry and on Possible Policy Mechanisms, pp. 1–461, December 2015
7. Kumar, A., Ranjan, S., Singh, M.B.K., Kumari, P., Ramesh, L.: Electrical energy audit in residential house. Procedia Technol. **21**, 625–630 (2015). http://linkinghub.elsevier.com/retrieve/pii/S2212017315003023

8. Magrini, A., Gobbi, L., D'Ambrosio, F.R.: Energy audit of public buildings: the energy consumption of a university with modern and historical buildings. Some results. Energy Procedia **101**, 169–175 (2016). https://doi.org/10.1016/j.egypro.2016.11.022
9. Marinosci, C., Morini, G.L., Semprini, G., Garai, M.: Preliminary energy audit of the historical building of the School of Engineering and Architecture of Bologna. Energy Procedia **81**, 64–73 (2015)
10. Qin, H., Jin, X., Zhang, X.: Research on extract, transform and load (ETL) in land and resources star schema data warehouse. In: Proceedings of the 2012 5th International Symposium on Computational Intelligence and Design, ISCID 2012, vol. 1, no. 4, pp. 120–123 (2012)
11. Quintero, A.S.d.V.: Inventario, consumo de energía y potencial de ahorro del alumbrado exterior municipal en España. Tech. rep., Instituto para la Diversificación y Ahorro de la Energía (IDAE) (2017)
12. Ruebsamen, T., Reich, C.: Supporting cloud accountability by collecting evidence using audit agents. In: Proceedings of the International Conference on Cloud Computing Technology and Science, CloudCom, vol. 1, pp. 185–190 (2013)
13. Shi, S., Wei, G., Bin, W., Wu, S., Wei, Z., Jun, Y., Jiaquan, Y., Guihai, L.: Design and development of electrical energy-saving diagnosis calculator APP. In: 2015 Sixth International Conference on Intelligent Systems Design and Engineering Applications (ISDEA), pp. 281–287 (2015). http://ieeexplore.ieee.org/document/7462615/
14. Thollander, P., Palm, J.: Industrial energy management decision making for improved energy efficiency-strategic system perspectives and situated action in combination. Energies **8**(6), 5694–5703 (2015)

Design and Implementation of Intelligent Agents and Multi-Agent Systems II

A Cooperative Agent-Based Management Tool Proposal to Quantify GHG Emissions at Local Level

Edgar Lorenzo-Sáez(✉), José-Vicente Oliver-Villanueva, Jorge E. Luzuriaga, Miguel Ángel Mateo Pla, Javier F. Urchueguía, and Lenin-G. Lemus-Zúñiga

Institute of Information and Communication Technologies (ITACA), Universitat Politècnica de València-Polytechnic University of Valencia, Valencia, Spain
edlosae@upv.es
https://ictvscc.webs.upv.es/en/

Abstract. Acting locally is essential to start working to limit the climate change impact and to reduce substantially and constantly GHG emissions. Indeed, the lack of GHG emissions quantification tools in real time and at local level, makes it difficult for public decision-makers to focus efforts and resources efficiently against climate change. In this paper, we propose an agent-based approach applied to an integral management tool to quantify GHG emissions in an accurate and up to date manner. The management tool ensures automatic and reliable management of large volumes of information minimising human interventions. The integral management tool is being tested in a pilot action in a medium-sized city in Eastern Spain (Llíria) to convert it into an innovative and sustainable Smart City.

Keywords: GHG emissions quantification · Climate change
Sustainable smart city · Balanced Scorecard · Agents

1 Introduction

Global warming is undeniable. Problems like droughts, forest fires, floods, soil erosions and desertification, among others, are affecting our planet, and specifically the Mediterranean region. Human influence is determined by the significant increase of the anthropogenic Greenhouse Gases emissions (GHG).

The transition to a low-carbon economy in cities is increasingly seen as a crucial contribution to limiting global warming [7]. As a basis for acting on climate change, cities need to quantify and report their GHG emissions [2,5,9,10,12,13]. Thus, after adopting the European Union's climate and energy package of measures in 2008, the European Commission presented the Covenant of Mayors for Climate and Energy initiative [3]. The commitments for Covenant Signatories have been proposed to exceed the European Union's goal of reducing CO_2

G. Jezic et al. (Eds.): KES-AMSTA-18 2018, SIST 96, pp. 243–252, 2019.
https://doi.org/10.1007/978-3-319-92031-3_24

emissions by 20% by 2020 and by 40% by 2030 [3]. Based on these quantifications and reports, cities must define specific policies and strategies to achieve their emissions reduction objectives. Therefore, public decision-makers at local administrations need appropriate tools that help them to evaluate objectively and transparently the activities and decisions carried out to reduce GHG emissions in quantitative terms. However, there are no such tools today.

Thus, from a multidisciplinary research group on Information and Communication Technologies against Climate Change (ICTvsCC) have developed an advanced management tool based on the quantification of GHG emissions within the integral SimBioTIC project. It facilitates decision-making optimizing the available resources to raise climate metrics and evidence-based objectives in real time according to environmental criterion.

The smart city concept integrates information and communication technologies (ICT), with different types of electronic data collection devices connected to a network, in order to supply information to optimize city operations and services, to manage assets and resources efficiently, as well as to empower their citizens [4]. SimBioTIC helps to add and improve services to a smart city as part of the strategy to face the climate change from a local scale. The first pilot action is being deployed successfully in the city of Llíria.

The developed tool was designed as an integral Balanced Scorecard [8] to be used by both public decision-makers and citizens, promoting transparency in decision-making processes. It also encourages participation and promotes massive engagement through creating a community of contribution, increasing the awareness and interest of the citizens to stay updated with the evolution of GHG emissions and the results of the decisions taken. The tool is in constant improvement, based on the results of experiences and the feedback from users. It is also envisaged to add new modules for the GHG emissions' management to the actual tool's version.

The rest of the paper is organized as follows. Section 2 is dedicated to present our current management tool system that process, organize and classify data from several number of sources with the aim of highlight the valuable data and transforms the way in which final users interact with the data. Section 3 focuses on materials used to adapt and construct our architecture and the methods used to structure the categorization of the modules. Next, Sect. 4 presents the proposed multiagent architecture to improve the performance of our current management tool system. Section 5 shows the preliminary results of the first successful pilot action in Llíria and Sect. 6 concludes the paper with the achievements and potentialities of our advanced management tool.

2 System Description

In this section, we describe the design of the platform architecture as a base for the development and subsequently deployed in Llíria as Smart City. The core mission of this platform is to address key topics of climate change problems with ICT infrastructure (technologies, frameworks, models) and services in communities, cities and regions. ICT become the structural enabler of change for addressing the problems mentioned in Sect. 1. In fact, we develop this platform in order to:

a. make the information transparent as possible to the end users,
b. take specific measures to reduce cost and consumptions,
c. monitor the consumptions information,
d. provide context from historical data,
e. allow a better understanding of cost and consumptions, i.e., when it is compared against same periods of past years or other neighbour communities and
f. adjust the terms and parameters to address efficiency to engage more effectively and actively with citizens.

The general architectural schema of the platform can be seen in Fig. 1. Where to meet previously defined functional requirements, it is separated through different logical layers: broker, data handling, and external services.

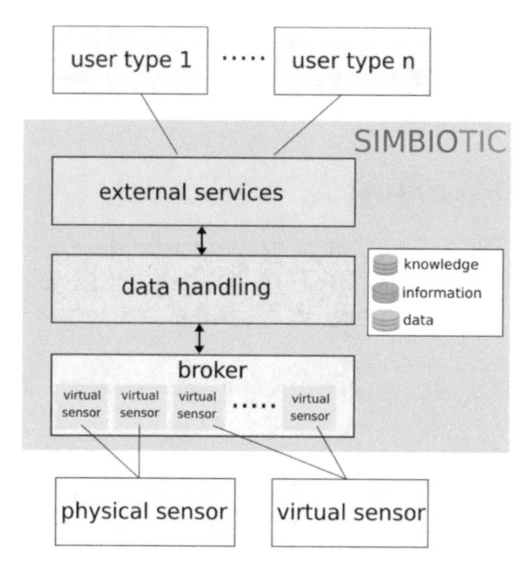

Fig. 1. SimBioTIC's general architecture.

The platform receives and gathers information from sensors. Two different kinds of sensors were defined and characterized depending on the source that provide data. On the one side, a physical sensor is considered as a simple Internet of Things (IoT) board with networking capabilities using either an Ethernet module or a wireless module. On the other side, a virtual sensor is an abstraction to heterogeneous data sources that produce measurements as if it would be a physical sensor (e.g. complex data calculated by the report [7] related to the GHG footprint measured and converted to its CO_2 equivalent in terms of global warming potential).

The broker layer is an intermediary between the external data sources and the system. It is in charge of getting data from registered sensors, regardless of their type: real or virtual. To take care of pre- and post- processing of data, this layer also includes a semantic adapter, which is used to add certain semantics to the received data.

Data handling layer manages the store of all the data that has passed through the broker layer. It will create a large non-relational database or data lake. It should be considered that the data volumes are very large since the granularity of the information is very fine. Thus, depending on the requirements from these data, it can be structured in tables and/or stored in relational databases. In the analysis, a variety of methods and approaches were used to generate information and certain knowledge, depending on location and different circumstances adapted to each case. Finally, the layer of external services is able to allow operations of visualization and management to different users or even other applications.

3 Material and Methods

This section is composed by two parts: the characteristics of the pilot city and the description of two modules of the advanced management tool.

3.1 City for the Pilot Study

The selection of the pilot city is important to test the tools and methodologies developed and to ensure results with high representativeness. To achieve this, a representative medium-sized city it is needed, with presence of all economic sectors.

For these reasons, the first pilot action was deployed in the city of Llíria, located 25 km NW of Valencia, as can be seen in Fig. 2. Llíria has approximately 25.000 inhabitants, placing itself in an average position between large cities and small towns. The city of Llíria has activity representation of the main economic sectors: in primary sectors (e.g. agriculture, forestry, livestock), in industrial sectors (e.g. ceramic, plastic, metal, electronic components, agro-food products) and in services (e.g. energy, transport, tourism, commerce), as well as a waste treatment plant.

Llíria's characteristics are ideal and representative to test the methodologies and tools developed.

3.2 Inventory of GHG Emissions

The first platform module is the basis of the tool. The inventory of GHG emission is structured in three levels of categorization (sectors, criteria and indicators) and their corresponding Key Performance Indicators (KPIs) to quantity the GHG emissions of each emitting source located in the city. The quantitative results are expressed in terms of CO_2-equivalent (CO_2-eq), calculated according to the

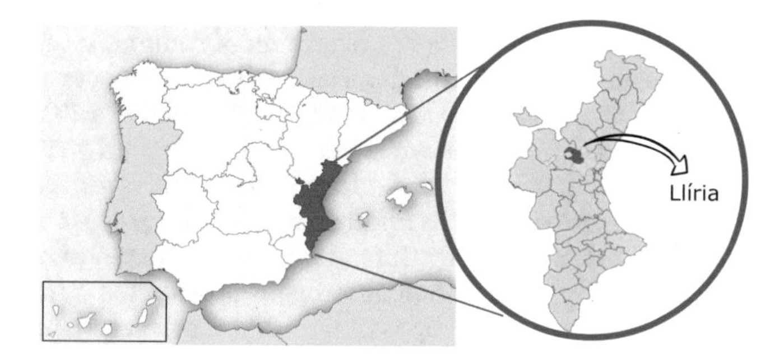

Fig. 2. Location of the pilot city Llíria.

global warming potentials of the 4th Assessment Report of the Intergovernmental Panel on Climate Change [6].

Different methodologies have been consulted and evaluated in order to elaborate this module. The more relevant are GHG PROTOCOL 2014 [1] as well as the IPCC guidelines [6]. The structure of the categorization of the inventory of GHG emission module facilitates the standardization of status reports generated by the second module.

3.3 Status Report Generator

The Status Report generator platform module can produce several report models, among them the normalized mandatory reports that each country must present at European level to comply with agreements and/or objectives at national and international level. This module helps in decision-making as well as the rectification of incidents or the modifications of inefficient measures detected, guaranteeing real-time monitoring of GHG emissions.

The modular structure of the tool allows to generate a specific report that complains with the Covenant of Mayors [3]. In order to achieve this, the standardization of the categorization of the IPCC guide [6] has been ensured, respecting the alphanumeric nomenclature between the structure of the categories in order to guarantee also the comparability between inventories.

4 Proposed Architecture

The decentralized approach of the SimBioTIC project (Fig. 1) allows to gain in flexibility, fault tolerance and scalability. These properties allow to model and implement a cooperative multi-agent system.

Following agents are considered in the multi-agent system: the sensors, actuators, smart objects (the IoT Things), the information adapters, the entity who stores the collected data in databases and the recipients of information (citizens, politics, institutions, among others). The objective is to develop a middleware based solution to allow that the agents could interact to each other and with the environment in a cooperative way, by the definition of communication protocols to the interaction, communication and interoperability made between them.

As shown in Fig. 3, the middleware architecture is composed by five main layers, at the bottom as our previous architecture, physical and virtual elements have identities, physical attributes, and virtual personalities in the sense that they are seamlessly integrated into the information network [14]. At the top is the user interface layer which presents by different dashboards an abstraction of the information in conjunction of fields and controls to provide the interaction with the final users. The layer called Composition and Management is where all the data are transformed to a common and manageable standard for all middleware architecture. Indeed, it is where most of the agents are introduced [11]. Following different agents' types are defined:

a. *PublisherAgents* embedded in IoT boards and in the addition of emission information of the IPCC reports.
b. *BrokerAgents* are responsible for routing information obtained by agents, coordinating and insuring the data delivery to rest of the agents.
c. *SubcriberAgents* embedded in all the agents that receives data, this type of agent is in charge of extracting and collecting all the information and leave it available for the data preprocessing techniques.
d. *PersistenceAgents* provide persistence to the information also perform backup tasks, as well as to ensure the correct consistency and storage of information.
e. *DisplayAgents* show real-time information about GHG emissions.
f. *ReportGeneratorAgents* this agent's type facilitate the retrieval of useful information from the database and provides custom reports.

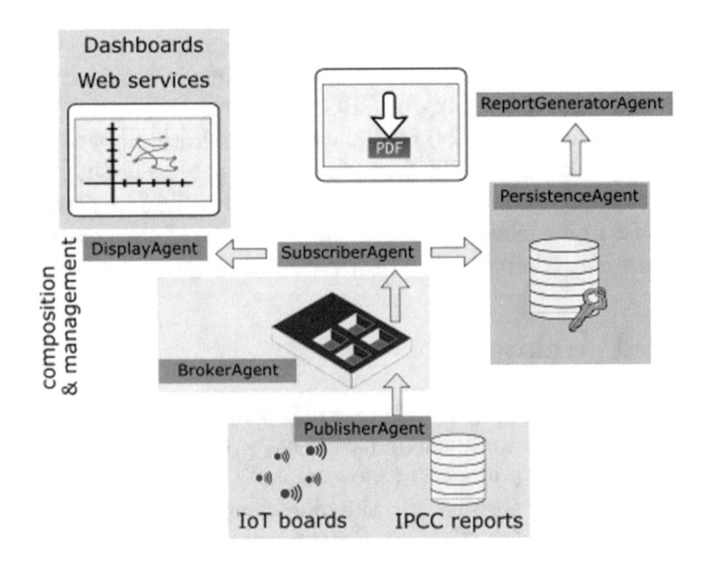

Fig. 3. SimBioTIC's general architecture based in cooperative agents.

5 Preliminary Results

5.1 Inventory of GHG Emissions

The GHG emissions inventory module applied to 2016 in the city of Llíria reflects a total of GHG emissions of 175.427 tonnes CO_2-eq. As it can be seen in Fig. 4 the total level of GHG emissions is considerably reduced by the "Forestry" sector. Due to its carbon sink effect, forestry is responsible for the absorption of a large amount of GHG. Thus, the total net value of emissions is reduced to 156.133 tonnes CO_2-eq.

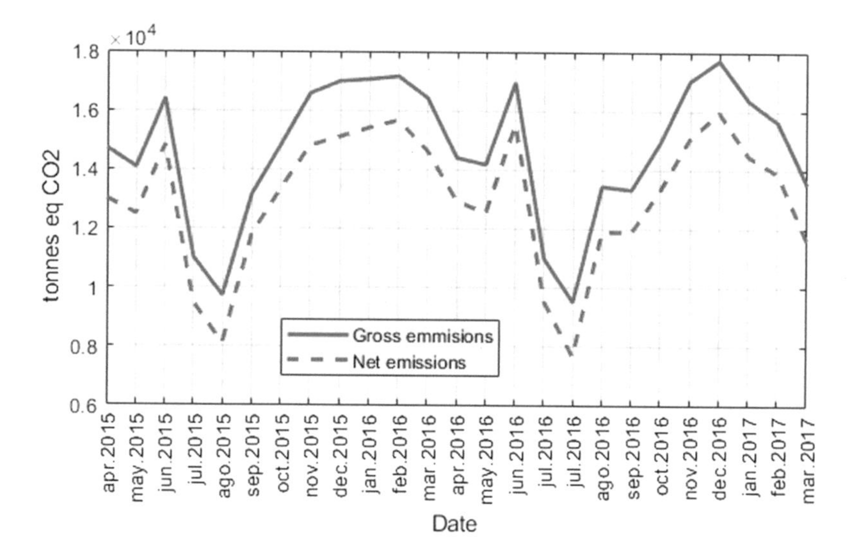

Fig. 4. Influence of forestry on net GHG emissions in Llíria in year 2016.

Disaggregating the GHG emissions by sectors (Fig. 5), close to 60% of total emissions were recorded in the "Energy (without Transport)" sector. Within this sector, the criterion with the highest emission levels is "Burning of fuel in manufacturing and construction industries" that represents 24% of the total, with the "Ceramic Industry" indicator as the main responsible with 18% of the total amount.

The next sector with the largest amount of GHG emissions is the "Transport" sector, with 25% of the total. The criterion "Private ground transport" dominates clearly with 24% of emissions, being the indicator "Automobiles" the most significant with 16% of the total GHG emissions.

The third highest GHG emissions sector is "Agriculture, Livestock and Other Land Uses", with 9% of total emissions, very far from the previous two. The sectors "Industrial Processes and Product Use" and "Waste" represent each of them, only the 3% of the emissions.

Concerning to the carbon sink effect, the forests of the municipality of Llíria calculated in the "Forestry" sector are responsible for an 11% reduction in gross GHG emissions.

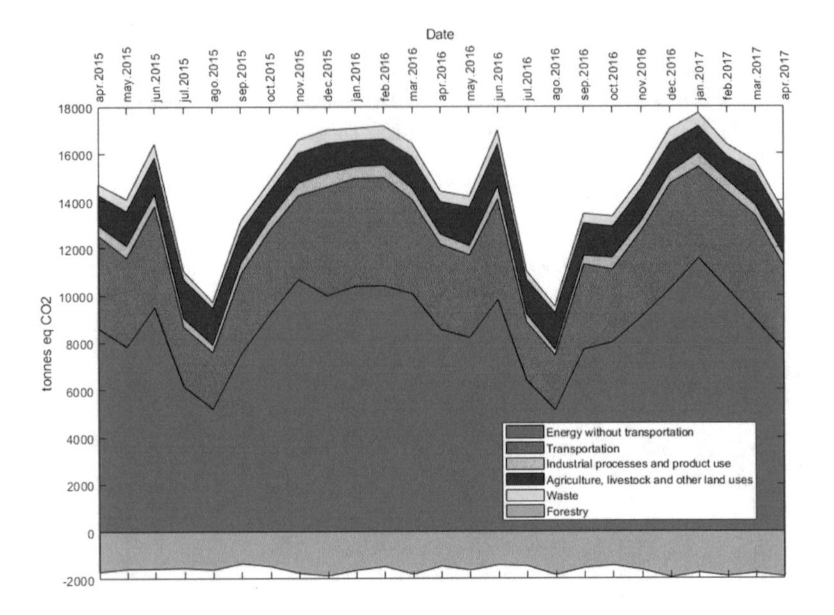

Fig. 5. Dynamic GHG emissions separated by sectors in Llíria in year 2016.

Figure 5 also represents the results obtained in its monthly distribution by sectors. A clear seasonal variability in the sectors "Energy (without transport)", "Industrial Processes and Use of products", "Transport" and "Waste" can be observed, due to the industrial activity that reflects a decrease in summer and two peaks of GHG emissions in June, before the break of August holidays and the months of December–January for industrial activity during the Christmas season. The sum of the seasonal variability observed mainly in the sectors "Energy (without Transport)" and "Transportation", determines the seasonality of the total GHG emissions in Llíria.

5.2 Status Report Generator

The results of the status report generator module consist in two standardized reports: a complete report of the 2016 emissions of all the emitting sources and with sink effect of the city of Llíria and a second report adapted to the requirements of the Covenant of Mayors for Climate and Energy that presents a comparative analysis with the reference year 2010.

6 Conclusions

An agent-based management tool has been developed and presented. This advanced tool allows to know accurately in real time the status and the evolution of GHG emissions from each source at local level. Thus, the tool helps local administrations to transit to Smart Cities using the potential of ICTs.

The emission sources could be showed in a quantitative way by sectors and by a disaggregated manner in criteria and indicators. With this systematic analysis and evaluation, the tool enables the decision-making process on the most efficient actions against climate change. Moreover, the tool generates automatically the mandatory reports to comply with the international agreements. The tool is in constant improvement, based on the results of experiences derived from the first pilot action carried out in a representative medium-sized city, being able to be transferred and extrapolated to other cities and regions with different socio-economic conditions.

Therefore, the standardization of reports and structure for categorization allow setting secondary objectives such as the development of carbon compensation mechanisms between sinks and emitting sources at local level, among others. This should be researched in near future.

Finally, based on the obtained results, the research group is developing new tool modules related to the integral management of GHG emissions at local and regional scale, e.g. carbon stocks management, risk management of large volumes of GHG emissions and simulation module of alternative scenarios to evaluate the impacts of mitigation measures using sustainable criteria.

References

1. Bhatia, P., Cummis, C., Brown, A., Rich, D., Draucker, L., Lahd, H.: Corporate Value Chain (Scope 3) Accounting and Reporting Standard. Supplement to the GHG Protocol Corporate Accounting and Reporting Standard, pp. 1–152 (2012). http://www.ghgprotocol.org/sites/default/files/ghgp/standards/Scope3_Calculation_Guidance_0.pdf
2. Dodman, D.: Forces driving urban greenhouse gas emissions. Curr. Opin. Environ. Sustain. **3**(3), 121–125 (2011)
3. European Commission: Covenant of Mayors: Greenhouse Gas Emissions Achievements and Projections. Publications Office of the European Union (2016)
4. Hamblen, M.: Just what IS a smart city? (2015). https://www.computerworld.com/article/2986403/internet-of-things/just-what-is-a-smart-city.html
5. Ibrahim, N., Sugar, L., Hoornweg, D., Kennedy, C.: Greenhouse gas emissions from cities: comparison of international inventory frameworks. Local Environ. **17**(2), 223–241 (2012)
6. IPCC: 2006 IPCC Guidelines for National Greenhouse Gas Inventories. Task Force on National Greenhouse Gas Inventories (TFI) (2006). http://www.ipcc-nggip.iges.or.jp/public/2006gl/pdf/0_Overview/V0_1_Overview.pdf
7. IPCC: Climate Change 2014: Mitigation of Climate Change: Contribution of Working Group III to the Fifth Assessment Report of the Intergovernmental Panel on. Cambridge University Press, 1132 pp. (2014). http://www.ipcc.ch/pdf/assessment-report/ar5/wg3/ipcc_wg3_ar5_full.pdf

8. Kaplan, R.S., Norton, D.: Using the balanced scorecard as a strategic management system. Harvard Bus. Rev. **74**, 75–85 (2007)
9. Kennedy, C., Demoullin, S., Mohareb, E.: Cities reducing their greenhouse gas emissions. Energy Policy **49**, 774–777 (2012)
10. Lin, T., Yu, Y., Bai, X., Feng, L., Wang, J.: Greenhouse gas emissions accounting of urban residential consumption: a household survey based approach. PLoS One **8**(2), e55642 (2013)
11. Padgham, V.L., Winikoff, M.: Developing Intelligent Agent Systems: A Practical Guide. Wiley, Chichester (2005)
12. Ramaswami, A., Bernard, M., Chavez, A., Hillman, T., Whitaker, M., Thomas, G., Marshall, M.: Quantifying carbon mitigation wedges in U.S. cities: near-term strategy analysis and critical review. Environ. Sci. Technol. **46**, 3629–3642 (2012)
13. Rauland, V., Newman, P.: Counting carbon in cities. Green Energ. Technol. **207**, 117–130 (2015)
14. Rogers, A.: Agent technologies for sensor networks. Comput. J. **54**(3), 307–308 (2011)

A Proposal to Improve the Usability of Applications for Users with Autism Considering Emotional Aspects

Ángeles Quezada[2(✉)], Reyes Juarez-Ramirez[2], Arnulfo Alanís Garza[1], Bogart Yail[1], Sergio Magdaleno[1], and Eugenia Bermudez[1]

[1] Department and Computer Systems, Tijuana Institute of Technology, Calzada Technological s/n Unit Tomas Aquino, Tijuana, Mexico
{Alanis,bogart,jmagdaleno,
maria.bermudez}@tectijuana.edu.mx

[2] Faculty of Chemical Sciences and Engineering, Autonomous University of Baja California, Calzada Universidad 14418, Parque Industrial Internacional Tijuana, 22390 Tijuana, B.C, Mexico
angeles.quezada@tectijuana.edu.mx,
reyesjua@uabc.edu.mx

Abstract. Emotions are an essential part of the functioning of a person needed for him to adapt to the environment and its changing conditions. Because of significant deficits in certain aspects of children with Autism Spectrum Disorder (ASD), it is necessary the development of new methodologies in order to treat the symptoms with software systems. In this article we present a proposal of a methodology for the development of software applications used for the teaching and recognition of emotions to children with Autism Spectrum Disorder.

Keywords: Emotions · Autism Spectrum Disorder · Emotional intelligence

1 Introduction

Emotions are a fundamental part of human nature, with the purpose of helping the adaptation process to the environment through his lifetime [1]. In human beings, the experience of an emotion involves a group of cognitive processes, attitudes and beliefs. Emotions are used the asses a situation, therefore influence on how said situation is perceived.

For a long time, emotions were considered unimportant emphasizing the rational part of human nature. But nevertheless, emotions, also called affective states, can indicate attitudes, motivations, desires, necessities, even goals [2] affecting everyday human activities.

Currently, there is research that shows that persons with Autism Spectrum Disorder (ASD) have difficulty processing social and emotional input [3] from the people around them.

G. Jezic et al. (Eds.): KES-AMSTA-18 2018, SIST 96, pp. 253–260, 2019.
https://doi.org/10.1007/978-3-319-92031-3_25

Research that shows the usefulness of software applications in the therapy and teaching of children with ASD [4]. The use of computers and consumer electronics in a teaching environment increased focused attention, attention span, fine motor function and generalization abilities [4].

Other developments include the use of virtual reality as a medium to offer a simplified interaction exploratory environment for children with ASD [6, 7].

Some of the main difficulties faced by persons with ASD are [5]:

- Recognition of facial expressions and emotions.
- Controlling and understanding his own emotions, and
- Understanding and interpreting the emotions of others.

In order to deal with the deficits presented in children with ASD it is of great importance the proper early intervention and therapy so as to improve prognosis. One way of helping with this is tools that help with dealing with the symptoms of ASD. Nevertheless, there is no characterization of what user abilities are needed for interact with electronic devices like computers, tablets, smartphones, among others and the user characteristics presented in children with ASD, which is needed that propose rulesets and methodologies for the design of software applications for the recognition of emotions.

This paper is organized in seven sections. In Sect. 2 we define the concept of emotions and their classification and we define the concept of Autism Spectrum Disorder and its characteristics. In Sect. 3 the concepts of emotional intelligence and ability is explained. In Sect. 4 the characterization of the user abilities needed to interact with a computer and persons with ASD. In Sect. 5 the conclusions and future work are presented.

2 Theoretical Foundations

2.1 Emotions

Ekman *et al.* [6] made popular the concepts of basic or pure emotions, which is still in use in the simulation of emotion expression. Popularized the concept of basic (or pure) emotions and proposed the existence of seven basic emotions, which are: surprise, fear, sadness, disgust, joy, anger.

According to Monica [3], emotions can be defined as a feeling and their singular thoughts, psychological and biological states and a variety of tendencies to act. She states that there are hundreds of emotions, and combinations of them, variables, mutations and refinements.

There exists a wide variety of models of basic emotions depending on what psychological point of view it is base on, each considering their own factors that determine or form the emotions. Some consider emotions universal, meaning that all cultures can recognize them or that in all cultures there is a name for each emotion, as shown in Table 1.

Table 1. Basic emotions

Theorist	Basic emotions
Plutchik [7]	Acceptance, anger, anticipation, disgust, joy, fear, sadness, surprise
Arnold [8]	Aversion, wrath, anger, sadness, desire, despair, hatred, fear, hope, love, sadness
Ekman, Friesen, and Ellsworth [6]	Anger, disgust, fear, happiness, sadness, surprise
Frijda [9]	Desire, happiness, interest, surprise, surprise, sadness
Izard [10]	Contempt, anger, disgust, anger, fear, guilt, interest, joy, shame, surprise
James [11]	Grief, fear, love, anger
McDougall [12]	Anger, disgust, joy, fear, submission, amazement
Mowrer [13]	Plan, pleasure
Oatley and Johnson-Laird [14]	Disgust, anger, anxiety, happiness, sadness
Tomkins [15]	Anger, interest, contempt, disgust, anger, fear, joy, shame, surprise
Weiner and Graham [16]	Happiness, sadness

2.2 Autism Spectrum Disorder

The word Autism was first used in 1911 by Eugene Bleuler [17], from the Greek word autós, meaning "self" to designate one of the most noticeable characteristics of schizophrenia in adults: the lost contact with reality, and in consequence, a great difficulty in the ability to communicate with others, evasion of reality, and retraction into one self. Later in 1943, Leo Kanner [18] countered the previous definition and described for the first time "Early Childhood Autism" as a differentiated personality disorder. Kanner used the term to define a segment of the population of children that had a tendency to retract into themselves and the inability to interact with the persons around them. The term was further defined, such as in [19–23] where it is described a list of the main affective and emotional disorders that define it. In a general way, the previously mentioned studies describe the essential characteristics that define the ASD, which are deficits in social communication and interaction, and limited repetitive behavior and interests [2]. Persons with ADS can be classified in three levels considering the level of independent functioning in their daily lives [8]:

1. Requiring support,
2. Requiring substantial support
3. Requiring very substantial support.

3 Emotional Intelligence

The term Emotional Intelligence was used first in 1990 by Peter Salovey of Harvard University and John Mayer of New Hampshire University, defined as the capacity to control and regulate the feelings of one self and others and use them as a guide for thinking and action.

Emotional Intelligence consists on a wide number of abilities and personality traits, such as: empathy, expression and comprehension of feelings, temper control, Independence, adaptation capacity, sympathy, interpersonal problem solving, social abilities, persistence, cordiality, kindness and respect [3].

3.1 Emotional Abilities of Emotional Intelligence

Practical proficiencies that derive from emotional Intelligence are five, classified in two areas [3]:

1. *Personal proficiencies.*
   (a) Knowledge of self. Capacity to know and recognize what is happening in one's body and what one is feeling.
   (b) Auto regulation. Emotional regulation and inhibition and mood modification and its exteriorization.
   (c) Motivation.
2. *Social Proficiency.*
   (a) Empathy. Understanding what other persons are feeling, seeing things from the perspective of others.
   (b) Social abilities. Abilities related to popularity among peers, leadership and interpersonal efficacy, that can be used to persuade and direct, negotiate and dispute solving, cooperation and team work.

4 Proposed Solution

4.1 Methodology for the Development of Emotion Expression and Recognition: Case Study of Children with Autism Spectrum Disorder

In order to develop a methodology, we must first understand the user characteristics necessary for the proper interaction with software systems.

Michel [4] describes some of the abilities needed for a typical user to be able to interact with a computer, such as:

(1) Three of the senses (Hearing, vision and touch)
(2) Voice
(3) Attention
(4) Fine motor ability
(5) Memory
(6) Generalization ability

For a typical user, it is not necessary to have all the previously mentioned attributes in order to interact with a computer system, meaning that if the user cannot interact in a particular way, it may be possible using other attribute.

In the case of users with ASD, they are impaired in some of those abilities, such as emotion recognition and certain cognitive processes, such as the case where they need to interpret, infer and socialize.

The visual ability is the most developed in a person with ASD [24] and the most used for everyday task by them. They also possess a developed hearing sense, but they are very susceptible to sudden and drastic changes in sound levels [5]. In the case of fine motor, vocal, and memory abilities depend greatly on the affliction level of ASD of the person [5]. In other words, persons with ASD possess those abilities but are greatly limited depending on the level of ASD [5].

Finally attention is a weakly developed attribute in persons with ASD, reason for that applications developed for persons with ASD in mind are primarily based on drawing as much attention possible unto them [5].

Figure 1 shows the relationship between abilities needed for a typical user in order to interact with a computer and the abilities that persons with ASD have.

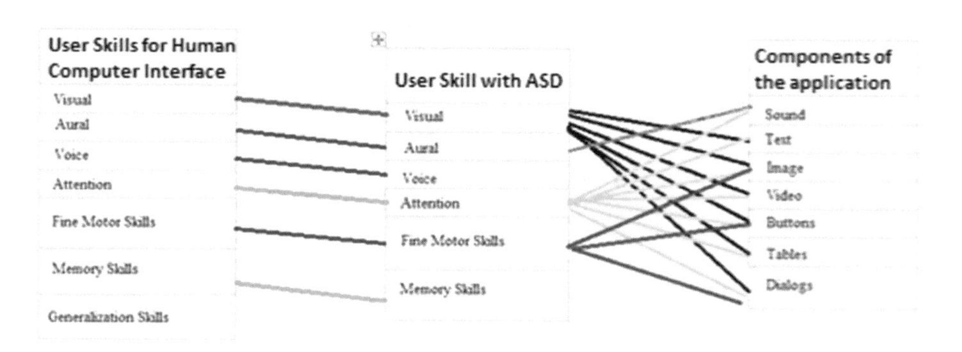

Fig. 1. Ability relationship.

Visual ability in persons with ASD is one of their core strengths since they are almost totally visual learners. Hearing, voice and motor ability are sufficiently developed to interact with electronic devices, such as a Tablet or smartphone, but usually not enough in order to properly use a desktop or laptop computer [4]. Attention and memory abilities vary greatly depending on the level of ASD, although attention is an ability can be at least partially developed with an adequately designed interface [5].

Due to the strength and weaknesses of some of the abilities presented in persons with ASD, we propose increasing the attention ability of this type of user, by emphasizing the visual sense and related interface components, such as text, images, buttons, tables and dialog boxes. Attentions can also be developed using sound, but in this case to be used in a subtle way because of the propensity to sudden changes in sound and volume in this type of user.

Finally, fine motor abilities are applied in order to interact with the interface components, such as images and buttons, via a touchscreen interface because of the difficulty presented to users with ASD with mouse and keyboard.

Improving attention levels can also help increase the motivation of use in users with ASD.

Emotional intelligence can be developed is we consider some of the emotional competencies and characteristics previously mentioned. This article centers on improving the motivation via a methodology for applications centered on helping with the emotional recognition aspect. We propose using an emotional intelligence test MSCEIT (Mayer-Salovey-Caruso test) [25] and determine what kind of emotional abilities does the user with Autism has. In order to try to establish a proper methodology for the development of therapeutic software for emotion recognition we are currently developing the following steps, as shown in Fig. 2:

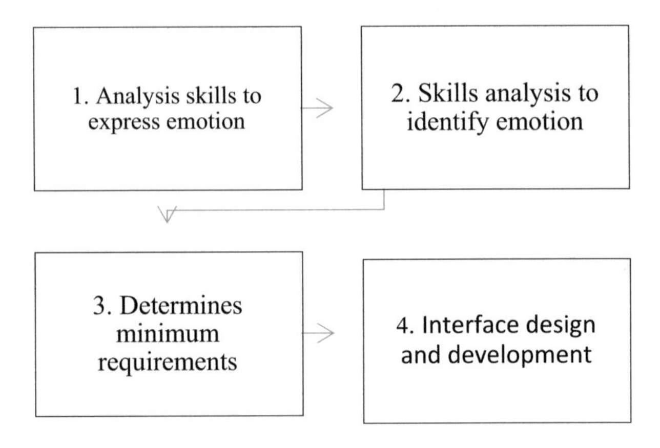

Fig. 2. Proposed methodology.

1. *Analysis skills to express emotion.* In this phase the necessary abilities for emotion recognition are determined considering the proposals by Goleman.
2. *Skills analysis to identify emotion.* In this phase it is necessary to identify the skills necessary to recognize an emotion
3. *Determine minimum requirements.* This phase depends on the results of the first for determine the minimum requirements needed for express and guarantee emotion recognition.
4. *Interface design and development.* With the minimum requirements, an interface in developed which will be submitted to testing to determine if the user with ASD can express and recognize an emotion with the interface.

5 Conclusions and Future Work

The presented article proposes a methodology for the development of software applications for emotion recognition and expression for children with ASD, considering some aspects necessary for interaction with computers and also necessary to recognize and express emotion. It is focused on the analysis of the necessary abilities needed for the expression and recognition of emotions and subsequently develop a user interface based off said abilities.

As we discussed, there is very little work concerning emotional intelligence and the abilities needed by children with ASD to recognize an emotion, this is why we need the development of new methodologies that consider those abilities for the development of specialized software.

Currently we are in the process of determining those abilities and what are the abilities that the child with ASD has with through usability testing of software applications and specialized emotional intelligence testing in schools for special needs users. It is necessary to identify the skills that are needed to interact with the technology and of which the child with autism possesses and in this way support the use of technology to the child with ASD to recognize an emotion, but for this it is also important to identify the skills needed to identify an emotion, which is the fundamental part of such research.

References

1. Fredrickson, E.V., Lourenço, B.L., La Cava, A.: Cytokines in systemic lupus erythematosus. Curr. Mol. Med. **9**(2), 242–254 (2009)
2. Díaz, R.R.T., Flores, R.: "Las Emodiones en el Niño Autista a Traves del Comic: Estudio de Caso," pp. 5–28 (2008)
3. Monica, M.P.: "El autismo y las emociones" (2013)
4. Michel, P.: "The use of technology in the study, diagnosis and treatment of autism". In: Final term Paper for CSC350, pp. 1–26. Autism Association (2004)
5. Association, A.P.: Diagnostic and statistical manual of mental disorders, 5th ed. (2014)
6. Ekman, P.: Facial expression and emotion. Am. Psychol. **48**, 384–392 (1993)
7. Plutchik, R.: 2001 plutchik natureo femotions.pdf (2001)
8. Arnold, M.B.: Emotion and Personality: vol. I Psychological Aspects. Columbia University Press, Oxford, England (1960)
9. Hamilton, N.H., Bower, V., Gordon, H., Frijda, N.H.: Cognitive Perspectives on Emotion and Motivation. Kluwer Academic Publishers, Dordrecht (1988)
10. Izard, C.E.: Human emotions (1977)
11. James, W.: What is an emotion? Mind **os-IX**, 188–205 (1884)
12. McDougall, W.: An Introduction to Social Psychology. Methuen & Co., London (1973)
13. Mowrer, O.H., Jones, H.: Extinction and behavior variability as a function of effortfulness of task. J. Exp. Psychol. **33**, 369–386 (1943)
14. Johnson-Laird, P.N., Oatley, K.: Towards a cognitive theory of emotions. Cogn. Emot. **1**, 29–50 (1987)
15. Tomkins, S.S.: Affect Imagery Consciousness: The Positive Affects. Springer, New York (1962)

16. Graham, S., Weiner, B.: Theories and principles of motivation. In: Berliner, D., Calfee, R. (eds.) Handbook of Educational Psychology, pp. 63–84. MacMillan, New York (1996)
17. Moskowitz, A., Heim, G.: Eugen Bleuler's Dementia Praecox or the Group of Schizophrenias (1911): A centenary appreciation and reconsideration. Schizophr. Bull. **37**(3), pp. 471–479 (2011)
18. Kanner, L.: Autistic disturbances of affective contact. Acta Paedopsychiatr. **35**, 100–136 (1968)
19. Asperger, H.: "Die" Autistisehen Psychopathen "im Kindesalter" (1943)
20. C.M.E.A.: "Schizophrenic syndrome in chilhood, pp. 945–947 (1961)
21. Lotter, V.: Epidemiology of autistic conditions in young children - II. Some characteristics of the parents and children. Soc. Psychiatry **1**, 163–173 (1967)
22. Rutter, M.: Concepts of autism: a review of research. J. Child Psychol. Psychiatry **9**(1), 1–25 (1968)
23. Wing, L.: Asperger's syndrome: a clinical account. Psychol. Med. **11**, 115 (1981). Published online on July 2009
24. Pierce, K.L., Schreibman, L.: Teaching daily living skills to children with autism in unsupervised settings through pictorial self-management. J. Appl. Behav. Anal. **27**(FAu), 471–481 (1994)
25. Mayer-Salovey-Caruso: MSCEIT, Test de Inteligencia Emocional

Towards a Model Based on Agents for the Detection of Behavior Patterns in Older Adults Who Start Using ICT

Consuelo Salgado Soto([✉]), Maricela Sevilla Caro, Ricardo Rosales Cisneros, Margarita Ramírez Ramírez, Hilda Beatriz Ramírez Moreno, and Esperanza Manrique Rojas

Facultad de Contaduría y Administración, Universidad Autónoma de Baja California, UABC, Tijuana, BC, Mexico
{csalgado,mary_sevilla,ricardorosales,maguiram, ramirezmb,emanrique}@uabc.edu.mx

Abstract. In this article, we present a proposal to develop a model based on agents, which allows the detection of the patterns that make up the behavior of older adults, who have been influenced by the use of new information and communication technologies. The model describes the iterations of four agents: Adult Agent, Detector Agent, Content Agent and Agent Activity; these interactions together with their knowledge base will result in the identification of possible risk factors to minimize them and increase the protective that allow mental activation and active aging. The protective derived from the correlation between energy and motivation, have guided the behavior of the adult towards better integration with society.

Keywords: Agent-based model · Protective and risk factors · Mental activation

1 Introduction

We live in the information age and the complex technological advances that gradually force people to interact with different tools and interfaces to the point of turning them into a necessity. Some of these new technologies have interaction with other systems that may involve human-computer activities, Artificial Intelligence is present in them, there are specific examples of this, we can mention a vacuum cleaner that runs its activity independently, homes with integrated technology, business applications for prediction, and devices for pattern recognition, among others. In the health area, there are also complex systems to support activities to diagnose diseases, monitor patients, etc. It is important to consider the mental health of a sector of the population, that of the elderly, who are affected positively or negatively by technological advances. Is it possible to think of a school environment that allows detecting patterns of behavior so that it can recommend content and activity to strengthen the protective factors and reduce the risk factors? With the simulation of an artificial environment and the characteristics that are planned in the question can be detected risk situations and those that support the generation of certain behavior in adults, this environment should be

G. Jezic et al. (Eds.): KES-AMSTA-18 2018, SIST 96, pp. 261–268, 2019.
https://doi.org/10.1007/978-3-319-92031-3_26

composed of agents with roles, rules and interactions defined, allow constant feedback and adapt them to new situations.

2 Background

Artificial Intelligence (AI) has been busy creating software to perform operations comparable to those performed by the human mind, as well as learning or logical reasoning [1]; In addition, interest in the development of this type of software has grown in recent years to achieve the incorporation into devices of a knowledge similar to that of a human being; On the other hand, AI provides a broad set of methods, techniques and algorithms that can be included in different applications from different fields and has extended research in knowledge directions such as medicine, biology, engineering, education, health, financial, videogames, among others, which are a clear example of the intervention that AI has in daily activities [2].

The area of health and medical care is a very favorable environment for the implementation of AI, through the technology of agents, given the knowledge that is required and at the same time has, where you can give a very precise interaction and attend technical as well as social aspects [3], for the acquisition and generation of knowledge. There are a variety of fields in the medical industry that involve medical support systems, some examples are systems that diagnose diseases, systems that recommend treatment, systems for examining patient history, and support from palliative care units [4]; The development of multi-agent systems has gradually become a key point for this type of health services and applications, particularly for assisted living, diagnosis, physiological tele-monitoring, intelligent hospital, and intelligent emergency applications [5].

On the other hand, the implementation of this type of systems ventures into more specific areas as proposed by [6], when the use of a multi-agent system dedicated to detect and examine the environmental components, to achieve and prevent a person is affected by diabetes, obesity or cholesterol by controlling the diet of people with health problems [6]; Another case is that presented by [7], which focuses on the use of agent technology to support human autonomy in recovery and improvement of cognitive and motor abilities.

As in the previous cases, there are other topics of interest in the health field, such as mental health in a specific sector of the population, that of the elderly. As a precedent, the aging process in them leads to a decrease in learning and memory, to the appearance of diseases such as depression and changes in attitude or lack of motivation, as well as loneliness and isolation [8]. It is worth mentioning that, somehow, little by little, the technological advance was isolating them, provoking negative behavior or suffering and diseases. Now, you can create an overview of the proposal that is the subject of this article, present a model based on agents that allows the detection of the patterns that make up the behavior of older adults, who have been influenced by the use of new technologies of Information and communication, supported by agent technology, the captured data can be analyzed to identify the factors that affect it at that moment and generate a particular behavior.

3 State of the Art

3.1 Agents

For the conceptualization of the agent, you can find descriptions in different bibliographies, such as the following one where you define it as anything capable of perceiving your environment based on sensors and acting in the same environment; makes decisions at a given moment based on the perceptions they have at that moment [9]; an agent is a software component that has autonomy and behaves like a human agent, working for some clients in, searching from their perspective, solving a certain problem. In a multi-agent system, agents may decide to cooperate for mutual benefit or compete to serve their interests; they are autonomous, social and reactive [10].

The term perception, applied in the context of agents, is used to indicate that they receive inputs at any moment, that is, it provides information to the agents about the environment where they live. Russell and Norvig et al. [9], indicates that perception originates in the environment through sensors that capture some aspect of it and is converted into an input for the agent itself to diagnose factors based on their decision.

3.2 Agent Based Model

A model can be visualized as a theoretical scheme of a system or a complex reality that is elaborated to facilitate its understanding and the study of its behavior [11]. On the other hand, simulation consists in representing something, pretending or imitating what it is not [12]. As for computer simulation, it is a part of modeling that tries to find analytical solutions where the prediction of behavior is allowed from parameters and conditions [13].

The modeling and simulation of complex systems from the point of view of the complexity sciences is focused on addressing the emergence of properties from the interaction between a large number of agents [14], the major advances of these systems are focused on Mathematics and computation, it should be mentioned that modeling and simulation from the perspective of thought, are known as artificial societies or agent-based modeling, as [14] cites [15].

The agent-based models (MBA) belong to a movement of social simulation, constitute an artificial society [16] composed of autonomous decision-making entities called agents [17], act in a self and sufficient manner in their environment to meet the objectives of its design [18]. MBA are identified by the following four assumptions: autonomy, independence, obey or follow simple rules, and are adaptive [19]. In the health scenario, the multi-agent system can have different roles to improve the physical and mental state of patients [20], due to their ability to react to situations detected in their environment.

3.3 Older Adults

As we age, our body undergoes various changes in its form, psychological and social functioning, which are generated by age-specific changes and accumulated wear. During the normal aging process, the speed of learning and memory decrease.

This deterioration is caused by disuse, disease, behavioral factors, psychological factors and social factors, rather than by aging itself [8].

The social environment that surrounds older adults has changed aggressively in recent decades. Technological advances have caused a radical change in the forms of production, dissemination, and acquisition of knowledge, [21]. Even, the forms of communication have changed influenced by these advances.

There are studies that reveal that a large percentage of this population needs help to carry out processes that were born in order to simplify and not complicate; in the absence of someone to support them, they are postponed by society [22], they can be affected by various factors that impair physical and mental health, they can be classified as risk factors and protective factors.

There are some characteristics that generate greater vulnerability in individuals, being potentially influential and/or favorable to provoke a behavior, they are called risk factors; On the other hand, there are characteristics that counteract the effects of risk factors, discouraging or preventing the appearance of problems, which have been called protective factors [23].

For the proposal, social isolation, fear, passivity, depression, low self-esteem, cognitive deterioration is observed as risk factors. On the other hand, the protective factors that can support the change in the physical and mental health of the elderly include motivation, better communication, greater social contact, mental activation, sense of belonging, and feeling useful.

A group of professors from the Faculty of Accounting and Administration, designed a course aimed at older adults, in which the activities are focused on students know the information and communication technologies (ICT). The course focuses on understanding the general concepts of computing, practicing with the different Windows accessories, preparing Word documents and using the Internet to make inquiries and communicate via email.

4 Modeling the Pattern of Behavior in Older Adults

Represent the real world through a model, simulate an artificial environment based on rules to obtain results and define actions, and in this case, through modeling and simulation of a school environment is intended to achieve the detection of behavior patterns in older adults.

4.1 Definition of the Agents of the Model

In the context of older adults in the classroom, the proposed model is composed of the following agents:

- AdultAgent represents the Elderly, who has different attitudes identified as protective factors and risk factors, required to determine the action to be followed.
- DetectorAgent, this agent personifies the Instructor, whose activity consists of detecting the risk factors and protective factors to assign the appropriate content, generating the one that will be stored in their knowledge base.

- ContentActivitiesAgent, represents the Activities and Content agent is responsible for receiving the content request, performs the search in the content and activities database to deliver it to the DetectorAgent.
- ActivityAgent receives is responsible for obtaining the requests and sending them to ContentActivitiesAgent, also, of these continually verifying the status of the environment.

4.2 Description of the Suggested Model

Agents can use their perceptions based on the extraction of characteristics, by detecting a small number of particulars in their input and send them directly to their main agent program to respond reactively to them or combine them with other information [10] to allow artificially experimenting on the system.

In this proposal, as shown in Fig. 1, it is described that from the detection of behavior patterns of older adults, who have been influenced by the use of ICT, an action will be taken through the agents. The interactions of the agents showed: AdultAgent, DetectorAgent, ContentActivitiesAgent and ActivityAgent Activity, the interactions in conjunction with the knowledge base will result in the identification of possible risk factors and increase the protective factors derived from the correlation between energy and motivation to guide the behavior of the adult.

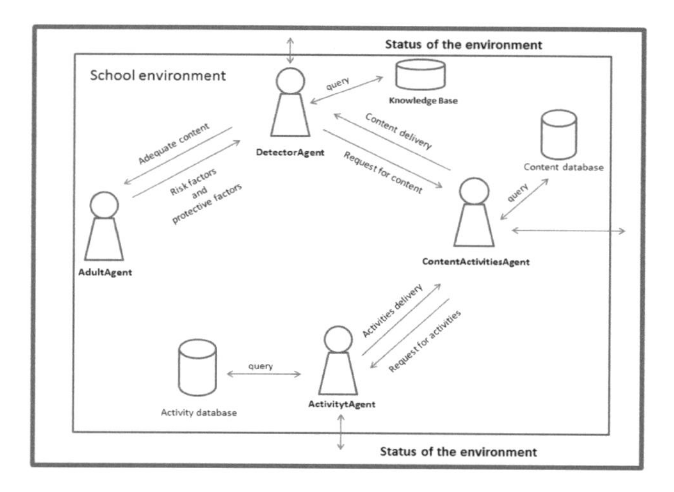

Fig. 1. Model proposal

4.3 Interaction Between Agents

Each one of the agents of the model has its activities defined, but within the environment, it is essential to establish the interactions that will be given to fulfill the objective. These interactions are listed below:

1. The AdultAgent contributes to the environment the risk factors and the protectors.
2. The Detector Agent detects risk factors and protective factors.

3. The Detector Agent consults the knowledge base and receives results.
4. The DetectorAgent, with the results obtained, makes a content request to the ContentActivitiesAgent.
5. The ContentActivitiesAgent receives the content request.
6. The ContentActivitiesAgent queries the content database, once it has the result based on the request of the DetectorAgent, it asks the ActivityAgent for the activity correlated with the content.
7. The ActivityAgent queries the activity database, once it has the result, sends the activity to ContentActivitiesAgent.
8. The ContentActivitiesAgent sends the content and activity to Detector Agent.
9. The DetectorAgent receives results and updates the knowledge base based on the correlation.
10. The DetectorAgent delivers the content, the activity to the AdultAgent, according to the factors detected in the environment.

The agents DetectorAgent, ContentActivitiesAgent and ActivityAgent updated at all times when reviewing the state of the environment in order to have a constant feedback.

5 Conclusions

The use of this systems in complex situations such as the one that is deepened can be used as tools to conceive artificial environments to act and make decisions, since the characteristics of agent technology are attractive and ideal to model and simulate in an environment where it is possible to predict, there is learning, and there is constant feedback.

The proposed aims to find solutions that allow the detection of an action based on certain rules and parameters to achieve the prediction of behavior. This model contains the interactions between the agents, the knowledge base, the database and constant feedback of the state of the environment to allow guiding the behavior of the elderly.

To be updated through courses, in this case, the designed for older adults, allows participants to know the information and communication technologies, also, to focus on establishing a communication link with the current society and enabling mental activation and the increase in a positive attitude.

With the future development will result in the identification of attitudes and possible risk factors in order to minimize them, and increase the protective factors that allow the physical and mental activation derived from the correlation between energy and motivation, guiding the behavior of the older adult towards a better integration with society, an improvement in health.

6 Future Works

The presented work is a proposal designed to detect patterns of behavior. The future works contemplated include the formalization of each of the agents to understand their properties, represent beliefs, desires and intentions of the agents, define the tuples of the agents, implement the knowledge base based on fuzzy logic Type 1 and Type 2 and define the rules of fuzzy inference.

References

1. Artificial intelligence: Spanish dictionary. Association of the Spanish Language, Spain (2018)
2. Furmankiewicz, M., Sołtysik-Piorunkiewicz, A., Ziuziański, P.: Artificial intelligence systems for knowledge management in e-health: the study of intelligent software agents. In: Mastorakis, N., Psarris, K.. Vachtsevanos, G., Dondon, P., Mladenov, V., Bulucea, A., Rudas, I., Martin, O. (eds.) 2014 Proceedings of the 18th International Conference on Systems (part of CSCC 2014) latest trends on systems - volume II Santorini Island. Institute for Natural Sciences and Engineering (INASE), Greece, pp. 551–556 (2014)
3. Cortés, U., Annicchiarico, R., Urdiales, C.: Agents and healthcare: usability and acceptance. In: Annicchiarico, R., Cortés, U., Urdiales, C. (eds.) Agent Technology and e-Health. Whitestein Series in Software Agent Technologies and Autonomic Computing. Birkhäuser Basel, Boston (2007)
4. Soltysik, A., Furmankiewicz, M., Ziuzianski, P.: E-health artificial intelligence system implementation: case study of knowledge management dashboard of epidemiological data in Poland. https://www.researchgate.net/profile/Anna_Soltysik-Piorunkiewicz/publication/271014656_E-health_artificial_intelligence_system_implementation_case_study_of_knowledge_management_dashboard_of_epidemiological_data_in_Poland/links/54bc20ed0cf253b50e2d1619.pdf. Recuperado en 24 de enero de 2018
5. Gallardo, E., Ávila, M., Ávila, R.: Aplicaciones de la inteligencia artificial en la Medicina: perspectivas y problemas. ACIMED, **17**(5) (2008). http://scielo.sld.cu/scielo.php?script=sci_arttext&pid=S1024-94352008000500005&lng=es&tlng=es. Accessed 24 Jan 2018
6. González-Campos, S., Gonzalez Crespo, R.: Agentes inteligentes para controlar la dieta de personas con problemas de salud (2007). https://www.researchgate.net/publication/262487664_Agentes_inteligentes_para_controlar_la_dieta_de_personas_con_problemas_de_salud. Accessed 24 Jan 2018
7. Cortés, U., Annicchiarico, R., Urdiales, C., Barrué, C., Martínez, A., Villar, A., Caltagirone, C.: Supported human autonomy for recovery and enhancement of cognitive and motor abilities using agent technologies. In: Annicchiarico, R., Cortés, U., Urdiales, C. (eds.) Agent Technology and e-Health. Whitestein Series in Software Agent Technologies and Autonomic Computing. Birkhäuser Basel, Boston (2007)
8. Galarza, K.: Envejecimiento activo, mejor vida en la tercera edad. Consulted by Internet 26 January 2018. http://www.saludymedicinas.com.mx/centros-de-salud/climaterio/prevencion/envejecimiento-activo.html. Accessed April 19 2017
9. Russell, S., Norvig, P., Corchado Rodríguez J., Joyanes Aguilar, L.: Inteligencia artificial. Pearson Educación, Madrid, pp. 32, 37–40, 979–982 (2011)
10. Bellifemine, F., Caire, G., Greenwood, D.: Developing Multi-Agent Systems with JADE. Wiley, Chichester (2007)
11. Model: Spanish dictionary. Association of the Spanish Languagea, Spain (2014)

12. Simulate: Spanish dictionary. Association of the Spanish Language, Spain (2014)
13. Computer simulation: ScienceDaily (2017). https://www.sciencedaily.com/terms/computer_simulation.htm. Accessed 14 Jan 2018
14. Susatama, K., Ruíz, K., Arévalo, L.: Modelación y simulación basada en agentes como alternativa para el estudio de las organizaciones empresariales, Revistas.ucc.edu.co (2018). https://revistas.ucc.edu.co/index.php/in/article/view/1838. Accessed 17 Jan 2018
15. Axelrod, R.: The Complexity of Cooperation: Agent-Based Models of Competition and Collaboration. Princeton University Press, Princeton (1997). http://doi.org/10.1002/(sici)1099-0526(199801/02)3:3<46::aid-cplx-6>3.0co;2k
16. Rodríguez, L., Roggero, P.: Modelos basados en agentes: aportes epistemológicos y teóricos para la investigación social. Revista Mexicana de Ciencias Políticas y Sociales **60**(225), 227–261 (2015)
17. Cardoso, C., Bert, F., Podesta, G.: Modelos Basados en Agentes [MBA]: definicion, alcances y limitaciones. http://www.iai.int/wp-content/uploads/2014/03/Cardoso_et_al_Manual_ABM.pdf
18. Wooldridge, M.: Multi-agent systems. Wiley, Chichester (2002)
19. Macy, M., Willer, R.: From factors to factors: computational sociology and agent-based modeling. Annu. Rev. Sociol. **28**(1), 143–166 (2002)
20. Chan, V., Ray, P., Parameswaran, N.: Mobile e-health monitoring: an agent-based approach. IET Commun. **2**(2), 223 (2008)
21. Blázquez Entonado, F.: Sociedad de la información y educación Dirección General de Ordenación, Renovación y Centros, Mérida (2001)
22. Monzón, A., Stanislavsky, P., Urrutia, M.: Los ancianos y la tecnología: ¿Se quedan afuera? (2008). http://fido.palermo.edu/servicios_dyc/publicacionesdc/vista/detalle_articulo.php?id_libro=34&id_articulo=4371. Accessed 20 de febrero de 2017
23. Mosqueda-Díaz, A., Ferriani, M.: Factores protectores y de riesgo familiar relacionados al fenómeno de drogas, presentes en familias de adolescentes tempranos de Valparaíso, Chile, Revista Latino-Americana de Enfermagem, vol. 19, pp. 789–795 (2011)

Intelligent Agents as Support in the Process of Disease Prevention Through Health Records

Hilda Beatriz Ramirez Moreno(✉), Margarita Ramírez Ramírez, Esperanza Manrique Rojas, Nora del Carmen Osuna Millán, and Maricela Sevilla Caro

Facultad de Contaduría y Administración, Universidad Autónoma de Baja California, UABC, Tijuana, BC, Mexico
{ramirezmb,maguiram,emanrique,nora.osuna, mary_sevilla}@uabc.edu.mx

Abstract. The information generated by the clinical histories of the different public and private health institutions demonstrates the need for the proper management of the large volume of information generated by each individual within a society. Having, analyzing, measuring and evaluating all this information is of vital importance for the care and prevention of health. Multi-Agent Systems (MAS) have been characterized by offering a possible solution in the development of complex problems. The development of any type of software requires methods and tools that facilitate the obtaining of a final product. With the implementation of Intelligent Agents, it will be possible to analyze and manipulate the information generated in the health area. This article presents only one agent of a Multi-Agent Systems model where its function is to provide support in the disease prevention process.

Keywords: Agent intelligent · Technological tools · Clinical records

1 Introduction

The information that is generated by the different clinical histories of each person is essential for the care and prevention of diseases. Scientific studies have shown that most of the diseases that are currently affecting humanity are susceptible to be avoided, modified or controlled [1]. In order to achieve this, we need appropriate technological tools, such as information systems, internet, big data, intelligent systems, intelligent agents, multi-agent systems, to mention just a few.

2 Background

Advances in science and technological trends in today's world have led society to migrate to the digital world in most of its daily activities. The health sector is no exception; medical records went from paper to electronic, but that's not all. To make use of all that information coming from different data sources with heterogeneous formats you need software and hardware; but if our intention goes beyond adding or

G. Jezic et al. (Eds.): KES-AMSTA-18 2018, SIST 96, pp. 269–274, 2019.
https://doi.org/10.1007/978-3-319-92031-3_27

deleting registers or making a query, obviously we need tools and applications for the collection, extraction and manipulation of information with the aim of creating intelligent and artificial means for a specific purpose, and to be able to support decision making [2].

The use of information and communication technologies helps us in the automation of activities, one of the main ones is the information systems used by companies or organizations to achieve operational excellence, develop new products and services or improve their process of taking decisions [3]. Having these systems will allow the management of information. At present, we can find different types of information systems and classifications; but its objectives are the same: automate processes, support decision-making and achieve competitive advantages. One of them is Multiagent Systems (MAS) that are based on Artificial Intelligence (AI), a discipline that involves computational processes and approximate models of human thinking, [4] Intelligent Systems (IS) and telecommunications.

The use of Artificial Intelligence is increasingly common in systems: it is the time when hundreds of products derived from it begin to be marketed worldwide. The AI seeks unification through the convergence NBIC -nano-bio-info-cogno-, has all the potential to improve the quality of life of individuals around the world [5].

Achieving an electronic database of patient information would allow health systems to strengthen their operation [6], but with the support of different technological tools, a detailed analysis of the data will be achieved to find patterns and establish the basis of the knowledge that allow us to provide better support in the prevention of diseases.

3 Intelligent Agents

An intelligent agent is a system capable of perceiving its environment, processing perceptions and responding to its environment in a rational manner, that is, correctly. By means of sensors it perceives its environment and responds with actions, according to the result, as shown in Fig. 1.

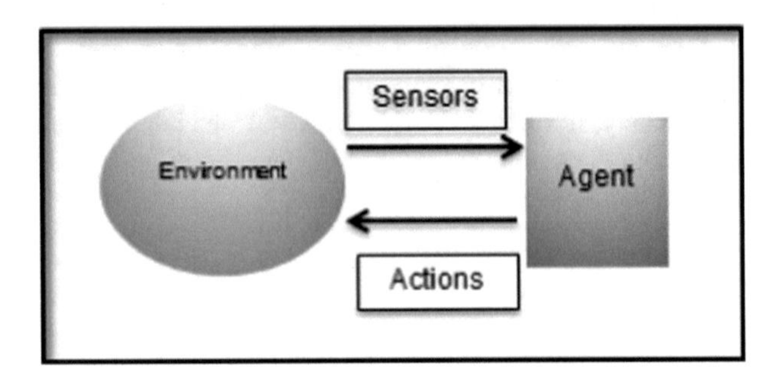

Fig. 1. Representation of an intelligent agent

3.1 Features of Intelligent Agents

The features that an intelligent agent must have are:

- Autonomous: It must act on its own, have control over its actions and can formulate goals.
- Sociability: Ability to communicate with other agents and with humans in a coherent manner.
- Reaction capacity: perceives its environment and reacts to adapt to it.
- Initiative: must be able to perform necessary actions for the objectives to be met.

Intelligent agents are used in the development of systems because they are friendly, flexible, adapt to users, allow greater functionality, integrated and compatible with other technologies [7].

An agent can learn from its environment through sensors that capture the environment to, based on the data that is captured and analyzed, establish a knowledge base to respond with actions to its environment.

We can use intelligent agents in different activities, such as:

- Capture, analysis and extraction of information
- Information services on the Internet
- Mobile apps
- E-commerce
- Process management

3.2 Agent Taxonomy

There are different ways to establish the taxonomy of intelligent agents, the two most powerful are described below [8].

The first is by researchers Wooldridge and Jennings [9], agents are divided into two classes that they call notions:

1. Weak notion of agent: With weak notion they refer to those agents capable of exchanging messages using a communication language, this is the most commonly used notion in software engineering. To be considered of weak notion, an agent must fulfill at least the following properties: autonomy, sociability, reactivity and initiative.
2. Strong notion: This notion is also called restrictive, considers an agent as a system that, in addition to the properties already mentioned, is defined using concepts that we normally apply to humans, such as belief, knowledge, obligation, etc.

The second, by Nwana, proposes an agent topology that identifies other classification dimensions, and distinguishes 7 types [10]:

1. Collaborative agents
2. Interface agents
3. Mobile agents
4. Information/Internet agents
5. Reactive agents

6. Hybrid agents
7. Intelligent agents

In the next section we will explain the agents to be used based on the function to be carried out for the management of the information of the clinical records and thus be able to support in the prevention of the diseases.

- **Collaborative agents:** They are characterized by their autonomy, cooperation with other agents, social capacity and response to achieve their objectives and the tasks of their users. For their operation they must have a negotiation coordination to achieve their acceptable agreements. Most of the collaborative agents currently used do not perform any complex learning. They can be used to solve problems such as: support for expert systems, decision support systems offering modularity, speed, flexibility among other things.
- **Mobile agents:** Mobile agents are computational software processes capable of navigating wide area networks such as the Internet, interacting with equipment and obtaining information from users. They communicate and cooperate with another agent for the exchange of data without necessarily providing all the distance information.
- **Information agents:** They arise because of the great demand for tools that exist for the management of the information that we have in our days on the WEB and its growth day by day. The role of information agents is to collect, integrate information from different distributed sources for handling and administration.

At present, there are systems and applications that use and combine agents of different categories to obtain better information and system performance. Intelligent systems (IS) look for problems that already exist or improve the way of working or the quality of life. The Intelligent System tries to generate representations, inference procedures and learning strategies that solve problems that until now have been solved by humans [11].

Another way to manipulate the large volume of information could be the incorporation of fuzzy logic to intelligent systems. Fuzzy logic extends from classical logic. Fuzzy logic seeks to create mathematical approximations in the resolution of certain types of problems; aims to produce accurate results from inaccurate data, the diffuse can be understood as the possibility of assigning more values of truth to classical statements [12].

Just as fuzzy logic can be applied, simulations provide a powerful alternative to complement, replace and expand traditional approaches in the field of social sciences. The field is intrinsically interdisciplinary, linked to the science of complexity, systems theory, data mining, neurodiffusion and distributed agencies [13].

4 Disease Prevention Agent

As mentioned above, the volume of information obtained from a medical record is large, its data comes from multiple sources of information and for its manipulation it requires specialized technological tools. This section describes the role of the Disease

Prevention Agent that is part of a Multi-Agent System model. Figure 2 describes how the Prevention Agent will work.

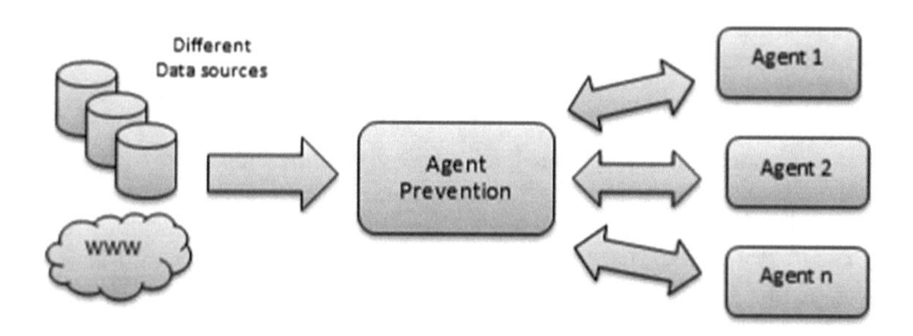

Fig. 2. Prevention agent

The Prevention Agent realizes its objective based on its knowledge obtained from the different sources of data that may be in different places, such as hospitals, doctors' offices, laboratories, X-rays, or on the Internet. It can also obtain information from Public Health Institutions.

The data that will be obtained from the agent's environment will be a universe of data by which the agent can identify and thus learn from the generated patterns of the information analyzed.

All the volume of information can be handled in a detailed way in a personal, social, regional, national or international way. These patterns will serve as support for the decision making that allows establishing a knowledge base to support in the prevention of human diseases.

5 Conclusions

The alternatives offered by the use of Artificial Intelligence with Intelligent Agents and Multiagent Systems is a different way of solving problems or situations in the traditional way, but the advantages offered by these technological tools are learning, modularity, speed, flexibility, all this in order to improve the way of working or quality of life.

With Multiagent Systems it is possible to have a combination of agents but the question is: Which is better? All have specific characteristics, the importance is for the agent to be used and to select the appropriate one and thus be able to have a functional Multiagent System.

To get to have a better quality of life can start with the care and prevention of diseases knowing social patterns, regional and not only nationally at the international level, the Intelligent Agent can learn from their environment such as suffering, susceptible people, nationality, work etc. and will respond with actions according to your objective or request of your user.

References

1. Secretaria de Salud de Baja California. Tu Salud. http://www.saludbc.gob.mx/tu-salud. Accessed Jan 2017
2. Organización Mundial de la Salud. La Salud. http://www.who.int/es/. Accessed Jan 2017
3. Laudon, K.C., Laudon, J.P.: Sistemas de Información Gerencial, decimocuarta edición 2016, pearson (2016)
4. Daza, C.: El mundo de los Agentes Inteligentes y su utilización en el mundo real (2014). https://www.seas.es/blog/e-learning/el-mundo-de-los-agentes-inteligentes-y-su-utilizacion-en-el-mundo-real/. Accessed 15 Jan 2018
5. Cairo, O.: El Hombre Artificial, El Futuro de la Tecnología. Alfaomega, Mexico (2011)
6. Ramírez, R.M., Ramírez, M.H.B., Osuna, M.N.C., Salgado, S.M.C., Vázquez, N.S.O., Alanis, G.A.: Big Data and health clinical records, In: Chen, Y.W., Tanaka, S., Howlett, R. J., Jain, L.C. (eds.) Innovation in Medicine and Healthcare 2017. Springer (2017)
7. Romero, M.: Sistemas MultiAgente (MAS), Departamento de Tecnología Electrónica Universidad de Sevilla (2009). http://www.dte.us.es/personal. Accessed 5 Jan 2018
8. González, R., Mediero, S.: Agentes inteligentes para controlar la dieta de personas con problemas de salud, ResearchGate GmbH (2014). https://www.researchgate.net/publication/262487664_Agentes_inteligentes_para_controlar_la_dieta_de_personas_con_problemas_de_salud. Accessed 8 Jan 2018
9. Wooldridge, M., Jennings, N.R.: Intelligent agents: theory and practice. Knowl. Eng. Rev. **10**(2), 115–152 (1995). http://www.cs.ox.ac.uk/people/michael.wooldridge/pubs/ker95.pdf. Accessed 10 Jan 2018
10. Nwana, H.: Software agents: an overview. Knowl. Eng. Rev. **11**(3), 205–244 (1996). http://teaching.shu.ac.uk/aces/rh1/elearning/multiagents/introduction/nwana.pdf. Accessed 10 Jan 2018
11. Schalkoff, R.J.: Intelligent Systems Principles, Paradigms, and Pragmatics. Jones and Bartlett Publishers (2011)
12. Perez, I., León, B.: Logica difusa: Teoria y Practica, Universidad Catolica Andres Bello, primera edicion (2007)
13. Bogart, Y., Castanon, M., Castro, J., Suarez, D., Magdaleno, S.: Fuzzy models for complex social systems using distributed agencies in poverty studies. In: Software Engineering and Computer Systems, pp. 391–400. Springer (2011)

Agent-Based Model as a Provider of Medical Services in Tijuana Mexico

Ricardo Rosales[1(✉)], Nora Osuna-Millan[1], Consuelo Salgado-Soto[1], Carlos Flores-Sanchez[1], Juan Meza-Fregoso[1], and Arnulfo Alanis[2]

[1] Facultad de Contaduría y Administración, Universidad Autónoma de Baja California, UABC, Tijuana, BC, Mexico
{ricardorosales,nora.osuna,csalgado,cflores,juan.meza70}@uabc.edu.mx

[2] Instituto Tecnologico de Tijuana, ITT, Tijuana, BC, Mexico
alanis@tectijuana.edu.mx

Abstract. The search for medical services to maintain family health remains a difficult goal to achieve because in addition family budget constraint is presented to desire to access the best possible medical services. The purpose of this paper is to present a model based on agents that analyzes a series of symptoms and sufferings, the results of this analysis allow provides a preliminary diagnosis of a patient. This model can be listing best options for medical service providers in Tijuana, Baja California, Mexico to prevent the erroneous self-diagnosis in the population of scarce resources. The model can help to create simulations of patient symptoms using computational tools and can predict in real time results with possible options of medical services to be attended. The model contains three main agents: Patient Agent, Medical Services Provider Agent and Content Medical Service Agent, the interaction of these agents supported by their Knowledge Base allow a finite and specialized search of suppliers in the health area. The use of agent-based models can reduce emerging costs derived from tests and errors in the real world.

Keywords: Modelling · Agents · Medical services

1 Introduction

Globally, main research lines of intelligent environments focus on systems operated by the end user; others focus on systems powered by autonomous agents, minimizing the user's cognitive load. This research represents the patient-user the option of knowing how much autonomy it maintains or how much it delegates to intelligent agents. Likewise, a Model Based on Agents is proposed that improves the ability to offer the adequate medical service based on the symptoms and conditions of the patient, allowing for adequate medical care.

G. Jezic et al. (Eds.): KES-AMSTA-18 2018, SIST 96, pp. 275–283, 2019.
https://doi.org/10.1007/978-3-319-92031-3_28

1.1 Medical Services

Nowadays, medical care plays an essential role in well-being and health of individuals. Medical Services promote and maintain the welfare (social, physical and mental) of people regardless of their social status. The medical services anticipate and take care of damages caused to health, maintaining protection against the risks that may be harmful to health.

Health requires constant maintenance by providing, diagnosing and treating diseases, medical services respond to this necessity, offering continuity of care by providing and understanding the specific needs of each patient. Proper medical attention allows patients to benefit from adequate attention promptly based on their symptoms or conditions. This article addresses in a different way how medical services can provide to patients based on their demands, as well as offering researchers an alternative to model medical care through multi-agent systems allowing abstraction of reality with the order to provide adequate medical services for the patient [1].

1.2 Agent Based Modelling (ABM)

The ABM is applied incrementally to empirical situations [2]. Its methodological advantage lies in the ability to explicitly simulate human decision-making processes considering a high degree of heterogeneity [3, 4].

The ABM has capabilities to deal with uncertainty in real-world actions using fuzzy logic techniques, approximate sets, Bayesian networks, etc. [5]. An agent-based on ABM can think and act like a human, can operate under autonomous control, perceiving their environment and adapting to changes to achieve specific objectives or goals [6]. In decision-making behavior, ABM outperforms simple if-then rules allowing for agents to learn and change behaviors in response to their experiences [7]. Even at the simplest level, an ABM consists of agents and the relationships between them and may have valuable conclusions about the system as a whole [8].

1.3 BDI a Practical Reasoning

The BDI theory tries to model the rationality of those actions taken by human beings in certain circumstances; this theory was developed by the philosopher Bratman [9]. A practical reasoning directed towards actions: towards process of deciding what to do. Unlike theoretical reasoning that is directed towards beliefs. This type of reasoning includes two activities: (I) Deliberation, that is, deciding which are the goals to be met, and (II) Analysis, related means, that is, deciding how the agent will achieve those goals. Both activities can be seen as computational processes executed by agents with bounded rationality.

2 Related Work

The existence of different research lines and applications using Systems Based on Agents in the realization of searches in different fields, can learn and perform searches based on established patterns [10]. The creation of intelligent systems allows performing specific activations of a specific context, allowing a comparison between the real world and the simulated world [11].

The autonomy feature offered by the agents allows to create reactive and personalized search assistants according to the preferences of the users, this allows information that is available to be accessed by searches derived from the combination of effective algorithms and techniques for the recovery of information improving the performance in the search for information [12].

The understanding of the processes among the elements involved is key allowing raise expectations offering quality information if these elements are represented in a Multi-Agent System, the simulated effectiveness is maximized through a virtual environment that can be contrasted in the real world [13].

3 Case Study

Modern and developed societies, with rare exceptions, have devoted considerable sums to the structuring of universal health service systems, with a strong public component, focused on the production and distribution of medical care. This dynamic was fostered in close connection with the technological fever of the "golden years," configuring so-called health industry, one of the most prosperous branches of economic activity in the world [14].

The goal of obtaining medical services is to maintain health. Achieving this objective depends in part on the services offered and the way they are organized in a region.

Specialized medical services have become an important activity for the private sector, ranking third regarding economic units registered in the city of Tijuana. Proper diagnosis, treatment, and prevention are promoted through the appropriate management of different conditions. Obtaining information quickly stimulates the efficient management of the process of obtaining medical attention, translating into a reduction in the duration of the process, thus promoting an efficient prognosis of the disease.

The importance of giving access to the population of these specialized services is important since in Mexico there has been a tendency of the population towards prevention, for which the demand for these services has been increasing, particularly in Baja California, besides the Foreign market adds to domestic demand.

The proliferation of establishments that provide medical care in the northern border of the country has been attributed to the development of a medical management model observed in Tijuana and Mexicali, becoming a case of success and an example for the rest of the country. During 2014 the economic spill in the state of Baja California for medical tourism was 673 million dollars, estimating an influx of visitors from the neighboring country in search of attention by order of 4.7 million people [15].

4 Modelling Medical Care Process

The design of a conceptual model depends importantly on knowledge of the study context, as well as on the perception of the modeler. We know that a model allows us to reduce the complexity of the study context and serves as a guide to represent difficult scenarios emerging on reality, likewise, it allows the generation of hypothetical cases helping to analyze and contrast different possibilities, to give the response to the problems raised. The model presented in this research is based on MAS, allowing elements involved to have autonomy in their interactions, beliefs, desires, and intentions, helping to solve problems.

4.1 Agents Modelling

We represent an abstraction of the real world using an agent; we motivated on this paradigm because the agents are entities that allow us to emulate rational behavior based on humans. Additionally, we used de Belief, Desire and Intention Paradigm (BDI) [9], in each agent, so that has more reasoning to do actions. Also, the agent can help the patient to accomplish a task such as find the best hospital and doctor for his correct treatment.

In our research we define agents based on the context of Medical Care, the proposed model is composed by three main agents Patient Agent (PA), Medical Services Provider Agent (MSPA), Content Medical Service Agent (CMSA). In the next part, each agent is defined and its interactions.

4.1.1 Formalization of Medical Care Process Agents

The formalization of the elements that compose the model helps to understand in details its properties. The formalization of methods is used to understand the wide variety of complex behaviors that an agent can exhibit. Each agent has three main elements Beliefs, Desires an Intentions, allowing them to have to reason similar to humans. In the next Fig. 1 is represented by its elements.

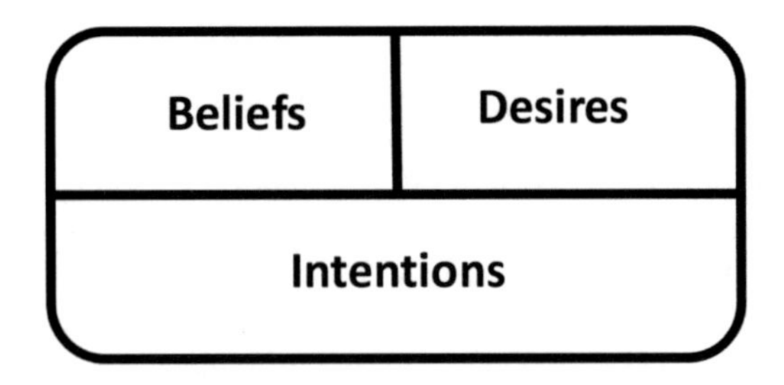

Fig. 1. A representation of BDI Agent

The following is a formalism, which is, defined a representation for BDI agents.

The Patient Agent (PA), is a Patient representation, is who has pain, symptoms, discomfort, this agent manifests its condition and requires medical care.

El PA is defined by a tuple of 3 elements:
A PA agent is a tuple of 3 elements:

$$\Pi = <\Omega, \Gamma, Y> \quad (1)$$

where:

1. Ω is the finite set of base beliefs. Each PA belief is represented by the emerging Symptoms
2. Γ is the finite set of base desires; Each PA desire is represented by find cure of Symptoms
3. Y is the finite set of intentions. Each PA intention is a stack of plans to execute requests or searches in order to find adequate medical services according its symptoms (Fig. 2).

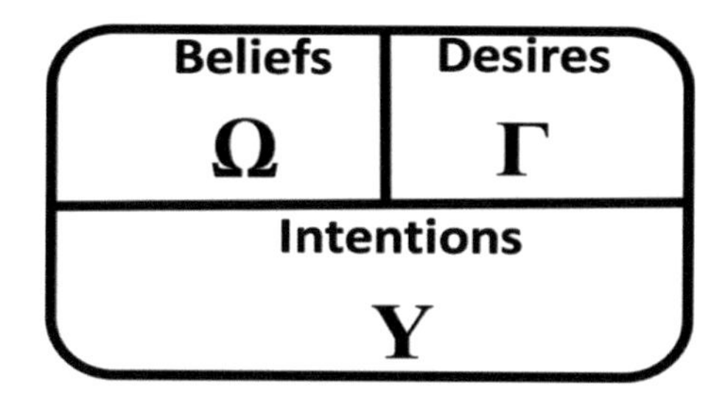

Fig. 2. A representation of Patient Agent (PA)

The Medical Services Provider Agent (MSPA). Is the Medical Services expert representation, analyze the symptom form filled, after this analyses using its Knowledge Base search the best results, based on Patient' Symptoms.
A MSPA agent is a tuple of 3 elements:

$$\Sigma = <\Theta, \Delta, \Phi> \quad (2)$$

where:

1. Θ is the finite set of base beliefs. Each MSPA belief is a selection and analysis of the kinds of symptoms that have the PA.
2. Δ is the finite set of base desires; Each MSPA desire is represented by the search of adequate medical service based on symptom belief of PA.
3. Φ is the finite set of intentions. Each MSPA intention is analyze based on the Knowledge Base to find the best option of medical services, request to CMSA the service a deliver to PA the adequate result (Fig. 3).

The MSPA knowledge Base is based on a Fuzzy Logic paradigm in the Patient Symptoms and Suffering to handle the emerging uncertainty to offer the adequate medical services. We had identify the involved linguistic variables for the input 'Symptom' (Strong, Medium, Weak), for the input 'Suffering' (Strong, Medium, Weak)

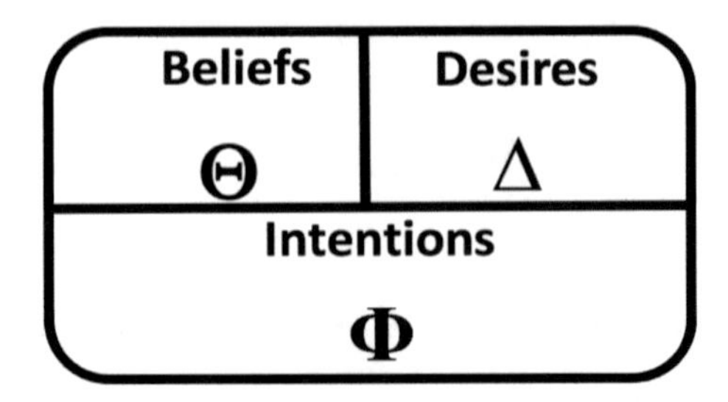

Fig. 3. A representation of Medical Services Provider Agent (MSPA)

and for the output 'disease Type' (RequiredMedicalService1, RequiredMedicalService2, RequiredMedicalService3, RequiredMedicalService4, RequiredMedicalService5, RequiredMedicalService6, RequiredMedicalService7, RequiredMedicalService8 and RequiredMedicalService9).

We defined 9-inference IF-THEN rules covering all linguistic variables. The proposed FIS is flexible and permits the addition or deletion of rules; this can be seen as an advantage as it can be adapted to different contexts or, if different variables exist, can be increased (Table 1).

Table 1. Inference fuzzy rules of MSPA.

Nu	Inference rules
1	If (Symptom is Strong) and (Suffering is Strong then (diseaseType is RequiredMedicalService1)
2	If (Symptom is Strong) and (Suffering is Medium then (diseaseType is RequiredMedicalService2)
3	If (Symptom is Strong) and (Suffering is Weak then (diseaseType is RequiredMedicalService3
4	If (Symptom is Medium) and (Suffering is Strong then (diseaseType is RequiredMedicalService4)
5	If (Symptom is Medium) and (Suffering is Medium then (diseaseType is RequiredMedicalService5)
6	If (Symptom is Medium) and (Suffering is Weak then (diseaseType is RequiredMedicalService6)
7	If (Symptom is Weak) and (Suffering is Strong then (diseaseType is RequiredMedicalService7)
8	If (Symptom is Weak) and (Suffering is Medium then (diseaseType is RequiredMedicalService8)
9	If (Symptom is Weak) and (Suffering is Weak then (diseaseType is RequiredMedicalService9)

The Content Medical Service Agent (CMSA). Is the container of Medical Services having a Data Base with content related to Medical Services constantly upgrade its

database to have the Medical Services upgraded, this agents response to MSPA requests.
A CMSA agent is a tuple of 3 elements:

$$\Lambda = <\vartheta, \mathrm{K}, \Psi> \qquad (3)$$

where:

1. ϑ is the finite set of base beliefs. Each CMSA belief is represented by MSPA requests based on the results of medical service required.
2. K is the finite set of base desires; Each CMSA desire is represented by search in its data base best medical services based on MSPA requests.
3. Ψ is the finite set of intentions. Each CMSA intention is find and response to the MSPA with best Medical Services found in its Data Base, this agent build a stack with results of medical services, the top of the stack is the first option and so on in order to cure the disease.

The following Fig. 4 illustrates the Model in general as well as the interactions of the agents involved within the Medical Care Process.

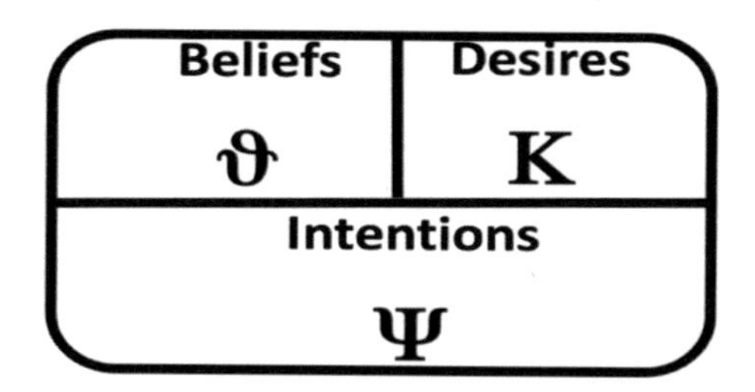

Fig. 4. A representation of Content Medical Service Agent (CMSA)

The Fig. 5. represents, all agents embedded in a context of Medical Care Process. This process is defined by a series of steps described below:

1. Patient Feels the Symptom
2. Patients seek Medical Care
3. Patient fill Symptom Form Provided by MSPA, "PA Provides the needs"
4. Patient brings Symptoms Form to MSPA
5. MPSA analyze the symptom form filled using its Knowledge Base, once have the best results, starts the request to CMSA
6. CMSA receive the MSPA Request a search in its Data Base, the best Medical Services based on MSPA request. (Constantly the CMSA check the Medical Service Environment (MSE) for Medical Service Updates) and upgrade its databases.
7. CMSA response to the MSPA with the best results and secondary results

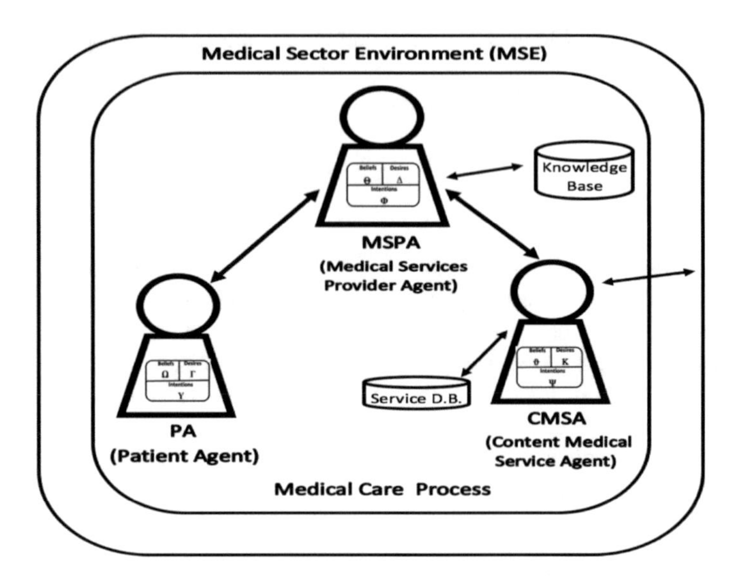

Fig. 5. Representation of Medical Care Process

8. MSPA receive the CSMA results and register these results to make historical user register
9. MPSA send to the PA the best secondary results of Medical Services
10. PA receive the results of Medical Services according to its symptoms

5 Conclusions

Many models have focused on building architectures with reasoning capabilities; the proposed article offers to the model based on agents, representing abstractly the provision of medical services based on the symptoms and sufferings of the patient. The proposed model defines an environment of medical care, where three main agents are considered, the interaction and the integration of these agents allow to study possible scenarios that may arise in the real world allowing offering the adequate average service based on the symptoms and sufferings of the patients.

Additionally, the use of agents motivates this research, because these agents can be intelligent, these entities emulate or simulate rational mental processes, behaviors such as the patient, doctor, and the expert in medical services.

6 Future Work

Derived from the complexity of medical care in Tijuana, Mexico, it is required that the model have more intelligent agents, the interaction between agents can be maximized adding different scenarios of the emergency level, the model will be manipulated to be attended allowing a closer approximation to reality. The agents will be programmed in a technology platform and will include artificial intelligence in its reasoning. Additionally, based on the proposed BDI aided by the Fuzzy Logic Type-2 we will create fuzzy perceptions allowing an adaptation to patients needs. The model will have patterns of perception, processing, and representation of knowledge to act in consequence of the emergence of the system.

Likewise, the use of Multi-Agent Systems will allow supporting the personalized, intelligent interaction and an active environment allowing to create complex adaptive systems configured according to the patient.

References

1. Rajeev, K., Prakash, N.: Elsevier Comprehensive Guide to Combined Medical Services (UPSC)-E-Book. Elsevier Health Sciences (2015)
2. Smajgl, A.: Challenging beliefs through multi-level participatory modelling in Indonesia. Environ. Model Softw. **25–11**, 1470–1476 (2010)
3. Gilbert, N.: Agent-Based Models. SAGE Publications, London (2008)
4. Parker, D., Manson, M., Janssen, M., et al.: Multiagent system models for the simulation of land-use and land-cover change: a review. Ann. Assoc. Am. Geogr. **93**(2), 314–337 (2003)
5. Ramos, A., Augusto, J., Shapiro, D.: Ambient intelligence-the next step for artificial intelligence. IEEE Intell. Syst. **23**, 15–18 (2008)
6. Russell, S., Norvig, P.: Artificial Intelligence. A Modern Approach, 3rd edn. Pearson Education International (2009)
7. Macal, C., North, M.: Tutorial on agent-based modelling and simulation. J. Simul. **4**, 151–162 (2010)
8. Bonabeau, E.: Agent-based modeling: methods and techniques for simulating human systems. Proc. Natl. Acad. Sci. U.S.A. **99**, 7280–7287 (2002)
9. Bratman, M.: Intention, Plans, and Practical Reason. Harvard University Press, Cambridge (1987)
10. Garvey, F., Sankaranarayanan, S.: Intelligent Agent based flight search and booking system. Int. J. Adv. Res. Artif. Intell. **1**(4), 12–28 (2012)
11. McTavish, C., Sankaranarayanan, S.: Intelligent agent based hotel search & booking system. In: 2010 IEEE International Conference on Electro/Information Technology (2010)
12. Jansen, J.: Using an intelligent agent to enhance search engine performance. First Monday, vol. 2, no. 3 (1997)
13. Wang, Y.: Multi-agent based logistics coordination system. Adv. Mater. Res. **433–440**, 3106–3111 (2012)
14. Almeida, C.: Reforma de sistemas de servicios de salud y equidad en América Latina y el Caribe: algunas lecciones de los años 80 y 90. Cad. Saude Publica, pp. 905–925 (2002)
15. Vargas-Hernández, J.G.: An exploration of Tijuana-San Diego marketing environment and marketing border of health service in Tijuana. Polish Assoc. Knowl. Manag. **55**, 55 (2011)

Business Informatics

Understanding the Potential Value of Digitization for Business – Quantitative Research Results of European Experts

Christopher Reichstein[1](✉), Ralf-Christian Härting[1], and Pascal Neumaier[2]

[1] Business Administration, Aalen University of Applied Sciences, Aalen, Germany
{christopher.reichstein, ralf.haerting}@hs-aalen.de
[2] Competence Center for Information Systems, Aalen University of Applied Sciences, Aalen, Germany
pascal.neumaier@kmu-aalen.de

Abstract. There are many drivers known for implementing aspects of digitization in business. While some firms are already successfully integrating the opportunities of digitization into their business processes, others still lack understanding and expertise. The impression is created that some firms rely on the bandwagon effect, i.e. they are doing it because everybody else is doing it! For that reason, the authors design a quantitative study based on a former qualitative research to prove main drivers of successful digitization aspects. Interviews with European experts, collected within the framework of this research, give empirical evidence of six factors (efficiency, innovation, data privacy, mobility, new business models and human integration) influencing the potential value of digitization in business. The research results show significantly that improvements of efficiency and mobility as well as the generation of new business models are main drivers of digitization for business success.

Keywords: Potentials · Digitization · Business · Strategies · European experts Quantitative study · Empirical results

1 Introduction

Digital transformations have caused a tremendous change of social and business structures [1]. The expansion of internet technologies enabled an emergence of new information and communication possibilities. In the course of this, the digital process affects individuals, as well as companies [2]. Because of the digital transformation, companies have to adapt to the far-reaching changes. Digital technologies such as Big Data, Internet of Things, Cloud and Mobile Computing, are changing value creation processes that in turn enables an emergence of new business models [3]. By new competitors pushing into the markets, the market situation gets more and more competitive. In order to remain competitive in future, it is inevitable for companies to scrutinize their own business model and adapt to new digital requirements [4].

G. Jezic et al. (Eds.): KES-AMSTA-18 2018, SIST 96, pp. 287–298, 2019.
https://doi.org/10.1007/978-3-319-92031-3_29

Digitization is not a new term since it has already appeared in the course of the third industrial revolution as a data conversion process from analog to digital. However, the understanding of the term digitization has changed. Nowadays, the term is associated with smart value creation processes using capable information and communication technologies. Although the term digitization is widely used, there is no consistent definition in literature. Moreover, main potentials of digitization for business strategies are not clearly defined yet in order to understand how to use driver of digital processes for better firm performances.

Thus, this study seeks to verify main potentials of digitization from an enterprise's (mostly employees with an IT background) perspective based on a former qualitative study. These six hypotheses were checked within this quantitative research by in total 216 experts in order to quantify main drivers of digitization potentials for business [5].

The paper is structured as follows: Sect. 2 defines the terms of digitization and its potentials. Hereafter, the authors are going to describe the research design and methods in Sect. 3 in order to understand among others how data were collected. The study results are presented in Sect. 4 followed by a conclusion of the paper in Sect. 5.

2 Definition and Potentials of Digitization

The digital revolution is not a new phenomenon. It has presented new challenges for companies and society for decades. An example is the electronic information exchange. It evolved from simple types of notification (e.g. symbols) to complex communication relations in the form of digital data over the course of time [6].

Along with the change of digital technologies, the understanding of digitization has been adjusted. As with the industry 4.0 case, the term "digitization" now stands for intelligent business and value-added processes by using powerful information and communication technologies such as Big Data, Cloud and Mobile Computing, Internet of Things or Social Software [7, 8]. Thus, digitization is not only the provision of information, but also the (partial) mapping of value-added processes in electronic form. In addition to industry 4.0, the conception of digitalization implies:

- new approaches to digitization found throughout the industry, particularly in the consulting and services sector [9, 10];
- digitization as a driver for the development of new business models with innovative products [11].

Digitization, according to Kagermann [11], describes the continuous convergence of the real and virtual world and can be regarded as the main driver of innovations and changes in all sectors of our economy. According to a former study [12], all operational as well as functional areas can benefit from digitization. Digitization has an impact on the existing analogue and/or manual work environment, for instance [13]. The change in a digital factory or privacy issues exemplifies this. There is an additional influence on already digitally implemented work areas such as the IT organization [14, 15].

Besides the economy, several other areas make use of digitization aspects. In the past, cultural goods have undergone a series of digitizing for the sake of protection of future purposes [16]. Teaching content at universities is increasingly driven by digital approaches [17]. The public administration is gaining an increasing amount of digital access to its stakeholders under the slogan "e-government", too [18].

A consideration of the concept of digitization cannot simply be reduced to technologies and their potentials for business processes, but must also include all factors from the social and legal environment. Hence, digitization should be viewed from a holistic, systemic perspective to be able to consider these factors within its near entirety [6]. A system, in this context, is a set of elements that consists of interrelationships [19, 20]. Core elements like the infrastructure of new digitization approaches result from new technologies as well as services and applications. Under the term "technology", all new developments, i.e. in the areas of "Big Data or Mobile Computing", can be subsumed. An example for digital services are cloud services, in particular the "software as a service" model. Digital services offer a wide range of business-oriented applications from the industrial as well as the service sector [21].

3 Research Design and Methods

Within the framework of a qualitative research in 2016 [5], different factors of the potential value of digitization in business were identified. This study was based on empirical data from 72 European experts using the methodology of Grounded Theory. In this former study, IT experts mentioned six main driver for a successful usage of digitization aspects for business predominantly. The results of the qualitative survey enable to extend the current scientific view of the potential value of digitization for further business strategies. In the following, the study design, research method and data collection are explained.

3.1 Study Design

To explore the potential value of digitization, the authors designed a quantitative research study based on the former qualitative research [5].

Interviews with European experts, collected within the framework of this research, give empirical evidence of six factors (efficiency, innovation, data privacy, mobility, new business models and human integration) that positively influence the potential value of digitization for business strategies. Further, all influencing factors are encouraged through specialized literature. Figure 1 shows the developed research model including six hypotheses.

The potential value of digitization can be defined as the individually perceived capability of the implementation of digital technologies. Digital technologies enable the acceleration of processes along the entire value chain, which leads to a reduction of costs [5, 22]. Therefore, hypothesis 1 was created:

H1: An improvement of efficiency positively influences the potential value of digitization.

In addition to efficiency improvements, digital technologies also enable companies to improve their innovation processes [5, 23]. Networking along the value chain

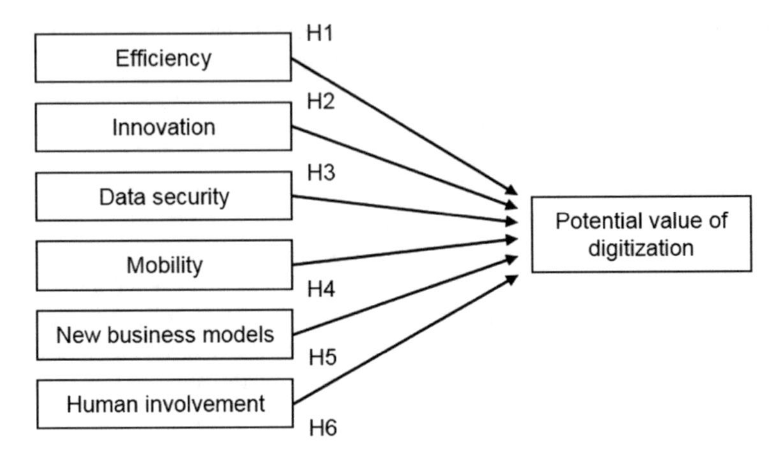

Fig. 1. Research model

improves the flow of information and shortens innovation cycles [24]. This leads to the creation of hypothesis 2:

H2: An improvement of innovation processes positively influences the potential value of digitization.

With an increasing degree of digitization, companies are becoming more and more dependent on reliable information and communication systems [25]. Due to several risk through the growing number of cyber-attacks, such as data theft [39] and sabotage [40], data security will play an increasingly important role in the future [5, 25]. Therefore, hypothesis 3 was created:

H3: Data security positively influences the potential value of digitization.

Nowadays, more and more companies are using mobile devices, such as smartphones and tablets [26]. The combination of mobility applications and mobile internet provides employees a flexible access to information. This can not only increase productivity, but also enhance employee satisfaction [5, 27, 28]. Consequently, this leads to hypothesis 4:

H4: An improvement of mobility positively influences the potential value of digitization.

Digital technologies like cloud computing and social media lead to a completely new set of business models and a change in consumer's behaviors and expectations [5], [11]. This phenomenon leads to an obsolescence of convectional business models. Adapting business models allows companies to gain competitive advantage and customer loyalty [4] that leads to hypothesis 5:

H5: The generation of new business models positively influences the potential value of digitization.

In addition to changing business models, digitization changes requirement profiles for employees [29]. The automation degree will rise due to an increase in networking.

On the one hand, this leads to a loss of lower qualified jobs and on the other hand, it can result in the gain of higher qualified jobs [30]. While potentially causing a loss in lower qualified jobs, digitization can increase the number of higher qualified jobs [30]. Hence, companies need to provide further education for employees [29]. Changes within a company can lead to uncertainties among employees. To avoid resistance, employees need to be involved in the process of change. To create acceptance among employees, good communication is inevitable [5] resulting in hypothesis 6:

H6: Human involvement positively influences the potential value of digitization.

In order to examine the impact of the defined determinants on the potential value of digitization, respondents had to answer on a Likert scale [31] of one to six (1: low to 6: high) within the online survey. All questions are conceptualized in accordance with general guidelines of quantitative studies [35]. The supported literature review is summarized in the following table (Table 1):

Table 1. Summary of literature review

Determinant	Item	Author
Efficiency	Single item	[5, 22]
Innovation	Single item	[5, 23, 24]
Data security	Sabotage & data theft	[5, 25, 39, 40]
Mobility	Single item	[5, 26–28]
New business models	Single item	[4, 5, 11]
Human involvement	Single item	[5, 29, 30]

3.2 Research Methods and Data Collection

The quantitative research study is based on a web survey, conducted within German-speaking countries (Germany, Austria and Switzerland). The study was implemented via the open source LimeSurvey [32]. A pre-test was carried out to ensure high quality standards. The main study started in June 2016 and ended in August 2016. The questionnaire was mainly distributed via email and personal contact to IT manager and IT experts with instructions to participate in the survey. The interviewers asked that the survey should only be attended if there was expertise in digitization aspects within the company. In total, 303 answers were collected. After data cleaning, 216 answers remained for evaluation. The majority of the participants came from Germany (78.24%), followed by Switzerland (16.20%) and Austria (5.56%). Most of the interviewed experts are working in the information and communication sector (26.85%), followed by those in professional, scientific and technical activities (20.37%) and manufacturing (18.52%). The classification of sectors was based on the European Classification of Economic Activities (NACE Rev.2). Table 2 shows the number of employees from participating firms. More than 87% of all interviewed experts are working in small or medium sized companies with less than 500 employees.

The study was evaluated by using a structural equation modeling (SEM) with Partial Least Squares (PLS) approach to visualize the relationship between different

Table 2. Number of employees per firm

Employees	abs.	in %
0–4	72	33.33%
5–9	27	12.50%
10–19	26	12.04%
20–99	39	18.06%
100–499	24	11.11%
500–4,999	15	6.94%
5,000–10,000	5	2.31%
>10,000	8	3.70%

variables [33]. In general, the tool Smart PLS is used to calculate the direct effects of the found dependent variables in literature on the dependent variable "potential value of digitization" as it is particularly suitable for smaller sample sizes [33]. In addition, SEM is a suitable method for this research to test the fit of a causal model with empirical data and new developed scales [34, 35]. The Smart PLS approach focusses on a partial least squares regression based on sum-scores and significances are calculated via bootstrapping. The non-parametric procedure called bootstrapping allows testing the statistical significance of various PLS-SEM results like path coefficients, Cronbach's Alpha and R^2 values [36]. It has to be noticed that there is no need to calculate metrics like Cronbach's Alpha (CA) or Average Variance Extracted (AVE) in case of single item sets. Furthermore, single item sets are not only allowed but also common in information systems research [36].

For multi-item measures (like the determinant "data security") common quality criteria such as Cronbach's Alpha (CA), Average Variance Extracted (AVE) and Composite Reliability (CR) are calculated to ensure good psychometric properties of the model. Accordingly, the authors examined reliability and validity to ensure a high quality of the measurement model and its constructs by considering the recommended threshold of .50 for Cronbach's Alpha (for new and developed scales), .50 for Average Variance Extracted, and .60 for Composite Reliability [38, 41].

4 Results

Figure 2 shows the results after analyzing the data via SmartPLS. Further details to measures, measurement, composite reliability (CR), Average Variance Extracted (AVE), and Cronbach's Alpha (CA) are shown in Table 4 (see Appendix).

With a coefficient of determination (R^2) of 0.385 the model can be regarded as sufficient according to Chin [34]. All constructs exceeds the recommended threshold for the common quality criteria like CR, AVE and CA as mentioned in Sect. 3.

Regarding Hypothesis 1, (**An Improvement of Efficiency Positively Influences the Potential Value of Digitization**) the results of the SEM show that efficiency positively affects (+0.456) the potential value of digitization on a high significance level ($p = 0.00 < 0.01$). Therefore, the hypothesis can be confirmed. This underlines that companies can increase their efficiency by digitizing their value creation processes.

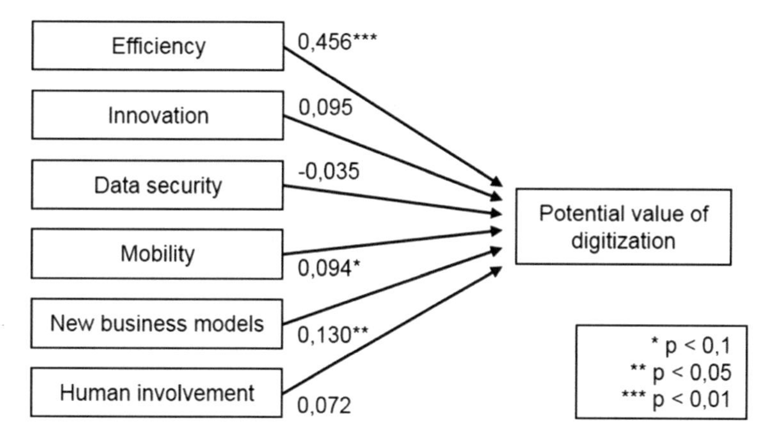

Fig. 2. Structural equation model with coefficients

The influence of enhanced innovation processes on the potential value of digitization was discussed in Hypothesis 2 (**An Improvement of Innovation Processes Positively Influences the Potential Value of Digitization**). With a value of +0.095, a positive influence could be proven. However, the influence is not significant ($p = 0.153 > 0.1$) which leads to the rejection of the hypothesis.

Hypothesis 3 (**Data Security Positively Influences the Potential Value of Digitization**) investigates the influence of data security on the potential value of digitization. Based on the results, the hypothesis had to be rejected due to a negative value (−0.035). A possible reason for the negative value could be attributed to companies estimating their data privacy adequately and, as a result, not feeling threatened by the emerging risks of digitization.

The evaluation of Hypothesis 4 (**An Improvement of Mobility Positively Influences the Potential Value of Digitization**) shows a positive (+0.094) and significant ($p = 0.098 < 0.1$) effect of enhanced mobility on the potential value of digitization. Hence, the hypothesis can be confirmed. In addition, the survey revealed that more than 89% of the companies are using mobile computing applications. Furthermore, about 70% are using cloud computing. Companies are able to increase their mobility by using these technologies.

The influence of new business models on the potential value of digitization was discussed in Hypothesis 5 (**The Generation of New Business Models Positively Influences the Potential Value of Digitization**). In accordance to the SEM results, a positive (+0.130) and significant effect ($p = 0.048 < 0.05$) could be detected. Therefore, the hypothesis can be confirmed as well. The survey further showed that more than 50% of the participants have already changed their own business model because of the digital transformation while others (22%) are planning to change their business model in the future.

The results of Hypothesis 6 (**Human Involvement Positively Influences the Potential Value of Digitization**) show a positive effect (+0.072) of human involvement on the potential value of digitization. The influence, however, is not significant ($p = 0.272 > 0.1$). Therefore, the hypothesis must be rejected. More than 63% of the participants thought that employees show a medium or high level of acceptance regarding the emerging digital technologies. More than 51% estimated that employees will not lose their jobs because of the digitization.

Important values of the SEM regarding path coefficients, t-statistics and p-values are shown in Table 3. For single-item sets, which are widely used in research, there is no need to calculate metrics like Cronbach's alpha [36]. The potential of digitization is shortened as PD.

Table 3. SEM coefficients

Hypothesis	SEM-Path	Path coefficient	T Statistic	Significance (P Values)
H1	Efficiency → PD	+0.456	6.108	0.000
H2	Innovation → PD	+0.095	1.430	0.153
H3	Data Security → PD	−0.035	0.626	0.531
H4	Mobility → PD	+0.094	1.657	0.098
H5	New business models → PD	+0.130	1.982	0.048
H6	Human involvement → PD	+0.072	1.099	0.272

5 Conclusion

In future, the digital transformation will continue to change business in all areas and sectors worldwide [37]. For companies, it is inevitable to adapt to the far-reaching changes caused by digitization. The term digitization already appeared in the course of the third industrial revolution and was seen as a conversion process from analog to digital. It is associated with smart value creation processes using capable information and communication technologies. Thus, digitization aspects become far more complex and definitions regarding the term of digitization have changed at a tearing pace.

In the course of this, firms have to decide on their own how to benefit from new technologies caused by digitization. The results within this research, however, empirically show six hypotheses (*efficiency, innovation, data privacy, mobility, new business models and human integration*) which can be used to underline the potential value of digitization aspects within firms. Thus, digitization can help companies to significantly increase their *efficiency*. In addition, *new business models* arise through digitization and enable new possibilities to satisfy changing customer requirements. Furthermore, companies can enhance their *mobility* by using digital technologies like mobile computing and cloud computing. Despite the increasing numbers of cyber-attacks, like data theft and sabotage, it is surprising that *data security* has no significant influence on the potential value of digitization within this examination. Although the improvement of *innovation* processes, as well as *human integration* both do not have a significant influence on the potential value of digitization within this study, the overall research results show tremendous arguments to integrate digitization technologies within business processes.

As with all studies, there are some limitations to consider. The sample size, the investigation period from June to August 2016 and the geographical area of German-speaking countries limit the research. In addition, the research only investigated the direct effect of determinants found through literature on the cross-industry potential value of digitization. The authors recommend to extend the model for future research by operationalizing control variables in order to shrink possible disturbing errors. For instance, it might be possible that firm and/or industry specifics effect the potential value of digitization.

Regarding the theoretical implications of this study, scientists should finally extend these parameters for further investigations in order to optimize the measurement model. In particular, it would be useful to focus on further quality criteria as mentioned to evaluate the model with respect to reliability and validity. Further exploration of this study may assist in the intention of a better understanding regarding the term digitization and the main potentials of digitization for business. With respect to practical implications, managers as well as (IT) consultants can use these study results to argue why to integrate digitization aspects in order to optimize business processes and to enhance business strategies.

Appendix

See Table 4.

Table 4. Measures, measurement, composite reliability (CR), Average Variance Extracted (AVE), and Cronbach's Alpha (CA)

Measures	Measurement	CR	AVE	CA
Efficiency *How do you estimate the potential benefits of digitalization with regard to increased efficiency in companies generally?*	Single item	n/a	n/a	n/a
Innovation *How do you assess the potential benefits of digitization for optimizing innovation processes in companies in general?*	Single item	n/a	n/a	n/a
Data security How high do you estimate the following risks of digitization in companies in general? • sabotage • data theft	Formative measurement	.93	.87	.86
Mobility *How do you estimate the potential benefits of digitalisation with regard to increase the mobility of companies in general?*	Single item	n/a	n/a	n/a

(continued)

Table 4. (*continued*)

Measures	Measurement	CR	AVE	CA
New business models *How do you estimate the potential benefits of digitalization in terms of the emergence of new business models?*	Single item	n/a	n/a	n/a
Human involvement *How do you estimate the acceptance of employees with regard to the emerging digital technologies in your company?*	Single item	n/a	n/a	n/a
Potential value of digitization *How do you generally assess the potential benefits of digitization in companies?*	Single item	n/a	n/a	n/a

References

1. Prins, C., Broeders, D., Griffioen, H., Keizer, A.-G., Keymolen, E.: iGovernment. Amsterdam University Press, Amsterdam (2011)
2. Boes, A.: Dienstleistung in der digitalen Gesellschaft: Beiträge zur Dienstleistungstagung des BMBF im Wissenschaftsjahr 2014. Campus, Frankfurt am Main (2014)
3. Bundesministerium für Wirtschaft und Energie: Zukunftschance
4. Digitalisierung – Ein Wegweiser. http://www.mittelstand-digital.de/MD/Redaktion/DE/PDF/broschuere-zukunftschance-digitalisierung,property=pdf,bereich=md,sprache=de,rwb=true.pdf. Accessed 21 July 2017
5. Härting, R., Reichstein, C., Jozinović, P.: The potential value of digitization for business – insights from german-speaking experts. In: Eibl, M., Gaedke, M. (eds.) Informatik 2017, vol. 47, Jahrestagung der Gesellschaft für Informatik. Lecture Notes in Informatics (LNI), Gesellschaft für Informatik, Bonn (2017)
6. Härting, R.: Elektronischer Geschäftsverkehr aus Sicht privater Haushalte. Gabler, Wiesbaden (2000)
7. Schmidt, R., Möhring M., Härting, R., Reichstein, C., Neumaier, P., Jozinović, J.: Industry 4.0 – potentials for creating smart products: empirical research results. In: Abramowicz, W., Kokkinaki, A. (eds.) 2015 18th International Conference on Business Information Systems. Lecture Notes in Business Information Processing, vol. 208, pp. 16–27. Springer (2015)
8. Härting, R., Schmidt, R., Möhring, M., Reichstein, C., Neumaier, P., Jozinović, P.: Nutzenpotenziale von Industrie 4.0: Einblicke in aktuelle Studienergebnisse. Books on Demand, Norderstedt (2015)
9. Loebbecke, C., Picot, A.: Reflections on societal and business model transformation arising from digitization and big data analytics: a research agenda. J. Strateg. Inf. Syst. **24**(3), 149–157 (2015)
10. Kagermann, H.: Change through digitization – value creation in the age of industry 4.0. In: Albach, H., Meffert, H., Pinkwart, A., Reichwald, R. (eds.) Management of Permanent Change, pp. 23–45. Springer, Wiesbaden (2015)
11. Tilson, D., Lyytinen, K., Sørensen, C.: Digital infrastructures: the missing is research agenda. Inf. Syst. Res. **21**(4), 748–759 (2010)
12. Schröder, C., Schlepphorst, S., Kay, R.: Relevance of the Digitalization for the German Mittelstand. (IfM), Bonn (2015)

13. Kuhlmann, M., Schumann, M.: Digitalisierung fordert Demokratisierung der Arbeitswelt heraus. In: Hoffman, R., Bogedan C. (eds.) Arbeit der Zukunft: Möglichkeiten nutzen-Grenzen setzen, pp. 122–140. Frankfurt, New York (2015)
14. Brettreich-Teichmann, W., Grötecke, J.: Wohin treibt die Digitalisierung die IT-Organisation? HMD Praxis der Wirtschaftsinformatik **53**(1), 1–2 (2016)
15. Weber, A.: Die Digitalisierung des Kulturerbes der Deutschen aus dem östlichen Europa. https://opus4.kobv.de/opus4-bib-info/frontdoor/index/index/docId/1579. Accessed 23 July 2017
16. Uskov, V., Howlett, R.J., Jain, L.C.: Smart Education and e-Learning. Springer, Wien (2016)
17. Chun, S.A., Shulman, S., Sandoval, R., Hovy, E.: Government 2.0: Making connections between citizens, data and government. Inf. Polity **15**(1), 1–9 (2010)
18. Bertalanffy, L.V.: General System Theory. George Braziller, New York (1968)
19. Roland, K.: Modelling ontology use for information systems. Springer, Wien (1973)
20. Härting, R., Mohl, M., Bader, S.: Digitalisierung als Treiber für Marketing 4.0. In: Härting, R. (ed.) Industrie 4.0 und Digitalisierung – Innovative Geschäftsmodelle wagen! Tagungsband, 8. Transfertag, Aalen 2016, pp. 134–150. Books on Demand, Norderstedt (2016)
21. Agarwal, R., Gao, G., DesRoches, C., Jha, A.K.: The digital transformation of healthcare: current status and the road ahead. Inf. Syst. Res. **21**(4), 796–809 (2010)
22. Nambisan, S.: Information technology and product/service innovation: a brief assessment and some suggestions for future research. J. Assoc. Inf. Syst. **14**, 215–226 (2013)
23. Dellarocas, C.: The digitization of word of mouth: promise and challenges of online feedback mechanisms. Manag. Sci. **49**(10), 1407–1424 (2003)
24. Lusch, R.F., Vargo, S.L., Tanniru, M.: Service, value networks and learning. J. Acad. Market. Sci. **38**(1), 19–31 (2010)
25. Keuper, F., Hamidian, K., Verwaayen, E., Kalinowski, T., Kraijo, C.: Digitalisierung und Innovation: Planung – Entstehung – Entwicklungsperspektiven. Springer Gabler, Wiesbaden (2013)
26. McAfee, A., Brynjolfsson, E., Davenport, T.H., Patil, D.J., Barton, D.: Big data. Manag. Revol. Harvard Bus. Rev. **90**(10), 61–67 (2012)
27. Knoll, M., Meinhardt, S.: Mobile Computing: Grundlagen – Prozesse und Plattformen – Branchen und Anwendungsszenarien. Springer Vieweg, Wiesbaden (2016)
28. Sabbagh, K., Friedrich, R., El-Darwiche, B., Singh, M., Koster, A.: Digitization for economic growth and job creation: Regional and industry perspective. The global information technology report, pp. 35–42 (2013)
29. Frey, C.B., Osborne, M.A.: http://www.oxfordmartin.ox.ac.uk/downloads/academic/The_Future_of_Employment.pdf. Accessed 24 July 2017
30. Babbie, E.: The Practice of Social Research. Cengage Learning, Boston (2012)
31. LimeSurvey GmbH: https://www.limesurvey.org/de/. Accessed 21 May 2017
32. Hooper, D., Coughlan, J., Mullen, M.: Structural equation modelling: guidelines for determining model fit. Electron. J. Bus. Res. Meth. **6**(1), 53–60 (2008)
33. Wong, K.K.-K.: Partial least squares structural equation modeling (PLS-SEM) techniques using SmartPLS. Mark. Bull. **24**, 1–32 (2013)
34. Chin, W.W.: The partial least squares approach to structural equation modeling. Mod. Meth. Bus. Res. **295**, 295–336 (1998)
35. Hewson, C., Yule, P., Laurent, D., Vogel, C.: Internet research methods: a practical guide for the social and behavioural sciences. Sage, London, Thousand Oaks, New Dehli (2003)
36. Ringle, C.M., Sarstedt, M., Straub, D.: A critical look at the use of PLS-SEM in MIS Quarterly. MIS Q. **36**(1), iii–xiv (2012)

37. PricewaterhouseCoopers AG: Die Digitalisierung verändert Unternehmen welt-weit und branchenübergreifend. http://www.pwc.de/de/digitale-transfor-mation/die-digitalisierung-veraendert-unternehmen-weltweit-und-branchenueber-greifend.html. Accessed 21 Jan 2017
38. Homburg, C., Baumgartner, H.: Beurteilung von Kausalmodellen: Bestandsaufnahme und Anwendungsempfehlungen, Marketing – Zeitschrift für Forschung und Praxis, **17**(3), pp. 162–176 (1995)
39. Sen, R., Borle, S.: Estimating the contextual risk of data breach: an empirical approach. J. Manag. Inf. Syst. **32**(2), 314–341 (2015)
40. Kagermann, H.: Change through digitization—Value creation in the age of Industry 4.0. In: Albach, H., Meffert, H., Pinkwart, A., Reichwald, R. (eds.) Management of Permanent Change, pp. 23–45. Springer, Gabler, Wiesbaden (2015)
41. Bagozzi, R.P., Yi, Y.: On the evaluation of structural equation models. J. Acad. Market. Sci. **16**(1), 74–94 (1988)

A Method of Knowledge Extraction for Response to Rapid Technological Change with Link Mining

Masashi Shibata(✉) and Masakazu Takahashi

Graduate School of Sciences and Technology for Innovation, Yamaguchi University, Yamaguchi, Japan
g501wc@yamaguchi-u.ac.jp

Abstract. This paper proposes an efficient clustering method of technology fields for future technical trend prediction from the public information. The speed of product development is steadily improving due to the spreading of ICT as social infrastructure and the rapid progress of machine learning. However, in the process of finding a new solution to the problem, the developer's capability is still an important. Referring to the problems and solutions to other technical fields with similar technology structure is one of the effective ways to find new solutions. However, selection of comparative fields often depends on the technical preferences and experience of developers. Thus, important signals might be overlooked. In this paper, we focus on the classification codes of patent for extracting the technology structure from the patent data. The link mining method is employed for visualizing the structure and extracting the feature. The structure is visualized as the graph of classification codes, and the feature is extracted as the features of the graph. From the result of the proposed method, we succeeded to reveal the cluster of the technology fields with similar technology structure.

Keywords: Patent · Patent analyses · Graph mining · Clustering Technological structure analyses

1 Introduction

There are various methods to capture the technological structure of the technological field, such as the market data analyses, gathering information from the technical magazines and the patent information, and comparing with closer technological fields. Since these methods are based on their personal heuristics knowledge, some important signals might be overlooked. Various kinds of technologies exist, even if they aim to the same function. Therefore, it is difficult to estimate potential technology.

A patent information is usually used for analyzing the technology. The patent information not only contains of the technological principles, but is public opened for easy to get. Furthermore, Patent information is also making use of formulating a corporate strategy, such as analyzing technological trends of competitors or evaluating market relevance of new businesses. Thus, Patent information makes possible to generate a wide range of knowledge.

G. Jezic et al. (Eds.): KES-AMSTA-18 2018, SIST 96, pp. 299–310, 2019.
https://doi.org/10.1007/978-3-319-92031-3_30

The patent electronic library was set up by INPIT (The National Centre for Industrial Property Information and Training) in Japan. Technical analysis can be achieved using the information in the collections from INPIT. However, the information stored on this site, the following restrictions are applied; (1) Data is stored from 1993, (2) Since there is no thesaurus, selection of a technical term is difficult. (3) The search indexed by company name is also difficult. Since above the dataset condition, it is difficult to extract the appropriate target search results. Therefore, we propose an efficient method for extracting knowledge from patent information in this paper.

The rest of the paper is organized as follows; Sect. 2 discusses the backgrounds of the research and related work; Sect. 3 briefly summarizes the gathered data on the target patent sector; Sect. 4 describes the analytics of the data and presents analytical results; and Sect. 5 gives some concluding remarks and future work.

2 Related Work

Many technical analysis methods with the patent were made so far [1–3]. They are performed for the business solution.

TRIZ (Teoriya Resheniya Izobretatelskikh Zadatch) is one of the method making use of the patents. This aims for technical development based on the structure of problem solving which appears repeatedly in the patent [4, 5]. Kawakami et al. proposed a support system for the idea gain aid of TRIZ [6].

KT (Kepner-Tregoe) method also aims for the problem solving, not achieved compared to similar successful case in the past. The successful examples were analyzed from the case of NASA, extracting the common principle was found out, then systematized problem solving process and decision-making. The merits of KT method are enumerated can be formally causative factor of the problem. On the other hand, the disadvantage is needed the expertise for solutions creation [7, 8].

Then, we focus on the patent analytical skills. One of the techniques for patent information analysis is the patent mapping. Kiriyama made content analyses with this method [9]. Shide et al. performed finding the change of the positioning for customer of research and development activities of the company with the patents analyses [10]. Kimura proposed technology evaluation method based on patent analysis for technology strategy planning [11].

As for the analysis of important information derived from patent analysis, Carpenter analyzed for important cited patents. Muguruma showed the validity of the patent citation analysis to propose the FCA (Forward Citation Applicant) map [12, 13]. Sato et al. proposed the importance calculation method of the patent document based on the citation information [14]. Ogawa et al. also proposed a basic patent extraction based on the citation information [15]. For citations, Albert conducted a validation of citation for important patent among the industry [16].

As for the patent classification in place of manpower, Tanaka proposed method of extracting the feature automatically [17]. Yamashita proposed a method of surveillance technology and specific method of patent classification with text mining [18]. Yamamoto et al. proposed a method to enhance the compatibility of the search by applying the information of related patent documents in search of academic papers.

Yamamoto, proposed a method to find the scientific papers with a variety of further information [19, 20]. Kleinberg extracted the topic and description of the relationship with graph theory [21]. Eto proposed a measure of co-citation based on structural units of the paper [22]. Ueda proposed the technical analysis with an active mining method focuses on the cognitive processes of the patent examiner, with patent classification such as IPC (International Patent Classification), FI (File Index), and F-term (File Forming Term) [23].

Then, we focus on the mining technology for knowledge extraction. The relationship of analytical methods and technologies, technology analysis method with patent information was described [24–26].

Graph theory as a structuring method was applied to many fields. For example, Chemical formula, WWW (World Wide Network), Social network, Statements with grammatical structure and dependency. Since the graph is made for expressing the structure of the object, the relationship can be described. Gettor proposed a method of the partial graph occurs frequently from the set of graphs [27]. Therefore, with the application of the technology of intelligent informatics, knowledge extraction was performed from patent information by graph mining method in this paper.

3 Experimental Configuration

Patent information consists of metadata, such as application date, applicant, classification codes, literal information and graphic information. In addition to the IPC, FI and F-term are used as the classification codes domestically in the Japanese patent classification system.

IPC indicates the technological fields of the main topics of the patent's claims. FI is used to indicate patent's technological fields more detail than IPC by adding extension symbol and/or file discrimination symbol to the IPC.

F-term is used to indicate patent's technological fields more detail than FI by adding the viewpoint of the problem and the solution. F-term is separated to a theme code and a viewpoint [28]. Figure 1 shows an example of the F-term's notation.

The theme code represents the technological field. The view point analyses the theme, such as material, purpose, operation, and manufacturing. The figure subdivides the viewpoint. A theme code is composed of 1-digit number, 1 alphabetic character, and 3-digit number. The first digit number takes the value of 2, 3, 4, and 5, and they represent residual technology, mechanics, chemistry, and electricity respectively. The first 2 letters of a theme code form a theme group, that group similar technological field.

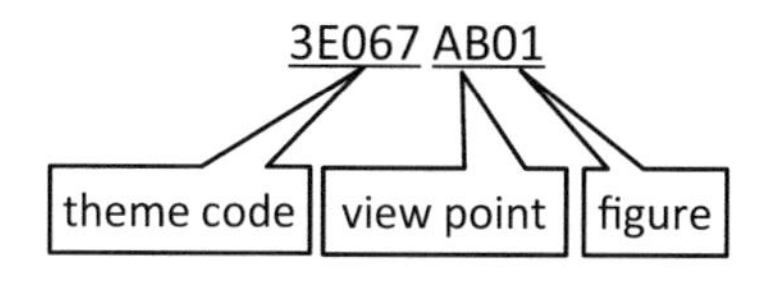

Fig. 1. F-term's notation

We regarded the 38 categories classified as theme groups as technological fields. We have selected one theme code from each theme group as the subject of the analysis. Table 1 shows the excerpted list of selected theme codes and their description. Since the F-term is reviewed annually, the theme code of a certain technological field may change during the experiment period. Thus, some theme groups have more than one theme code. The numbers of the theme codes for the analysis by each first digit number are 9 for "2", 10 for "3", 11 for "4", and 9 for "5". We gathered the target data from the patent information database [29].

We focused on patent publication since it is ensured inventive step and inquiry by inventive examining. We gathered the data submitted from 2007 to 2009. This is because we also focus the technological structure effect on the global financial crisis in 2008. Based on the above condition, 26,705 patent data are gathered from the database. Table 2 shows the application number for each theme group by the year of application.

Table 1. Theme codes list (Excerpted)

	Theme group description		Theme code description
2B	Natural resources	2B022	Cultivation of plants
2F	Measurement	2F003, 2F103	Optical transform
2G	Applied physics	2G001	Analyzing materials by the use of radiation
2K	Light-device	2K103, 2K203	Projection apparatus
3B	Textile-packing machine	3B006	Electric suction cleaners
3C	Production machinery	3C001	Automatic control of machine tools
3D	Transportation	3D001, 3D301	Chassis suspension devices
3F	Conveyance	3F001, 3F343	Sheets, magazines, and separation thereof
3G	Power machinery	3G001, 3G301	Electrical control of air or fuel supplied to internal-combustion engine
3J	General machinery	3J001	Connection of plates
4B	Biotechnology	4B004	Coffee makers
4C	Medical care	4C001, 4C301, 4C601	Ultrasonic diagnostic devices
4F	Plastics engineering	4F040, 4F041, 4F042	Coating apparatus 1 (contact or immersion)
5C	Video system	5C001	Electron microscopes
5D	Electric equipment	5D002	Stringed musical instruments
5L	Electronic commerce	5L096	Image analysis

Table 2. Application number for each theme group by year

Theme group	Theme code	The number of application		
		2007	2008	2009
2B	2B022	148	164	185
2B	2B104	81	88	87
2C	2C001	561	614	578
2D	2D001	61	58	54
2E	2E013	3	11	13
2F	2F003, 2F103	127	97	95
2G	2G001	295	269	273
2H	2H006	183	192	189
2K	2K103, 2K203	284	156	614
3B	3B006	61	60	48
3C	3C001	43	34	51
3D	3D001, 3D301	290	295	237
3E	3E001, 3E040	374	441	435
3F	3F001, 3F343	316	314	308
3G	3G001, 3G301	796	783	706
3H	3H001	5	5	6
3J	3J001	140	172	155
3K	3K001, 3K092	108	131	114
3L	3L015, 3L018, 3L345	128	112	90
4B	4B004	106	135	105
4C	4C001, 4C301, 4C601	501	499	502
4D	4D001, 4D006	484	474	486
4E	4E001, 4E078, 4E079	96	118	152
4F	4F040, 4F041, 4F042	590	516	535
4G	4G001	87	80	61
4H	4H001	209	205	212
4J	4J029	271	286	321
4K	4K001	223	268	297
4L	4L031	158	152	125
4M	4M118	523	524	606
5C	5C001	67	78	71
5D	5D002	9	13	10
5E	5E001	239	248	201
5F	5F001, 5F040, 5F140	321	303	273
5G	5G001	10	10	8
5H	5H001	17	13	10
5J	5J001	60	52	41
5L	5L096	568	708	692
5M	5M024, 5B024	251	177	110

In this study, the technological structure of a category is defined by the contained technological fields and their relations. It is represented by graph structure. The theme codes which are given to patent represent the technological fields and are used as nodes of the graphs. First, for representing the technological structure of a patent, a complete graph is created for each patent.

The graph is unweighted and undirected graph. Both the number of the node appearance and the number of the link appearance are counted one for each existence node and link in the graph. Figure 2 shows an example of the complete graph of each patent. Then, annual graphs for each theme group are created. First, overlaying the complete graphs of the patents of the theme group in each year of application. The number of link appearance is accumulated each time they appear. Thus, the graphs are weighted undirected graph, and are ego-centric graph. "Ego" means a node connecting the all other nodes in the graph. In this experiment, the selected theme code of the theme group becomes ego. Then, the ego and its links are excluded from the annual graphs. This is because the links connected to other than ego are regarded as representing the technological structure. Figure 3 shows an example of the excluding ego and its links. As a result, 117 graphs are created. They represent annual technological structure for each technological field.

As elements of the feature vector of the created graphs, the number of links between the first digits 2, 3, 4, and 5 are extracted. As a result, each graph's feature vector is ten-dimensional. Table 3 shows the list of the feature vector of the graphs. K-means method is used as the classifier. First, the graphs are classified into 8 classes. Then, the biggest cluster is further classified into 12 classes. Finally, 117 graphs are classified into 19 classes.

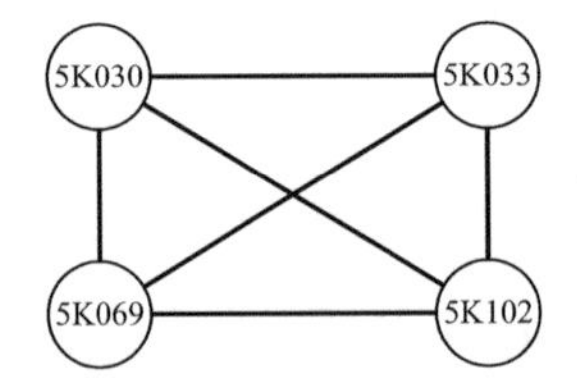

Fig. 2. Complete graph of each patent

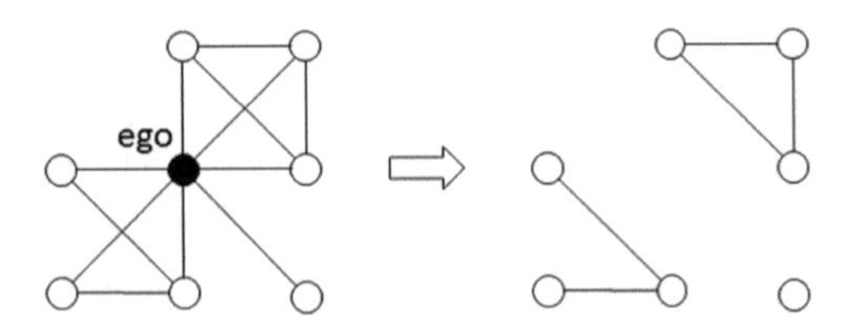

Fig. 3. Exclude ego and its links

Table 3. List of feature vector (Excerpted)

Year	Theme code	2-2	2-3	2-4	2-5	3-3	3-4	3-5	4-4	4-5	5-5
2007	2B022	66	16	74	0	6	14	0	128	0	0
2007	2D001	10	0	10	8	0	0	0	14	0	12
2007	2G001	114	2	98	100	2	0	0	30	54	34
2007	3B006	0	0	0	0	8	8	26	2	0	4
2007	3H001	0	0	0	0	6	0	0	0	0	0
2007	4B004	0	0	0	0	14	130	0	56	0	0
2007	4H001	84	12	32	72	0	10	86	58	86	768
2007	4L031	18	0	50	0	34	148	6	776	32	2
2007	5C001	20	6	18	66	4	0	14	0	30	102
2007	5G001	0	0	0	0	0	0	0	0	0	4
2007	5L096	42	16	14	276	22	2	82	54	64	1,140
2008	2B022	80	4	62	8	10	0	20	114	0	20
2008	2D001	4	2	6	4	0	6	2	20	6	0
2008	2G001	144	2	98	140	0	6	6	38	60	60
2008	3B006	0	0	0	0	20	0	16	0	0	4
2008	3H001	0	0	0	0	8	0	6	0	0	0
2008	4B004	2	0	4	0	34	178	2	120	4	0
2008	4H001	50	8	68	16	0	0	92	46	56	740
2008	4L031	2	2	50	0	14	134	4	722	38	4
2008	5C001	16	4	18	70	2	0	2	8	20	56
2008	5G001	0	0	0	0	0	0	0	0	0	6
2008	5L096	66	36	22	306	30	4	164	46	146	1,458
2009	2B022	142	24	124	0	14	6	4	148	0	2
2009	2D001	12	10	2	4	0	4	2	0	2	2
2009	2G001	162	2	76	78	0	2	0	28	30	42
2009	3B006	0	8	2	0	16	4	8	0	0	2
2009	3H001	0	0	0	0	2	0	6	0	0	0
2009	4B004	0	0	0	0	28	102	0	112	2	0
2009	4H001	62	10	66	78	2	28	52	106	68	402
2009	4L031	8	0	34	0	2	64	2	376	10	4
2009	5C001	12	0	6	48	0	0	0	0	14	76
2009	5G001	0	0	0	0	0	0	0	0	0	2
2009	5L096	50	24	18	204	26	8	120	10	60	1,188

4 Experimental Results

This chapter describes the result of the technological structure clustering of the target data using link mining.

Table 4 describes the result of the classification. In the theme groups which is assigned by multiple theme codes, only representative theme code is described. Overall, the most of the graphs of the same theme group are clustered into the same

Table 4. Clustering results

Cluster	Year	Theme code	Cluster	Year	Theme code	Cluster	Year	Theme code
0	2007	2K103	9	2008	2H006	12	2007	4L031
	2008	2K103		2007	4J029		2008	4L031
	2007	4M118		2008	4J029	13	2007	4C001
	2008	4M118		2009	4J029		2008	4C001
	2009	4M118	10	2007	2C001		2009	4C001
2	2007	3G001		2008	2C001	14	2007	5E001
	2008	3G001		2009	2C001		2008	5E001
	2009	3G001		2009	4H001		2009	5E001
3	2007	4F040		2009	5M024	15	2009	2H006
	2008	4F040	11	2007	2D001		2007	4E001
	2009	4F040		2008	2D001		2008	4E001
4	2009	2K103		2009	2D001		2009	4E001
5	2007	4H001		2007	2E013		2007	4K001
	2008	4H001		2008	2E013		2008	4K001
	2007	5F001		2009	2E013		2009	4K001
	2008	5F001		2007	3B006		2009	4L031
	2009	5F001		2008	3B006	16	2009	3C001
	2007	5L096		2009	3B006		2007	3J001
	2008	5L096		2007	3C001		2008	3J001
	2009	5L096		2008	3C001		2009	3J001
	2007	5M024		2007	3H001		2007	3K001
	2008	5M024		2008	3H001		2008	3K001
6	2007	4D001		2009	3H001		2009	3K001
	2008	4D001		2009	4G001		2007	3L015
	2009	4D001		2007	5C001		2008	3L015
7	2007	3D001		2008	5C001		2009	3L015
	2008	3D001		2009	5C001	17	2007	2F003
	2009	3D001		2007	5D002		2008	2F003
	2007	3E001		2008	5D002		2009	2F003
	2008	3E001		2009	5D002	18	2007	2G001
	2009	3E001		2007	5G001		2008	2G001
	2007	3F001		2008	5G001		2009	2G001
	2008	3F001		2009	5G001	19	2007	2B022
	2009	3F001		2007	5H001		2008	2B022
8	2007	4B004		2008	5H001		2009	2B022
	2008	4B004		2009	5H001		2007	2B104
	2009	4B004		2007	5J001		2008	2B104
	2007	4G001		2008	5J001		2009	2B104
	2008	4G001		2009	5J001		2007	2H006

cluster even though the year of application is different. It is confirmed that the global financial crisis did not affect the technology structure. Cluster1 is the biggest cluster among the result of the first clustering. It contains 81 out of 117 graphs. Thus, it is further divided into 12 classes numbered 8 to 19. Table 5 describes the coordinates of the center of the clusters. It contains the coordinate of cluster1 for reference.

Cluster0 consists of 2K103 and 4M118. They have many connections within the first digit number 2: residual technology and 5: electricity, and between 2 and 5. This means these 2 theme codes' patents are deeply related to these 2 categories.

Cluster2 only consists of 3G001. It has many connections within the first digit number 3: mechanics. And it has a close relation to 4: chemical, and 5. Thus, it is closely related these 3 categories. And it is confirmed that it is actively developed by the number of the links.

Cluster3 only consists of 4F040. It has many connections between 2 and 4, and within 4. Thus, it is closely related to these 2 categories. And it is confirmed that it is actively developed by the number of the links.

Cluster4 only consists of 2K103 in 2009. It has many connections within 2 and 5, and between 2 and 5. Thus, it is closely related to these 2 categories. And comparing with cluster0, the number of connections increases.

Cluster5 consists of 4H001, 5F001, 5L096, and 5M024. It has many connections within 5. Thus, it is mainly related to electricity.

Cluster6 only consists of 4D001. It has many connections within 4. Thus, it is mainly related to chemistry and is not very related to other fields.

Cluster7 consists of 3D001, 3E001, and 3F001. It has relatively many connections within 3, but the number of the links is low, given the number of the applications. Thus, it is considered that the patent of this cluster covers a small area.

Cluster8 consists of 4B004 and 4G001. It has relatively many connections between 3 and 4, and within 4, but the number of links is very small. Thus, it is considered that these areas are not actively developed.

Cluster9 consists of 2H006 and 4J029. It has relatively many connections within 4. Thus, it is mainly related to chemistry and is not very related to other fields.

Cluster10 consists of 2C001, 4H001 in 2009, and 5M024 in 2009. It has relatively many connections within 5. Thus, it is mainly related to electricity and is not very related to other fields.

Cluster11 is the largest cluster. It consists of 2D001, 2E013, 3B006, 3C001, 3H001, 4G001, 5C001, 5D002, 5G001, 5H001, and 5J001. The number of the application of the member of this cluster is very small. This is why these categories fall into the cluster. Thus, it is considered that these areas are not actively developed.

Cluster12 only consists of 4L031. It has many connections within 4, and compared with it, it has quite small connections with other combinations. Thus, it is mainly related to chemistry and is not very related to other fields.

Cluster13 consists of 4C001. It has many connections within 4, and between 4 and 2, and 4 and 5. Thus, it has the relation with residual technology and electricity.

Cluster14 only consists of 5E001. It has many connections within 5, and between 5 and 4, and has quite small connections with other combinations. Thus, it is related to chemistry and electricity.

Table 5. Cluster centre coordinates

Cluster	2-2	2-3	2-4	2-5	3-3	3-4	3-5	4-4	4-5	5-5
0	930.40	61.60	92.40	812.00	14.80	1.20	28.00	9.60	76.40	607.20
(1)	34.70	5.76	31.47	16.53	27.57	26.10	8.22	131.06	36.55	57.69
2	1.33	40.00	0.00	0.00	5723.33	812.00	288.67	146.67	5.33	32.67
3	440.00	97.33	1252.67	280.00	55.33	471.33	54.00	1476.00	703.33	268.67
4	3980.00	600.00	46.00	3244.00	78.00	0.00	224.00	2.00	4.00	1432.00
5	31.60	11.60	18.20	117.60	7.80	3.00	59.60	25.20	287.80	1270.60
6	18.67	15.33	128.67	3.33	26.67	174.67	20.67	2412.67	328.67	96.67
7	54.22	156.89	1.56	24.89	639.11	16.22	173.11	2.89	0.89	50.89
8	1.20	0.00	1.60	1.60	28.80	136.40	0.80	119.60	12.40	19.20
9	160.50	3.00	187.50	18.50	28.00	32.50	1.50	557.50	51.00	16.00
10	27.20	3.60	16.40	53.20	1.60	6.00	12.80	22.00	22.00	491.60
11	3.67	1.13	2.60	8.20	7.53	2.13	3.60	2.93	3.07	19.67
12	10.00	1.00	50.00	0.00	24.00	141.00	5.00	749.00	35.00	3.00
13	20.00	1.33	173.33	42.67	0.00	0.67	1.33	658.00	332.67	49.33
14	34.67	0.00	0.00	25.33	1.33	0.00	3.33	62.67	348.00	322.67
15	13.00	0.50	29.75	2.25	7.25	39.25	0.50	354.25	22.25	10.75
16	3.40	28.20	2.80	2.40	161.60	59.00	44.00	18.00	12.80	15.20
17	266.00	16.67	8.00	68.67	4.67	0.00	0.67	2.00	0.00	20.67
18	140.00	2.00	90.67	106.00	0.67	2.67	2.00	32.00	48.00	45.33
19	63.71	9.43	73.14	1.14	8.00	9.14	3.43	153.71	0.57	3.14

Cluster15 consists of 2H006 in 2009, 4E001, 4K001, and 4L031 in 2009. It has relatively many connections within 4, and has small connections with other combinations. The number of the links is low, compared with the number of the applications. Thus, it is considered that the patent of this cluster covers a small area.

Cluster16 consists of 3C001 in 2009, 3J001, 3K001, and 3L015. It has relatively many connections within 3. And the number of the links is low, compared with the number of the applications. Thus, it is related to machinery, and is considered that the patent of this cluster covers a small area.

Cluster17 consists of 2F003. It has relatively many connections within 2. And both the number of links and the number of applications are small. Thus, it is related to residual technology and is considered that this area is not actively developed.

Cluster18 consists of 2G001. It has relatively many connections within 2, and between 2 and 4, and 2 and 5. Thus, it has relation to residual technology, chemistry, and electricity.

Cluster19 consists of 2B022, 2B104, and 2H006 in 2007. It has relatively many connections within 4. Thus, it has relation to chemistry.

5 Concluding Remarks

This paper presents the clustering of the 38 technological fields which are represented as the theme groups. We utilize the patent classification codes and link mining method to represent the technology structure of the technology areas. We create the graphs by patents' theme codes and reveal the feature of the technology areas. As a result, some different technological fields which have a similar technological structure are found. In addition, it was revealed that the global financial crisis had not influenced the technological structure.

Our future work is improving the clustering accuracy by optimizing the graphs' structure and their feature vector with appropriate classifier.

References

1. Rosenzweig, R.: The hazards of recombinant DNA, Stanford's patent application natural selection effects. Trends Biochem. Sci. **2**(4), 84 (1977)
2. Shuchman, H.L.: Engineers who patent: data from a recent survey of American bench engineers. World Patent Inf. **5**(3), 174–179 (1983)
3. Narin, F., Norma, E., Perry, R.: Patents as indicators of corporate technological strength. Res. Policy **16**(2–4), 143–155 (1987)
4. Altshuller, G.: The Innovation Algorithm: TRIZ, Systematic Innovation, and Technical Creativity. Technical Innovation Center, Worcester (1999)
5. Altshuller, G.: 40 Principles: Extended Edition. Technical Innovation Center, Worcester (2005)
6. Kawakami, H., Naito, K., Hiraoka, T., et al.: Idea generation support system for implementing benefit of inconvenience by employing the theory of inventive problem solving. Trans. Soc. Instrum. Control Eng. **49**(10), 911–917 (2013)

7. Kepner, C.H., Tregoe, B.B.: The Rational Manager: A Systematic Approach to Problem Solving and Decision-Making. McGraw Hill, New York (1976)
8. Kepner, C.H., Tregoe, B.B.: The New Rational Manager. Princeton Research Press, Princeton (1981)
9. Kiriyama, T.: IP information analysis (Patent information: analysis and effective utilization). J. Inf. Sci. Tech. Assoc. **60**(8), 306–312 (2010)
10. Shide, K., Ando, M.: The shift of positioning of Japanese general contractors' R&D activities. J. Archit. Plan. **76**(668), 1929–1935 (2011)
11. Kimura, H.: One approach of technology stocktaking and evaluation for corporate technology strategies: emphasizing future intentions and quantification through patent analysis. J. Sci. Policy Res. Manag. **26**(1/2), 52–61 (2012)
12. Carpenter, M.P.: Citation rated to technologically important patents. World Patent Inf. **3**(4), 160–163 (1981)
13. Muguruma, M.: The usefulness of patent forward citation analysis and its practical examples. J. Inf. Sci. Tech. Assoc. **56**(3), 114–119 (2006)
14. Sato, Y., Iwayama, M.: A study of patent document score using patent-specific attributes in citation analysis. J. Inf. Process. Manag. **51**(5), 334–344 (2008)
15. Ogawa, T., Watanabe, I.: Finding basic patents using patent citations. IPSJ SIG Notes, Inf. Process. Soc. Jpn. **35**, 41–48 (2005)
16. Albert, M.B., Avery, D., Narin, F., et al.: Direct validation of citation counts as indicators of industrially important patents. Res. Policy **20**(3), 251–259 (1991)
17. Tanaka, K.: Multi-viewpoint clustering of patent documents. IPSJ SIG Notes **4**, 9–14 (2008)
18. Yamashita, Y.: Text mining technology for patent analysis and patent search: patent search and patent analysis service patent integration. J. Inf. Process. Manag. **52**(10), 581–591 (2010)
19. Yamamoto, M., Maze, H., Yajima, T., et al.: A journal paper filtering using the profile revised by patent document information. IEEJ Trans. Electr. Inf. Syst. **130**(2), 358–366 (2010)
20. Yamamoto, M., Kinugawa, H.: A journal paper filtering using the multiple information. IEEJ Trans. Electr. Inf. Syst. **131**(6), 1250–1259 (2013)
21. Kleinberg, J.M.: Authoritative sources in a hyperlinked environment. J. ACM **46**(5), 604–632 (1999)
22. Eto, M.: A New Co-Citation Measure Based on Structures of Citing Papers, Database, 49 (SIG 7(TOD 37)), The Information Processing Society of Japan, pp. 1–15 (2008)
23. Ueda, I.: Active mining utilizing the patent classification IPC, F1, F term on the basis of the cognitive processes of the examiner. Proc. SIG-FAI, Jpn. Soc. Artif. Intell. **46**, 13–21 (2001)
24. Karamon, J., Matsuo, Y., Ishizuka, M.: Link mining from networks of academic papers. Tech. Rep. IEICE, KBSE **106**(473), 73–78 (2007)
25. Kashima, H.: Mining graphs and networks. J. Inst. Electr. Inf. Commun. Eng. **93**(9), 797–802 (2010)
26. Kajikawa, Y.: Utilization of citation information by link mining. J. Inf. Sci. Tech. Assoc. **60** (6), 224–229 (2010)
27. Gettor, L.: Link mining: a new data mining challenge. ACM SIGKDD Explor. Newsl. **5**(1), 84–89 (2003)
28. Outline of FI/F-term, by The Japan Patent Office. https://www.jpo.go.jp/torikumi_e/searchportal_e/pdf/classification/fi_f-term.pdf. Accessed 23 Jan 2018
29. YUPASS. http://www.yupass.jp. Accessed 23 Jan 2018

Agent-Based Gaming Approach for Electricity Markets

Setsuya Kurahashi(✉)

University of Tsukuba, Tokyo, Japan
kurahashi.setsuya.gf@u.tsukuba.ac.jp
http://www2.gssm.otsuka.tsukuba.ac.jp/staff/kurahasi/english.html

Abstract. The Electricity Market in Japan has been an oligopolistic market since the previous century, but it will be a liberalised competitive market soon due to a policy change. It is supposed to provide wholesale power markets. Therefore it has high possibilities to become two-sided markets with strong wholesalers. The two-sided markets have been researched using mathematical economics models recent years. The model, however, can only deal with one or two players on a market, therefore it has limitations to analyse more various players. On the other hand, many research projects of dynamic pricing and incentive mechanisms have been carried out on power markets. Some of the studies use agent-based modelling to analyse them as autonomous agents and optimised decision-making algorithms. These studies have shown interesting results, but they also have limitations to analyse further more complex markets and decision making processes of market players taking managing conditions into consideration. In this study, we adopt Agent-based gaming to analyse them on a two-sided electricity market.

Keywords: Electricity market · Two-sided market
Agent-based gaming

1 Introduction

The government of Japan has clearly announced that it would realise liberation for participation of power operators into small consumers such as general households in 2016. It would launch unbundling of power generation and distribution during the period around 2018 to 2020. These policies might bring about advancement of innovation with a wide variety of enterprises participating and increasing the use of renewable energy.

The purpose of this research is to achieve an efficient market while taking into consideration electricity market liberalisation. Additionally, this research studies incentive mechanisms for a competitive electricity markets for enabling energy transformation from fossil energy to renewable energy.

Through this research, by applying the agent-based gaming method, our goal is to propose an incentive design. It promotes innovation such as electricity

G. Jezic et al. (Eds.): KES-AMSTA-18 2018, SIST 96, pp. 311–320, 2019.
https://doi.org/10.1007/978-3-319-92031-3_31

supply and demand adjustments, stable supply, and dissemination of renewable energy through free decision-making by market participants including consumers and power operators. In a new liberalised energy market old and new energy companies will base their actions and plans on the behaviour of their competitors as well as on the (expected) responses of the consumer market. We propose using an agent based simulation of a market of consumers as a laboratory setting to study the behaviour of human decision-makers in an energy transition game.

2 Research Background

As clearly demonstrated by examples of the communications and Internet markets, the liberalisation of participation by many enterprises can create new markets while bringing about many benefits such as increase in business opportunities, diversification of services, and lowering of fees. On the other hand, leaving everything to free market competition prevents products and services with higher transaction cost from being transacted. It results in causing market failure.

On the other hand, in the ICT market which has two sides, consumers and suppliers, platform competitions are being developed on a global basis. These are attractive on the price side, the supply side, and the service side. This two-sided market mechanism has been analysed by using mathematical models [1,2]. In addition, recently, studies regarding real-time dynamic pricing based on agent modeling and studies regarding incentive mechanisms [3] have been made.

Smart grid is expected to gain profits from real-time dynamic pricing. This pricing system enables both power consumers and power companies to reflect changes in wholesale prices on the demand side [4]. Conversely, auction-based power pricing is not an uncommon concept. However, the demand side which participates in auction sessions is based on renewable energy such as solar energy. Therefore, electricity generated is very variable.

An important challenge here is the valid modelling of the population of agents in the model. Realistic agent behaviour is important to make an agent based game a tool that provides applicable insights [6].

In two-sided markets with consumers and suppliers, platform competitions are being developed on a global basis which are attractive on the price side, the supply side, and the service side. This two-sided market mechanism has been analysed by using mathematical models. However, mathematical models were applied to analyse market mechanisms with only one or two players [7–9]. Therefore, mathematical models have limitations in analysing mechanisms with multiple diversified players such as consumers. In addition, studies regarding ABM-based dynamic pricing and incentive mechanisms have been in progress. In these studies, however, the decision-making process of agents was controlled by an algorithm. For this reason, there are limitations in these studies to analyse complicated decision-making processes taking into account movements of actual environments, human behaviour and complex energy consumers markets, and corporate management conditions. Based on these traditional models, in this research, we made an attempt to build a two-sided market model for electricity markets by applying agent-based gaming.

Serious game sessions have been held in recent conferences regarding social simulation [5]. As for the traditional approaches of serious games, however, societies and environments which served as backgrounds were defined by game designers.

Afterward, chapter three describes the research objectives, and chapter four explains the outline of the energy conversion model. Chapter five gives a description of experimental environment and chapter six discusses the experimental results, while chapter seven summarises this research.

3 Research Objectives

The objective is to analyse what players can obtain market ascendancy under what kind of conditions in an electricity market. In order to achieve electricity platform design which maximises social welfare, this research focuses on aggregators and imbalance adjustment. Currently, utilisation of market functions associated with electricity supply and demand adjustment has been considered, with a proposal for establishing a new one-hour-ahead market and a real-time market in order for electricity distribution operators, power producers and retailers to procure the most efficient regulated power supplies from these markets (Ministry of Economy, Trade and Industry, 2013). Use of these market prices in imbalance settlement for renewable energy can secure transparency and fairness. This should have positive influence on the efficiency of electricity markets and the promotion of renewable energy dissemination (Fig. 1).

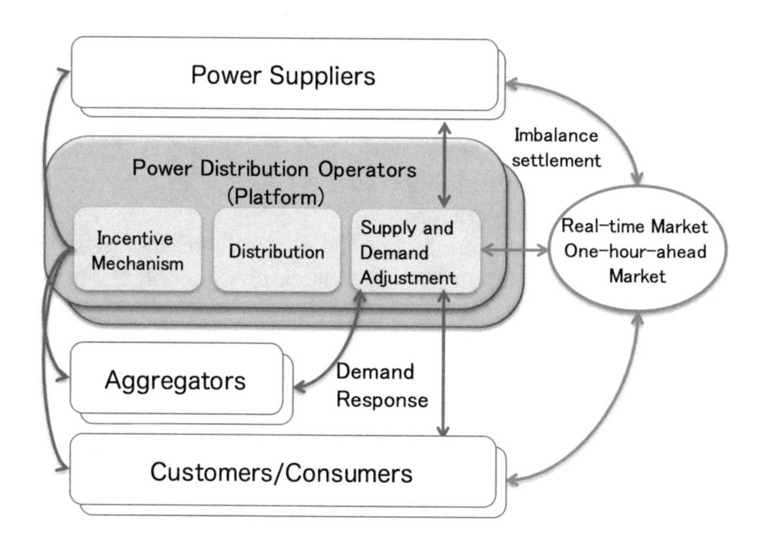

Fig. 1. Imbalance settlement and the electricity market

3.1 Analysis of Market Structure Which Brings About Energy Conversion

System design in electricity markets have a significant influence on generation of market rulers. Our additional goal is to design a system which is effective for energy conversion to renewable energy. Design of a mechanism for achieving stable electricity supply equilibrium based on utilisation of a wide variety of energy sources needs to play the role of a platform for maximising the utility for both electricity suppliers and consumers. In order to analyse these structures, we use ABM.

3.2 Comparative Analysis of Decision-Making Structures

While expanding electricity consumers and power producers to multiple agents, their behaviour is expressed by using a multi-agent model. With that, we conducted comparative analysis on the decision-making results obtained by introducing participatory agent-based gaming. By analysing differences brought by each individual agent, we evaluated strategies of imbalance adjustment incentives for electricity, and government subsidies and tax rate policies. In addition, observing the targeted phenomenon not only from a single viewpoint, but from several different viewpoints, in order that each phenomenon can be expressed accurately by using only one model [11].

4 Energy Conversion Gaming Model

In energy conversion gaming models based on agent-based gaming models (Fig. 2), in an electricity market where power producer players and aggregator players participate, power producers make their decisions based on electricity sale prices, advertising investments, and plans for power-generation facilities. Sale prices are adjusted based on imbalance settlement in supply and demand with electricity distribution operators.

The proposed agent-based gaming model is based on the government plan of energy market reform in Japan [10]. In this gaming model, the actual participants participate in the game playing the roles of power producers, electricity retailers, and aggregators. In addition, computer agents also participate in the market autonomously as a number of consumer agents. The government agents conduct imbalance settlement based on the predetermined market rules. Based on this gaming model, the game participants can experience the complexity of this market and they can design a market system while verifying the effectiveness of the system designed. Our ultimate goal is to verify whether real-time characteristics are satisfied by conducting simulation based on the actual climate data in order to develop further verification.

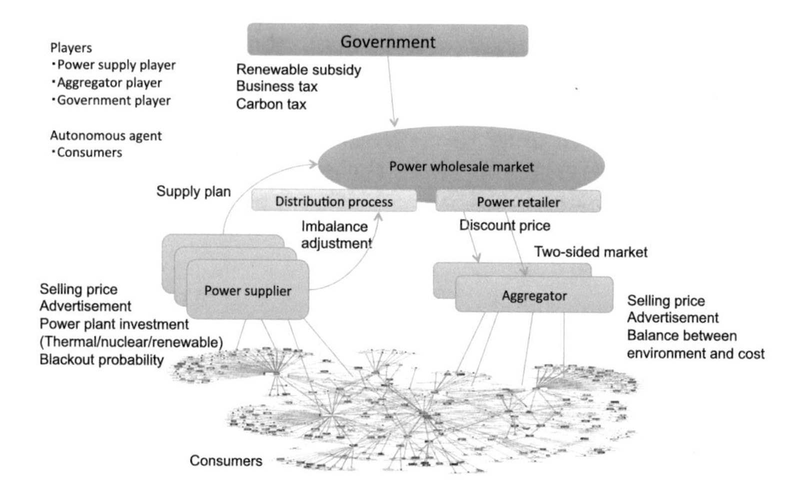

Fig. 2. Energy conversion gaming model

4.1 Model Outline

According to the ODD protocol, the section below describes the outline of the model. The ODD (Overview, Design concepts, and Details) protocol was proposed to standardise the published descriptions of individual-based and ABMs [11]. The primary objectives of ODD are to make model descriptions more understandable and complete, thereby making ABMs less subject to criticism for being irreproducible.

In this model, 'Entities' are electricity suppliers, aggregators, the government, and consumers. 'State variables' are defined as follows:

- Electricity suppliers
  Sale prices, discount rates for major clients, investments (advertising, thermal, nuclear, and renewable energy), costs (thermal, nuclear, and renewable energy), carbon generation rates (thermal, nuclear, and renewable energy), power generation amounts (thermal, nuclear, and renewable energy), operator attractiveness, carbon gas generated, and rate of power failure occurrences
- Aggregators
  Sale prices, advertising investment, the number of operators that purchase electricity, and energy proportions (thermal, nuclear, and renewable energy)
- Government
  Imbalance prices, business tax rates, carbon tax rates, and renewable energy investments
- Consumers
  Norm effect parameters, information effect parameters, network generation parameters, and the number of consumers

'Process overview and scheduling' are as below. Suppliers generate power, and sell it to consumers and aggregators. While taking into account the environment of consumers and their intentions toward prices, suppliers determine the power generation proportions of thermal power generation, nuclear power generation, and renewable energy, electricity prices (discounts for general/major clients), and advertising investments in order to maximise their own profits. Increase in the proportion of renewable energy increases the power failure probability, resulting in paying the imbalance cost. Additionally, their own competitiveness declines in proportion to the power failure probability.

Aggregators purchase electricity with discounts for major clients from suppliers, while re-selling the electricity to consumers. While taking into account the environment of consumers and their intentions toward prices, aggregators determine the power generation proportions of thermal power generation, nuclear power generation, and renewable energy, electricity prices (for general clients), and advertising investments.

While considering their own preferences for electric power and electric power charges, consumers purchase electric power from appropriate suppliers. Consumers are network-linked with their acquaintances receiving the norm effect. The government determines imbalance prices, business taxes, carbon taxes, and renewable energy subsidies. Based on these, the total amount of carbon gas generated and the entire probability of power failure are determined. The goal of the government is to optimise these variables.

The electric power proportions are determined in uniform random numbers expressed by the base proportion $\pm 10\%$. Based on the synthesised attractiveness of prices and electric power preferences, suppliers and aggregators are determined by using roulette selection. The power failure probability is an exponential function based on the renewable energy proportions.

5 Experiment

An electricity gaming model was implemented by the agent programming environments, NetLogo and HubNet, which was operated from each terminal connected to the local network. Figure 3 shows the screen displayed for players.

From any of these screens, the condition of consumers allocated on the network and the condition of suppliers could be confirmed. Consumers were able to identify order destinations in different colours, so that they could intuitively understand the current share condition of suppliers. Electric power prices, investment for power generation facilities, and management information including surplus funds could be confirmed as supplier conditions. From this player screen, each supplier player entered necessary information such as the electric power prices, advertising investments, investments for thermal power generation, investments for nuclear power generation, investments for renewable energy power generation, and the discount rates for major clients. Aggregate players determined the price and the energy source weight, in addition to the electric power price and advertising investments, as factors for deciding order destinations. In this experiment, four supplier players and one aggregator agent made their decisions for 18 periods.

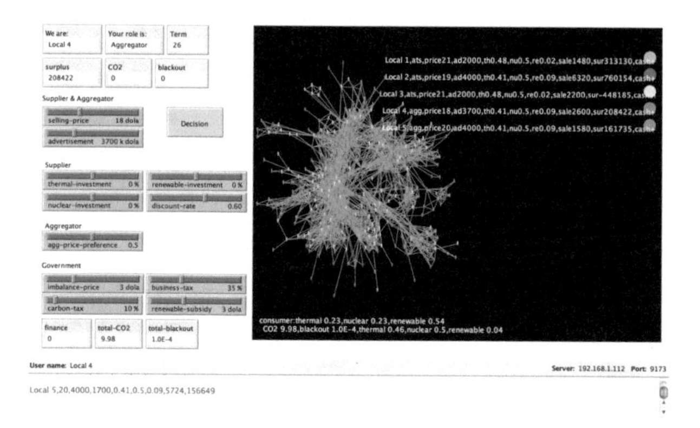

Fig. 3. Player panel: Networks of consumers and market shares are graphically observed. Decision-making and management conditions of other players, including the preferences of consumers, are observed on a panel. Decision-making and management condition of all players are able to be observed on a panel in every period.

The goal of this study is to clarify decisive factors for making decision of energy selection based on human competitive and collaboration behaviour to be helpful for an incentive design of energy markets. For the purpose, two hypotheses were set in the experiment. First is that energy transition to renewable source is achieved by players while keeping their profit. Second is that aggregators have ability to control the energy market through the share of consumers' power market as well as other two-sided markets.

6 Results and Discussion

The left chart of Fig. 4 shows the proportion of each energy source, the amount of CO_2 emissions, and the transition of the power failure probability. In the initial stage, the proportion of thermal power generation exceeded 60%; however, it declined gradually, finally going down to less than 40%. This also reduced the amount of carbon emissions (The right chart of Fig. 4). The first hypothesis, which energy transition to renewable source is achieved by players while keeping their profit, has been adopted with this result.

On the other hand, the proportions of nuclear power generation and renewable energy power generation increased. This is because of the influence given by the energy orientation of consumers. In particular, the proportion of renewable energy gradually increased in tune with the orientation of consumers, while it declined in the later stages. This result might be because whereas the power generation proportion of each electric power supplier was inclined toward the use of thermal power generation in the initial stage, the energy orientation of consumers was about 1/3. Therefore, there must have been an incentive that worked where the order volume increased by changing the power generation investment according to this proportion (The left chart of Fig. 5).

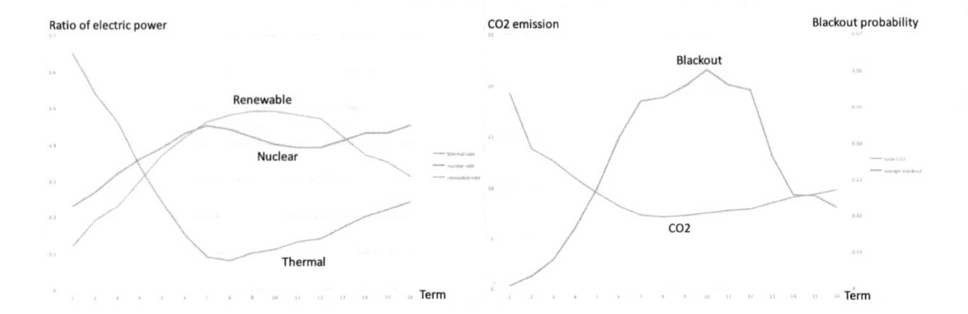

Fig. 4. Left: Trend of energy source rate, Right: Trend of CO_2 and blackout rate

However, the situation, which was originally expected that the proportion of nuclear power generation decreased, was not observed, while nuclear energy with lower cost and carbon gas emissions continue to be relied on. This result shows that the management of electric power suppliers gave the first priority to maximising their profits, while giving almost no consideration to risks of nuclear power generation accidents. On the other hand, the aggregator agent made profit as well as suppliers players, but it could not monopolise the electric consumer market because one possibility is that the supplier players learnt how to keep their market share in competition from the aggregator(The right chart of Fig. 5). The second hypothesis, which aggregators have ability to control the energy market through the share of consumers' power market as well as other two-sided markets, was rejected with the result.

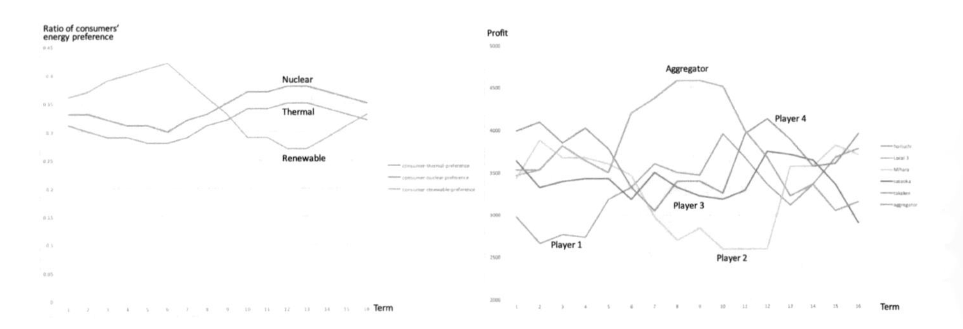

Fig. 5. Left: Trend of consumers' energy preference, Right: Trend of players' sales in the power market

All of the four players participating in this experiment were business people in their 30's, who might have had a custom to make decisions to maximise business profits as corporate managers. They were at the same time consumers, however, this experiment suggests that their concepts of accident risks might significantly change when they play a social role as entities to make corporate decisions.

7 Conclusion

In this research, based on agent-based models, serious games, design of electricity market platforms, and social network models, we built a model having the items below as purposes.

(1) Feature analysis on electric power imbalance adjustment for achieving new system designs
(2) Design of competitive electricity market platforms
(3) Design of incentive mechanisms for imbalance adjustment
(4) Evaluation and examination of mechanism design based on agent-based gaming models

The goal of this study is to clarify decisive factors for making decision of energy selection based on human competitive and collaboration behaviour to be helpful for an incentive design of energy markets. For the purpose, two hypotheses were set in the experiment. First is that energy transition to renewable source is achieved by players while keeping their profit. Second is that aggregators have ability to control the energy market through the share of consumers' power market as well as other two-sided markets.

Our experiment confirmed that the energy orientation of electric power consumers could give a significant influence on power generation investment of electric power suppliers, and the risk of nuclear energy was underestimated. And the first hypothesis was adopted and the second was rejected by the experiments through the agent-based gaming. These findings enabled us to analyse the decision-making process of people and operators, while being able to obtain effective knowledge regarding social ecosystems which disseminate renewable energy and adaptive behaviour. In the future, we are going to examine gaming models for system evaluation in liberalised electricity markets including aggregators as electric power wholesalers. Moreover, we will have to think about monitoring and reporting on the decision-makers traits and decision strategies they use in the game in relation to the performance of the simulated markets. A next challenge is the empirical validity of the agents behaviour in the model.

References

1. Boudreau, K.J., Hagiu, A.: Platform rules: multi-sided platforms as regulations. Platf. Market. Innov. **1**, 163–191 (2009)
2. Unno, M., Xu, H.: Opitmal platform strategies in the smartphone market. IEEJ Trans. Electr. Inf. Syst. **132**(3), 467–476,(2012)
3. Bacon, D.F., et al.: Predicting your own effort. In: International Conference on Autonomous Agents and Multiagent Systems (AAMAS 2012), vol. 2, pp. 695–702 (2012)
4. Samadi, P., Schober, R. Wong, V.W.S.: Optimal energy consumption scheduling using mechanism design for the future smart grid. In: 2011 IEEE International Conference on Smart Grid Communications, pp. 369–374, October 2011

5. http://www.essa.eu.org
6. Jager, W., Van der Vegt, G.: Management of complex systems: towards agent based gaming for policy. In: Janssen, M. (ed.) Policy Practice and Digital Science: Integrating Complex Systems, Social Simulation and Public Administration in Policy Research. Springer Science in the Public Administration and Information Technology series edited Christopher G. Reddick (2015)
7. Rochet, J., Tirol, J.: Platform competition in two-sided markets. J. Eur. Econ. Assoc. **1**(4), 990–1029 (2003)
8. Rochet, J., Tirol, J.: Two-sided markets : a progress report. Rand J. Econ. **37**(3), 645–667 (2006)
9. Sannikov, Y.: A continuous time version of the principal-agent problem. Rev. Econ. J. Stud. **75**(3), 957–984 (2008)
10. Agency for Natural Resources and Energy. http://www.enecho.meti.go.jp/en/category/electricity_and_gas/energy_system_reform/
11. Grimm, V.: Pattern-orinted modeling of agent-based complex systems: lessons from ecology. Science **310**, 987–991 (2005)
12. Delre, S.A., Jager, W., Janssen, M.A.: Diffusion dynamics in small-world networks with heterogeneous consumers. Comput. Math. Organ. Theory **13**, 185–202 (2007)
13. Toivonen, R., Onnela, J.P., Saramaki, J., Hyvonen, J., Kaski, K.: A model for social networks. Balt. Congr. Future Internet Commun. **371**(2), 851–860 (2006)
14. Kurahashi, S., Saito, M.: Informative and normative effects using a selective advertisement. SICE J. Control Meas. Syst. Integr. **6**(2), 076–082 (2013)
15. Yang, C., Kurahashi, S., Ono, I., Terano, T.: Pattern-oriented inverse simulation for analyzing social problems: family strategies in civil service examination in imperial China. Adv. Complex Syst. **15**(7), 1–21 (2012). 1250038

Finding the Better Solutions for the Smart Meter Gateway Placement in a Power Distribution System Through an Evolutionary Algorithm

Ryoma Aoki(✉) and Takao Terano(✉)

Department of Computer Science, School of Computing, Tokyo Institute of Technology, Tokyo, Japan
aoki.r.ae@m.titech.ac.jp, tterano@computer.org

Abstract. In this paper, we propose an evolutionary computation algorithm to find the better solutions for the smart meter gateway placement problem for electric power distribution system. In a given business situation, a smart meter at each household will be used to manage electricity charges, appliance controls, and so on, to make a house smarter. In order to manage the communication among such smart meters, we need expensive gateway devices. Among thousands of households, therefore, it is necessary to adequately place several dozen of gateways under complex distance, geometrical, and number constraints. Furthermore, smart meter gateway systems are gradually developed in a town. This requires temporal strategies for the placement. To solve the problem, we are developing an evolutionary computation algorithm, whose objective is to minimize the number of gateways within feasible computation time and scalable in the number of smart meters and gateways. The proposed algorithm is characterized by dynamic area decomposition and sophisticated crossover mechanisms. The experimental results of the proposed method have revealed that (1) the placement costs becomes half compared with the existing method and (2) the computation time is enough feasible against the increase of the problem scales.

Keywords: Smart meter/Gateway placement problem
Electric power distribution system · Dynamic area decomposition
Evolutionary computing

1 Introduction

New generation electric power distribution systems provide customers with both energy and information. For the purpose, each household or customer has smart meters [1]. Distribution systems requires gateways, or concentrators to send and/or receive information of smart meters to/from upper level electric facilities. As the cost of a gateway device is very expensive, they require its optimal placement, that is, the number of gateway devices should be as small as possible, and however, they must communicate with connected all the smart meters in a given area. Thus, our problem is to get the better solutions of smart meter and gateway placement under various

G. Jezic et al. (Eds.): KES-AMSTA-18 2018, SIST 96, pp. 321–330, 2019.
https://doi.org/10.1007/978-3-319-92031-3_32

constraints in a given area with several thousand households. In this paper, we apply evolutionary computation techniques and propose a new algorithm for them. The proposed algorithm is characterized by dynamic area decomposition and sophisticated crossover mechanisms. The experimental results in a real sample urban area have revealed that (1) the placement costs becomes half compared with the existing method and (2) the computation time is enough feasible against the increase of the problem scales. The model area is shown in Fig. 1, which includes 1,598 households, or smart meters in 16 km * 25 km wide area. Figure 1 is created by Open Street Map [6].

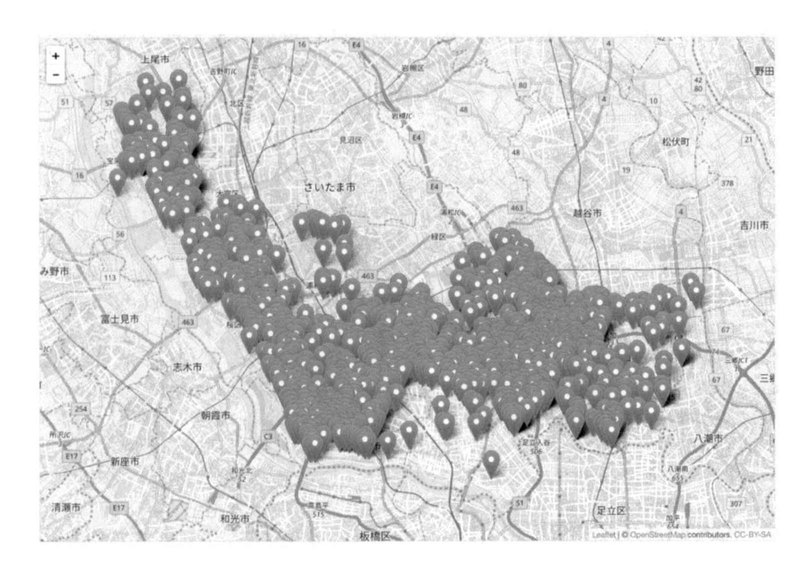

Fig. 1. Smart meter distribution in a sample urban area with 1,598 households

The rest of this paper is organized as follows: Sect. 2 describes the characteristics of the placement problem, Sect. 3 explains the proposed evolutionary computation technique. Section 4 conducts intensive experiments and related discussions. Finally, Sect. 5 concludes the paper and proposes some ideas for future work.

2 Descriptions of the Placement Problem

A gateway gets information from/to plural smart meters and sends/receives it to/from a power control center. From the given specifications of the system, a smart meter and a gateway have a short-range communication function within 950 m length. The communicable number of one gateway with smart meters is up to 200. Each gateway is equipped on an electric utility pole, which is set in 50 m length in the household area. Under the conditions, the objectives of the optimization problem are threefold: (1) to minimize the number of gateways, (2) to set all gateways in order to let every smart meter be communicable with them, and furthermore, (3) within the above two constraints, to increase the number of gateways proportional with the increase of households in the future development of the area. The third issue states that the solution method should be scalable with the number of smart meters. In the problem of the given area, there exists obvious feasible solutions, which extracts any gateways such that (1) the number of communicable smart meters is zero, or (2) another gateway exists to communicate with all the smart meters. To satisfy the conditions, we are able to get 4,762 candidate utility poles at most in the given area. To get optimal solutions, however, we must solve a typical combinatorial problem to select the places. Thus, we employ evolutionary algorithms to represent the candidate places as bit strings in the chromosome.

3 Proposed Evolutionary Algorithm

In this section, we propose an evolutionary algorithm to get optimal solutions for the problem. We have employed a rather standard genetic algorithm [2] to formulate the problem. The chromosomes of the placement problem are represented in the bit string forms. Each chromosome shows in which, the bit 1/0 means that the corresponding gateway is/is not selected among 4,762 gateway candidates places. However, the uniqueness of our proposal lies in the three points: (i) adding to usual mutation algorithm, we utilize a nearest neighbor method; (ii) instead of solving the total problem at once, we employ dynamic area decomposition algorithm to get feasible local solutions, which aims at gradually handling the feasible solution areas larger; and (iii) to cope with the scalability of the problem size, we apply backward method, which gradually decreases the number of households from current placement of them. In the following subsections, the outline of the algorithm, neighborhood mutation method, dynamic area decomposition method, and backward method are described in order.

3.1 Applied Genetic Algorithm

The genetic algorithm we applied is characterized in the following operations.

(1) Crossover: We select two parents from the population, then a usual two point crossover operation is applied to get eight children.
(2) Mutation 1: For selected two parents from the population, a usual mutation operation is applied to change the bit with uniform random probability to get two children.
(3) Mutation 2: Nearest neighbor mutation is applied, which will be described below, to get two children.
(4) Duplication deletion: Using the above three genetic operations, we get twelve children. Among them, we delete the duplicated ones.
(5) Generation Change: We apply the Minimal Generation Gap (MGG) method [3] to manage the generation change. The outline of MGG is as follows: (i) select two parents from the population, (ii) apply genetic operations to the two parents, (iii) among the generated children, the best one and the other random one are selected, and (iv) return the two to the population.

(6) Dynamic Area Decomposition: When the number of generations increases, we change the divided areas from smaller to larger. The details will be described below.

3.2 Nearest Neighbor Mutation

Nearest neighbor mutation is a new mutation method for the placement problem, which aims at searching another candidate place nearby in the physical distance from the current feasible gateway placement. Using the mutation, we expect to maintain the diversity of the population. The procedure is as follows:

(1) Among *n* gateways selected as feasible parents, apply steps (2), (3), and (4) operations with *1/n* probability to each of them;
(2) Set the feasible state into infeasible state;
(3) Select one gateway from the most five nearest ones from the one changed to infeasible;
(4) Set the gateway feasible.

3.3 Dynamic Area Decomposition

As the effect of gateway placement changes in each operation is considered local, to get initial feasible solutions, the genetic algorithm should be applied in the narrower areas. However, such local solutions may not globally generate thee better solutions. Thus, we dynamically change the boundary of the local areas, then the number of local areas gradually get smaller. This is a basic idea of dynamic area decomposition, which is introduced in the algorithm as yet another mutation method.

3.4 Backward Smart Meter Placement

We assume that the gateways have been placed in sparse at the initial phase of area development and then the numbers of gateways have gradually been increasing. The process is directed to forward as the time goes by. However, the forward process is hard to implement in our settings, because there are no information about the future plans of the area development. Thus, to simulate this process, we start from the current 1,598 smart meters, then 50.0%, 25.0%, 12.5%, 6.25%, and 3.125% of smart meter placement cases are considered. To select the places where smart meters are, using random numbers with normal distribution, we get places with each percentages. Then, we apply the proposed evolutionary computation algorithms without the dynamic area decomposition function in each placement.

3.5 Implementation of the Proposed Algorithm

The proposed method is developed and implemented on TSUBAME supercomputer system at Tokyo Institute of Technology, which is a cluster based system and easily used as a backend batch processes. We have implemented the algorithm in R language [4]. To generate and draw maps, we have used LeafletR interactive Web-Maps [5].

4 Experiments and Discussions

This section describes the results of the intensive computer experiments. In the following subsections, the changes of global feasible solutions, the results of backward smart meter placements, and the discussion on the experiments are described in order.

4.1 Getting Global Optimized Solutions

Applying the method in Sect. 3, we have obtained the solutions shown in Fig. 2. Each point in Fig. 2 represents the place the gateway placed. Each corresponding circle 2 means the area the gateway communication reached. From the figure, all the smart meters are covered with only 62 gateways.

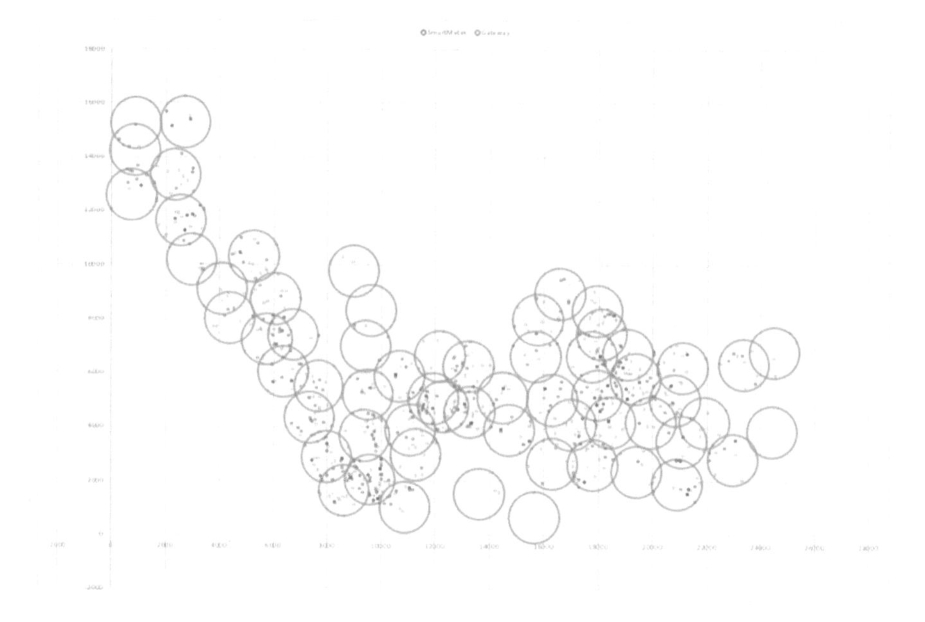

Fig. 2. Optimized solutions found by the proposed method in the given area

We have evaluated the objective function using the following equation:

$$Evaluated\_Value = \#\_GateWay\_Used * 70 + \#\_Smart\_Meters\_Not\_Connected * 10,000.$$

The convergence of the evaluated values are shown in Fig. 3. The X-axis of the graph represents of number of evaluation with genetic operations and the Y-Axis represents the evaluation values calculated with the above equation.

Figure 4 depicts the changes of the number of gateways needed (blue line) and the number of unconnected smart meters (orange line). From Figs. 3 and 4, we observe that

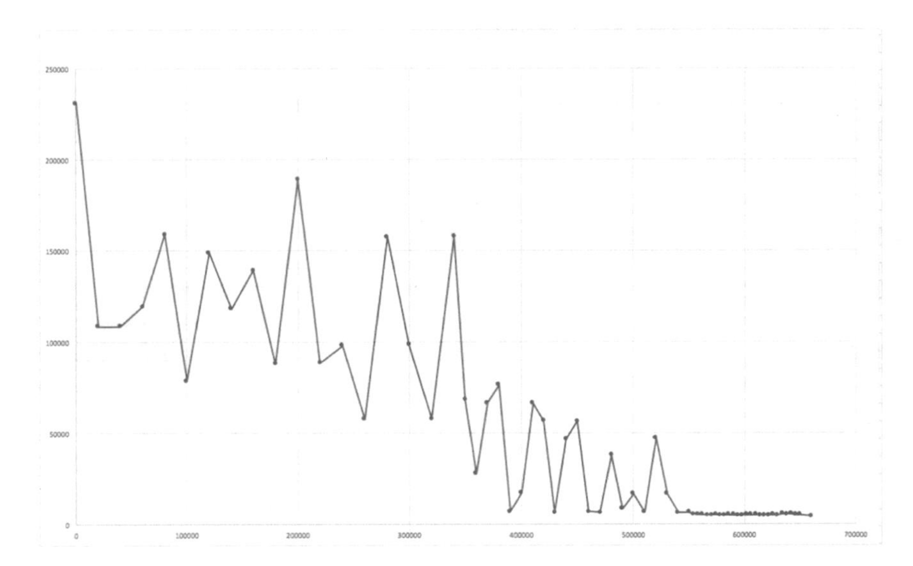

Fig. 3. Convergence of the evaluation values of the best individuals

we have got the better results during the iterations. Because the limit of the space, we have omitted the experimental results without some genetic operations shown in Sect. 3. However, as shown in Table 1, the results have suggested that all the operations we have designed work well in the problem.

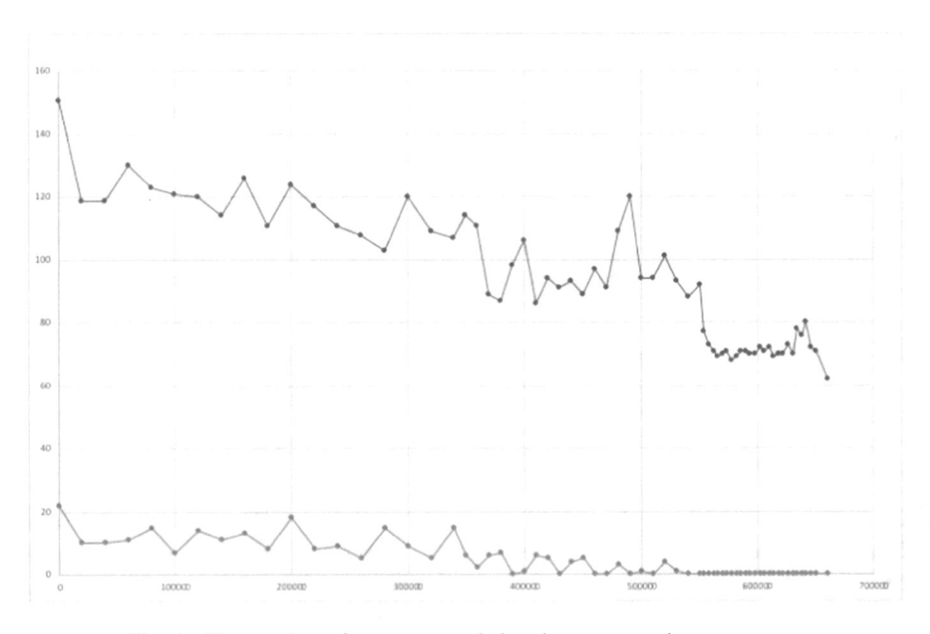

Fig. 4. The number of gateway needed and unconnected smart meters.

Table 1. Summary of the effects of the proposed procedures

Nearest neighbor mutation	Usual mutation	Number of function evaluations	Number of gateways needed	Number of unconnected smart meters
X	X	660,001	62	
X	–	1,310,001	70	24
–	X	1,310,001	65	0
–	–	1,310,001	76	10

Also, in Fig. 5, the effects of dynamic area decomposition are depicted by adding the decomposition information in Fig. 4. In the experiments, at first, we have set the given area into 10 * 10 decompositions, then, we have reduced areas into 5 * 5, 2 * 1, then, finally 1 * 1 to get the final results. From Fig. 5, we have found the decomposition strategy has worked well.

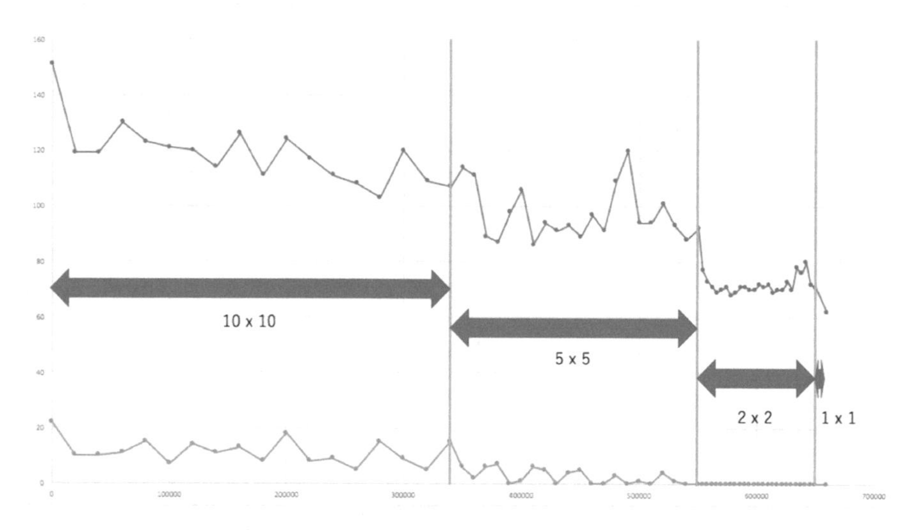

Fig. 5. The effects of dynamic area decomposition

4.2 Results on Backward Smart Meter Placements

We have tested the backward smart meter placement method to the given area. As described in the previous section, we have set five backward patterns, then have repeated 1,000 patterns experiments with 3.125%, 6.25%, 12.5%, 25.0%, and 50.0% smart meter placement cases by changing random seeds. The experimental summaries are shown in Table 2 and Fig. 6. Also, each one of the experimental results on 3.125% and 50.0% placement cases are shown in Figs. 7 and 8. From these results, we have observed that the proposed method is scalable in different situations and are able to get the better results on the combinatorial gateway placement problem.

Table 2. Summary of ratio of smart meter placements and gateways needed in 1,000 trials with different random seeds

	Ratio of setting smart meter numbers					
	3.125%	6.25%	12.5%	25.0%	50.0%	100.0%
Minimal gateway numbers	22	33	43	51	56	62
Maximal gateway numbers	38	49	58	59	61	62
Median gateway numbers	29	41	50	56	59	62
Average gateway numbers	29.3	40.9	50.3	58.8	58.8	62

Note: Only one run is executed in 100.0% placement case

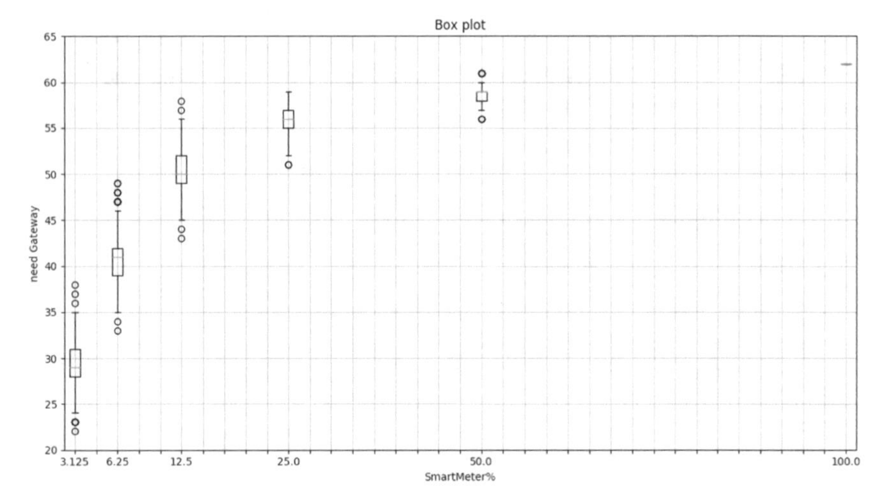

Fig. 6. Ratio of smart meter placements and gateways needed in the experiments

4.3 Discussion

They have reported that, with a conventional method to place gateways in the given area with heuristics, the number of gateways needed to cover the given area is over one hundred. However, from the intensive experiments for the given area, we have confirmed the effectiveness of the proposed method: (i) We are able to get near optimal solutions against various constraints; (ii) Each procedure in the proposed method is indispensable to apply the large scale problems; and (iii) the method is enough scalable against various constraints. On the other hand, in the given experiments, fortunately, there are no infeasible solution violating the limit of number of the smart meters to be handled by each gateway. For the one of the future issues, we should test the robustness of the method against the number of smart meters the gateway can handle. Furthermore, some additional conditions about gateway smart meter combination systems should be considered, which would cause the other combinatorial problems.

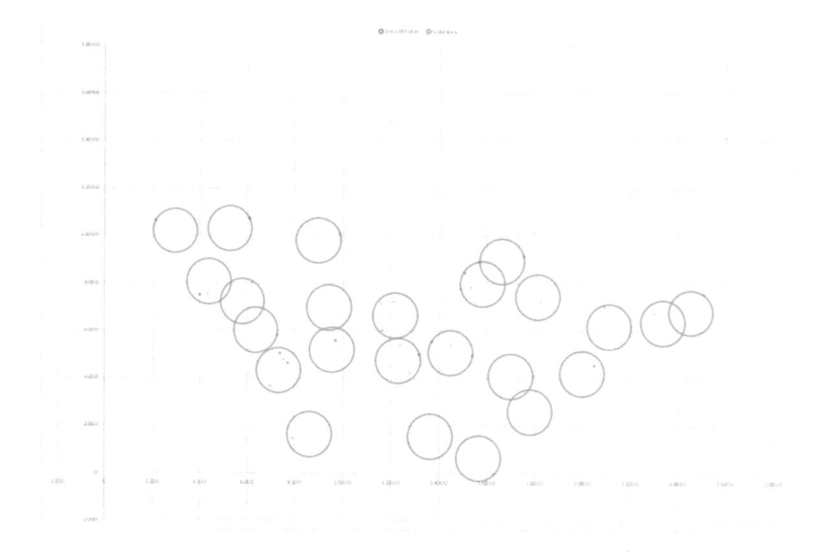

Fig. 7. A case of 3.125% smart meter placement

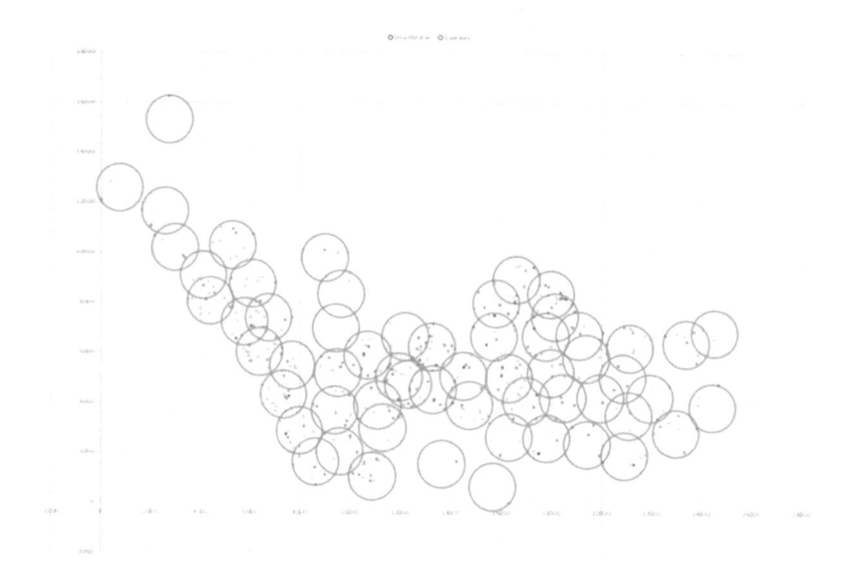

Fig. 8. A case of 50.0% smart meter placement

5 Concluding Remarks

In this paper, we have proposed an evolutionary computation algorithm to solve smart meter gateway placement problems, whose objective is to minimize the number of gateways within feasible computation time and scalable in the number of smart meters and gateways. The proposed algorithm is characterized by dynamic area decomposition and sophisticated crossover mechanisms. The intensive experimental results with TSUBAME Super computer system of the proposed method have revealed that (1) the placement costs becomes half compared with the existing method and (2) the computation time is enough feasible against the increase of the problem scales.

Our future work includes to adding other constraints about gateway functionalities with restricted communication ranges and/or boundary conditions about the specified areas. Also, further improvement for such combinatorial problems will be necessary.

References

1. Fang, X., Misra, S., Xue, G., Yang, D.: Smart grid—the new and improved power grid: a survey. IEEE Commun. Surv. Tutor. **14**(4), 944–980 (2011)
2. Mitchell, M., Forrest, S., Holland, J.H.: The Royal Road for Genetic Algorithms: Fitness Landscapes and GA Performance. MIT Press, Cambridge (1992)
3. Sato, H., Ono, I., Kobayashi, S.: A new generation alternation model of genetic algorithms and its assessment. J. Jpn. Soc. Artif. Intell. **12**(5), 734–735 (1997)
4. Graul, C.: leafletR: Interactive Web-Maps Based on the Leaflet JavaScript Library: R package version 0.4-0 (2016). http://cran.r-project.org/package=leafletR
5. R Core Team: R: A Language and Environment for Statistical Computing: R Foundation for Statistical Computing (2017). https://www.R-project.org/
6. OpenStreetMap contributors: Planet dump (2017). https://planet.osm.org, https://www.openstreetmap.org

Japanese Health Food Market Trend Analysis

Yoko Ishino(✉)

Yamaguchi University, 2-16-1 Tokiwadai, Ube, Yamaguchi 755-8611, Japan
ishino.y@yamaguchi-u.ac.jp

Abstract. So-called health food has attracted increasing attention for more than two decades so that the health food market share has continued to increase. However, definitions and regulations regarding health food differ among countries; thus, comparing multiple countries simultaneously is difficult. This study focused on Japan because it has established health food rules and regulations, and both macro and micro data are available for analysis. In this study, government statistics were first used to determine the Japanese health food market's macro trend. Then, a questionnaire was administered to identify consumer attitudes and behaviors, and important factors for consumers to purchase health food were determined by conducting chi-square tests using the survey data. The obtained factors corroborated the results of the regression analysis used to determine the macro trend. Then, the following hypothesis was posed: differences between the actual sales of health food and sales estimated by regression analysis are primarily due to advertising activities and government regulations. A straightforward agent-based social simulation (ABSS) was performed to investigate how advertising and government regulations influence health food sales. Several reasonable findings that can be utilized in a future ABSS were obtained.

Keywords: Market trend · Health food · Agent-based social simulation

1 Introduction

The concept of "functional food" or "health food" has been widely accepted in Asia, America, Europe, and elsewhere [1]. The phrase "functional food" was used first in Japan in the 1980s. Since then, the health food market has been expanding steadily in many countries. From a technology management perspective, development of health food represents product innovation that often involves an innovative material or ingredient. The development of innovative materials to achieve new functions has occurred frequently in the functional food area. For a product to survive in such an environment, businesses must precisely understand consumer needs.

Although the idea of functional food has been accepted in many countries, the types of functional foods vary widely depending on culture, religion, and environment [1]. For example, fortified cereals and various supplements are very popular in the USA, while ginseng powder is a favored product in South Korea. In addition, health food definitions and regulations differ among countries. Consequently, it is difficult to compare national health food markets uniformly.

G. Jezic et al. (Eds.): KES-AMSTA-18 2018, SIST 96, pp. 331–340, 2019.
https://doi.org/10.1007/978-3-319-92031-3_33

This study focuses on the Japanese health food market. In 1984, the Japanese Ministry of Education, Science and Culture classified the functions of food as nutritional (primary function), palatability (secondary function), and disease prevention by biological regulation (tertiary function). Among these functions, the third function received special attention, and new food products with this function were referred to as functional foods. The Food with Health Claims (FHC) regulatory system was established in 1991. Since then, the health food market in Japan has been expanding steadily. The market grew quickly from the 1990s to 2005. After 2005, the growth rate slowed due to labeling regulations enacted in the 2006 Pharmaceutical Affairs Act. However, the market continued to grow and reached 1.21 trillion yen in 2013 [2]. In addition, several institutional reforms were established and an additional category, i.e., Food with Function Claims (FFC), was established in 2015. Generally, the main factor influencing market growth is considered to be growing health awareness due to an increase in lifestyle-related diseases and Japan's aging population.

The objectives of this study were to identify Japanese consumer attitudes and behaviors regarding health food, model such behavior, and investigate factors that affect market scale. First, statistical multiple regression analysis was performed to determine the macro trend of the health food market. Then, a questionnaire was administered to identify consumer attitudes and behaviors. Important factors for consumers to purchase health food were determined by performing chi-square tests using the survey data. To bridge the macro and micro trends, this study focused on the differences between actual health food sales and the sales estimated by regression analysis. We hypothesized that such differences are primarily related to advertising activities and government regulations. Based on this hypothesis, a straightforward agent-based social simulation (ABSS) was performed to investigate how advertising activities and government regulations influence health food sales.

The remainder of this paper is organized as follows. Section 2 provides general information about health food in Japan. Our research process is described in Sect. 3, and the results are discussed in Sect. 4. Conclusions and suggestions for future work are presented in Sect. 5.

2 Definitions and Coverage of Health Food in Japan

The definitions of health food differ among countries. The definitions and coverage of health food in Japan used in this research are as follows.

According to the Ministry of Health, Labor and Welfare, there is no legal definition of so-called health food. The term health food indicates the range of products sold and used to improve health. However, health food that meets government safety and efficacy standards can be referred to as FHC and, according to government regulations, the efficacy and function of FHC can be advertised. FHC can categorized as Food with Functional Claims (FFC), Food with Nutrient Function Claims (FNFC), and Food for Specified Health Uses (FOSHC) [3]. FFC, which was established in April 2015, is the most recent category. This study targeted a broad range of health foods, including FHC, such as supplements, drinks, and other products (fortified biscuits, concentrated extracts, etc.).

3 Research Method

First, statistical multiple regression analysis was performed to determine the macroeconomic trend of the Japanese health food market. Second, a dataset derived from the questionnaire results was analyzed using chi-square tests to identify key consumer attributes related to the consumption of health food. Finally, a straightforward ABSS was performed to investigate how advertising activities (including word-of-mouth communication and a company's sales force) and government regulations impact health food sales.

3.1 Analysis from Macroeconomic Point of View

Multiple regression analysis is an extension of simple linear regression analysis [4]. Generally, regression analysis is a statistical process to estimate the relationships among variables. Multiple regression analysis is performed to determine the extent to which a dependent variable (or objective variable) changes when any one of independent variables (explanatory variables) is changed, while the other independent variables are fixed. The multiple regression analysis targets to estimate a function consisting of the independent variables called the regression function. Multiple regression analysis also reveals the overall fit of the model (i.e., variance is explained) and the relative contribution of each independent variable to the total variance is explained.

In this study, the market size of health food (JPY) of each year was set as an objective variable. A previous study [5] found that gross domestic product (GDP) and the population aging rate should be two major explanatory variables when interpreting the per-capita health food expenditures of eight Asian countries in 2011. Therefore, the following variables were considered explanatory variable candidates in this study: nominal GDP, real GDP, the senior population (65 or older), and average annual income.

We first computed correlation and partial correlation matrices using all variables to understand each relationship. Then, after removing the variable suspected of having high multicollinearity, multiple regression analysis was performed to obtain the regression function. Note that data from 1990 to 2015 were used in this study.

3.2 Analysis of Questionnaire Survey Data

Survey Design. An Internet survey regarding health food was performed by Marsh Co., Ltd. in Japan from July 9 to July 14, 2015. The panels organized by this research firm were used for the survey. The number of the valid responses was 1028, and the respondents were allocated evenly in terms of sex and age.

Pearson's Chi-square Test. Contingency tables, which are related to conditional probability tables, are useful when examining the association between two nominal variables. The Pearson's chi-squared statistic is typically calculated to test the independence of two variables. The chi-squared statistic is defined as follows:

$$\chi^2 = \sum_i \frac{(O_i - E_i)^2}{E_i} \quad (1)$$

where O_i is the observed number of cases in category i and E_i is the expected number of cases in category i [6]. Here the respondents were sliced into two groups, i.e., those who consume health food nearly every day and those who consume health food less frequently. Regarding health food intake, we searched for attributes that did not allocate respondents evenly (significance level of 5%).

3.3 Agent-Based Social Simulation

In an ABSS, the different elements of social systems are modeled using artificial agents [7]. To simplify the problem, we set a situation in which an agent should choose one of two alternatives. One case is where the agent consumes health food, and the other is where the agent consumes normal food (not health food). Here, an agent moves randomly and receives information about the good reputation of the food from neighbor agents and a health food company. The former represents word-of-mouth communication and the latter represents the company's sales promotions. If the merged recommendation value exceeds a predetermined threshold, the agent adopts the recommended food. The design of the ABSS was derived from the multiple regression analysis and chi-square test results. Therefore the details of ABSS settings are discussed in the next section.

4 Results

4.1 Multiple Regression Analysis Results

The correlations and partial correlations between two variables in all combinations were calculated (Tables 1 and 2, respectively). As shown in Table 1, "Market Size," "Real GDP," and "Senior Population" are very strongly correlated to each other (greater than 0.94). On the other hand, the partial correlation between "Market Size" and "Real GDP" is only 0.04 (Table 2). This is evidence that there is high multicollinearity between "Real GDP" and "Senior Population." Based on these results, stepwise multiple regression analysis was performed using "Nominal GDP," "Senior Population," and "Average Annual Income" as explanatory variables.

Finally, the best regression function was obtained using only two explanatory variables, i.e., "Nominal GDP (=X_1)" and "Senior Population (=X_2)," where the coefficient of determination (adjusted R^2) was 0.914, which demonstrates how well the model fits. In addition, this R^2 value was sufficiently high to explain the market size of health food (=Y).

Next, we examined how much the published statistical market size of health food (=Y) and the estimated market size (=$\widehat{Y}$) derived from the best regression function differed (Fig. 1). The difference between Y and $\widehat{Y}$ is shown in Fig. 2. Note that the values of Y and $\widehat{Y}$ in Fig. 1 are quite close.

Table 1. Correlation matrix between variables

	Market Size of Health Food (JPY)	Nominal GDP (JPY)	Real GDP (JPY)	Senior Population (# of people)	Average Annual Income (JPY)
Market Size of Health Food (JPY)	1	0.40	0.96	0.94	0.27
Nominal GDP (JPY)	0.40	1	0.46	0.25	0.42
Real GDP (JPY)	0.96	0.46	1	0.96	0.45
Senior Population (# of people)	0.94	0.25	0.96	1	0.39
Average Annual Income (JPY)	0.27	0.42	0.45	0.39	1

Table 2. Partial correlation matrix between variables

	Market Size of Health Food (JPY)	Nominal GDP (JPY)	Real GDP (JPY)	Senior Population (# of people)	Average Annual Income (JPY)
Market Size of Health Food (JPY)	1	0.22	0.04	0.61	−0.64
Nominal GDP (JPY)	0.22	1	0.39	0.13	0.37
Real GDP (JPY)	0.04	0.39	1	0.40	0.14
Senior Population (# of people)	0.61	0.13	0.40	1	0.23
Average Annual Income (JPY)	−0.64	0.37	0.14	0.23	1

From a macroeconomic perspective, these results indicate that the aging population and economic growth should be crucial relative to expanding the health food market. We then investigated whether a similar tendency could be found in the behavior of ordinary consumers.

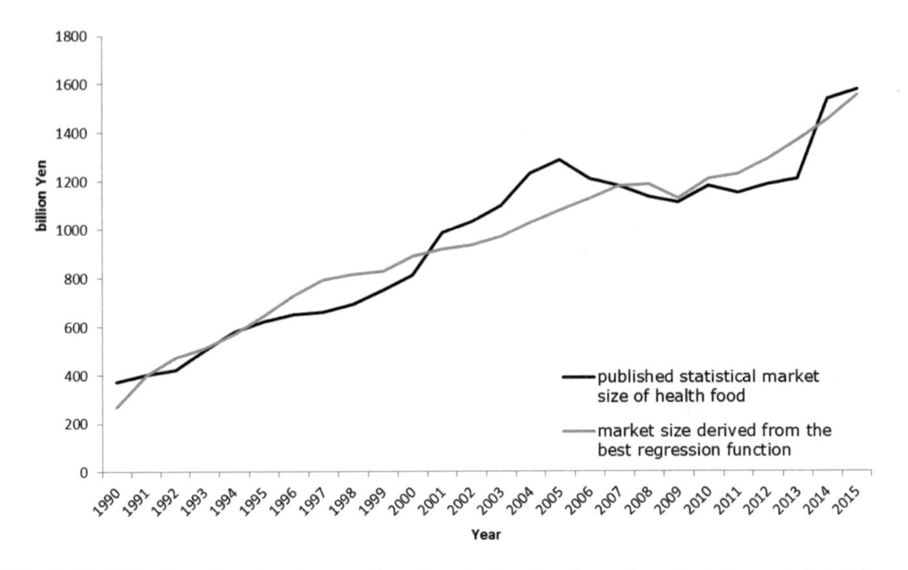

Fig. 1. Published market size (= actual market size) and estimated market size calculated from multi-regression

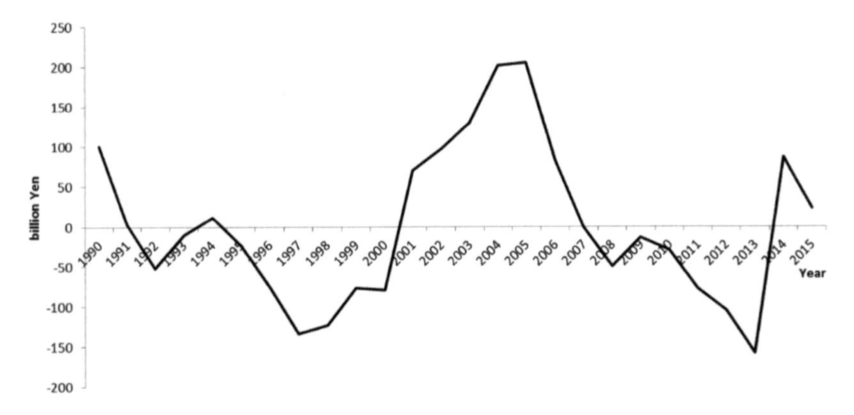

Fig. 2. Difference between published and calculated market sizes

4.2 Chi-square Tests Using Survey Data

Five attributes, i.e., age, sex, household income, marital status, and living with children, were selected to examine whether they influence the frequency of health food intake. The results of the chi-square tests indicated that only age and household income affect the frequency of health food intake. A significant relationship was found between health food intake and age group with a significance level of 5%, as shown in Fig. 3. In addition, a significant relationship was found between health food intake and household annual income with a significance level of 5%, as shown in Fig. 4.

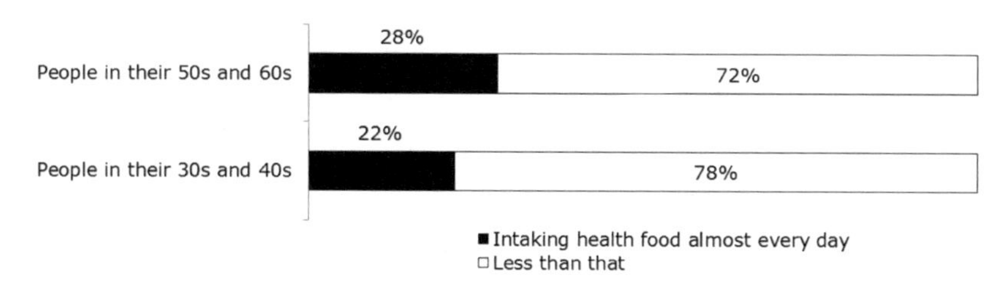

Fig. 3. Influence of the age group: $\chi^2(df = 1, N = 1027) = 5.74, p = 0.017$

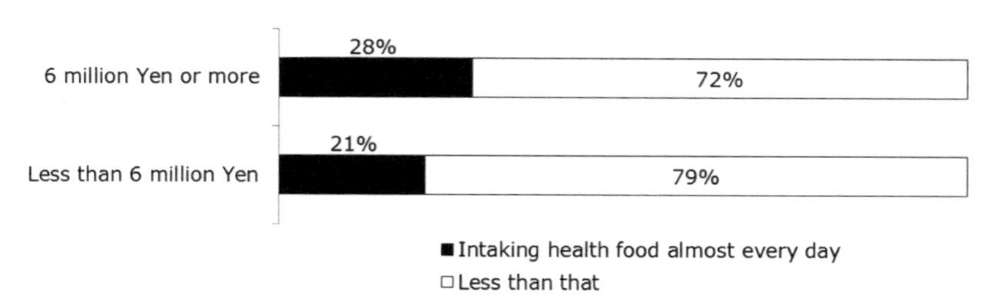

Fig. 4. Influence of annual household income: $\chi^2(df = 1, N = 900) = 5.55, p = 0.018$

The results show that older people and people with higher incomes consume more health food. Note that the same trend seen in the multiple regression analysis was obtained from the ordinary consumer survey.

4.3 Agent-Based Social Simulation Results

As described previously, it was found that the sales scale of health food in Japan can be predicted from the elderly population and nominal GDP. However, it is not possible to predict the sales scale of health food perfectly using only these two attributes (see the residual error in Fig. 2).

Past events regarding health food were also examined.

- In 2004, the effectiveness of a coenzyme Q10 supplement was introduced in television programs, and this resulted in significantly increased use of this supplement.
- In 2006, the Ministry of Health, Labor and Welfare indicated that some products that use Agaricus can drive oncogenesis. As a result, producers voluntarily recalled Agaricus-related products.
- In 2007, a television program introduced the effectiveness of some health foods; however, it was later revealed that the data used in the television program were fabricated. As a result, the program was canceled.

By superimposing these events onto the graph in Fig. 2, the author considered that positive and negative effects, such as a company's sales promotion, consumer word-of-mouth, laws and regulations, and corporate scandals, might result in fluctuations in the health food market. Therefore, the following hypothesis was formed: differences between the real sales of health food and sales estimated by regression analysis may primarily be due to advertising activities and government regulations.

Based on this hypothesis, we designed an ABSS as follows.

- Two types of agents were employed in this ABSS, i.e., a company agent and a consumer agent. The company agent did not move and remained stationery through the simulation. However, the consumer agent moved randomly and received information about the reputation of the food from neighbor agents, i.e., from the consumer and company agents. Information provided by a company agent represented the company's sales promotions and advertisements, and information conveyed by a consumer agent represented word-of-mouth communication.
- To simplify the problem, there was only one kind of health food and alternative normal food in the model. A consumer agent decided to purchase health food or alternative normal food at each time step.
- The consumer agent did not remember the previous food selection state and could only inform other agents about the current food selection state.
- The consumer agent randomly walked and gathered information about the food selection of neighbor consumer agents and the recommendation of a neighbor company agent. At each time step, a consumer agent determined food selection by majority vote based on the obtained information with a probability of 70%.
- At each time step, the model calculated the share of health food selected by consumer agents. If this share was greater than a threshold (65% by default), the balance of the food recommendation by company agents was readjusted so that the health food accounted for a small proportion of the total recommendation. In the same manner, if that share was less than a threshold (35% by default), the balance of the food recommendation by company agents was readjusted so that the health food accounted for a large proportion of the total recommendation.

The experimental conditions were as follows. The number of consumer agents was 1000, and the number of company agents was 50. In the initial state ($t = 0$), the probability of a consumer agent adopting health food was 30%. Also, the probability of a company agent advertising health food was 20%. Note that we did not have a sufficient amount of data to conduct calibration relative to the actions of a consumer agent to adopt health food. Therefore, we set the parameter values in a trial-and-error manner.

Figure 5 shows one of the resultant share transitions of the health food in the simulations. In Fig. 5, an x-axis and y-axis indicate the time steps in the simulations and the heath food share among consumer agents, respectively. As can be seen, the health food share fluctuated, similar to that shown in Fig. 2. The results of the ABSS can be interpreted as follows. If the spread of health food progressed too much due to advertising and word-of-mouth communication, etc., something, such as laws and regulations, would slow the spread (and vice versa). If market expansion was experiencing a slowdown, advertisements and new laws could encourage the spread of health food.

Although precise parameter settings are an issue that cannot be avoided, this simulation improved our understanding of the Japanese health food market.

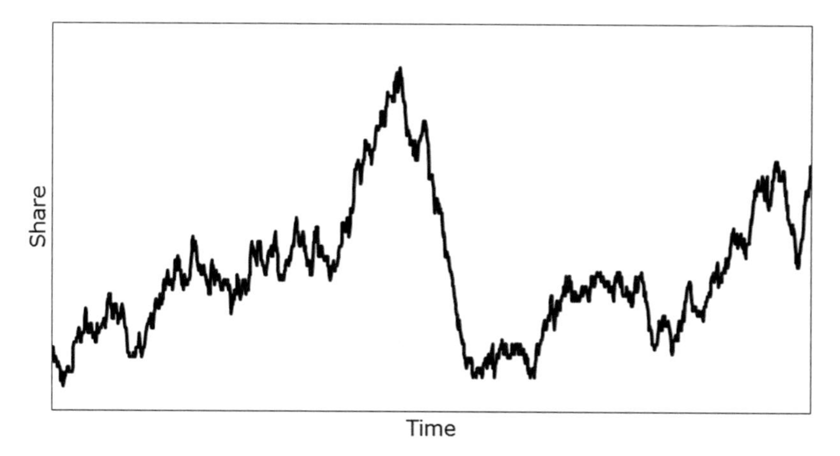

Fig. 5. Output example of ABSS: health food share transitions

5 Conclusions

In this study, growth in health food expenditure was explained using only two indices, i.e., nominal GDP (economic index) and the aging population (social index). The regression function obtained in this study allowed us to predict future health food market size quantitatively because the coefficient of determination (adjusted R^2) was sufficiently high. This can greatly contribute to creating health food marketing strategies. However, it was not possible to perfectly predict market size using only these two attributes; thus, residual error occurred. This residual error may have come from the influence of consumer communication (positive and negative) and/or government regulations. Before conducting the ABSS, the analysis of a questionnaire revealed that older people and people with higher incomes consume more health food. This corroborated the results of a multiple regression analysis using macroeconomic data. Finally, a straightforward ABSS was conducted to investigate how advertisements and government regulations influence health food sales. In future, we plan to perform a more precise ABSS using new consumer survey data.

Acknowledgment. This work was supported by JSPS KAKENHI Grant Number 26560121 and JP26282087.

References

1. Siró, I., Kápolna, E., Kápolna, B., Lugasi, A.: Functional food. Product development, marketing and consumer acceptance - a review. Appetite **51**(3), 456–467 (2008)
2. Editorial Department of Food Processing and Ingredients: Trends in the market for dietary supplement, health ingredients and production technology. Food Process. Ingred. **49**(3), 43–45 (2013). (in Japanese)
3. MHLW (Ministry of Health, Labour and Welfare) Homepage. http://www.mhlw.go.jp/stf/seisakunitsuite/bunya/kenkou_iryou/shokuhin/hokenkinou/. Accessed 2 Feb 2018

4. Freedman, D.A.: Statistical Models: Theory and Practice. Cambridge University Press, Cambridge (2005)
5. Aoki, H., Ishino, Y.: Essence of growing health food markets in Asian countries. J. Innov. Sustain. **8**(3), 108–117 (2017). ISSN 2179-3565
6. Sheskin, D.J.: Handbook of Parametric and Nonparametric Statistical Procedures. CRC Press, Boca Raton (1997)
7. Sun, R. (ed.): Cognition and Multi-Agent Interaction: From Cognitive Modeling to Social Simulation. Cambridge University Press, Cambridge (2006)

Simulation of the Effect of Financial Regulation on the Stability of Financial Systems and Financial Institution Behavior

Takamasa Kikuchi[1(✉)], Masaaki Kunigami[2], Takashi Yamada[3], Hiroshi Takahashi[1], and Takao Terano[2]

[1] Keio University, 4-1-1 Hiyoshi Kohoku-ku, Yokohama, Kanagawa, Japan
takamasa_kikuchi@keio.jp, htaka@kbs.keio.ac.jp
[2] Tokyo Institute of Technology, 4259 Nagatsuta-cho, Midori-ku, Yokohama, Kanagawa, Japan
mkunigami@gakushikai.jp, terano@dis.titech.ac.jp
[3] Yamaguchi University, 1677-1 Yoshida, Yamaguchi-shi, Yamaguchi, Japan
tyamada@yamaguchi-u.ac.jp

Abstract. This study focuses on the influence of financial institutions' behavior under financial regulation on financial systems. For this purpose, the authors propose simulation models of systemic risks expressing financial regulation and the financing and investment behavior of financial institutions. Using this model, scenario analysis is performed based on cases generated from a combination of various regulations. We then approach the issue of the trade-off between the stability of the financial system and a decline in liquidity. Our numerical experiment shows that the liquidity of marketable assets could be reduced by imposing a BS restriction.

Keywords: Agent-based simulation · Systemic risk · Financial regulations Asset liability management

1 Introduction

Following the financial crisis triggered by the Lehman shock, there began an international movement to strengthen financial regulations, a movement that has gained momentum in recent times [1]. While many consider that regulation strengthening has improved the stability of financial institutions, others are concerned that it may also lower the liquidity of the market and reduce the incentives of financial institutions [2, 3].

There have been many theoretical and empirical studies on financial crisis propagation and infection. Likewise, financial regulatory reform has been also studied. In addition, computational approach has been employed to analyze the dynamics of financial crisis and the effects of financial regulatory reform [4–7][1]. However, to the best of our knowledge, there are few studies to explicitly analyze the effects of financial regulatory reform of financial systems by agent-based approach.

[1] The complete list is given in [8–10].

G. Jezic et al. (Eds.): KES-AMSTA-18 2018, SIST 96, pp. 341–353, 2019.
https://doi.org/10.1007/978-3-319-92031-3_34

In this paper, we study the influence of financial institutions' behavior under financial regulation on financial systems. For this purpose, the authors propose simulation models of systemic risks expressing financial regulation and the financing and investment behavior of financial institutions [8–10]. Using this model, scenario analysis is performed based on cases generated from a combination of various regulations. We then approach the issue of the trade-off between the stability of the financial system and liquidity decline.

2 Model

2.1 Outline

Agents are financial institutions with balance sheets and financial indicators (e.g., capital adequacy ratio). The market value of securities fluctuates in response to price fluctuations of marketable assets. There are two kinds of networks: lending networks to businesses and interbank networks for short-term investment and funding involving financial institutions. Additionally, each financial institution is directly connected to the central bank and can access central bank deposits and lending facilities. Furthermore, each financial institution engages in (a) investment behavior (increasing or decreasing securities and/or lending to business corporations on its balance sheet) and (b) financing behavior (filling any difference between balance sheet debits and credits).

The authors proposed an extended model based on the model by May and Arinaminpathy (M-A model hereafter) [6] to analyze aspects of bankruptcy chains through changes in the balance sheets of individual financial institutions [8–10]. The proposed model has the following characteristics:

1. Each financial institution has a simplified balance sheet and engages in short-term lending and borrowing through the interbank network;
2. The collapse of each financial institution is permitted to observe the impact of such a collapse on the capital of other financial institutions with which the collapsed institution has lending relationships; and
3. An institution suffers from bankruptcy when its capital is insufficient to absorb a given financial shock, a feature that is important in representing the chain of failure.

Our model also expresses the following:

1. Deterioration in the financial and credit situation of financial institutions through fluctuations in the market prices of assets held (capital adequacy ratio);
2. Increase in cash flow shortfall and liquidity risk due to the deterioration of the financing environment in the I/B market; and
3. Central bank funding to prevent bankruptcy.

Thus, our model has become an agent-based model that focuses on the endogenous mechanisms of the financial crisis (Fig. 1)[2].

[2] Note that, since we look at the impact of surplus operating behavior and price fluctuations of market assets on the financial condition and cash flow of financial institutions, we do not deal with trading networks of non-marketable assets and trading networks between the central bank and commercial financial institutions.

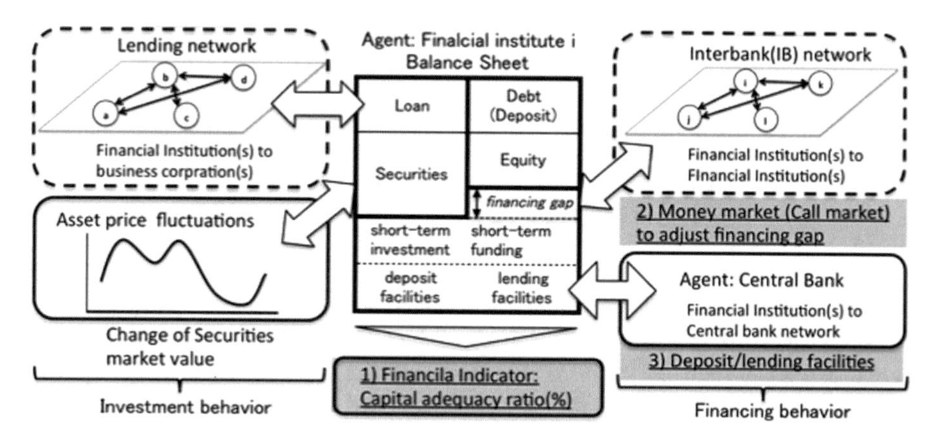

Fig. 1. Conceptual model: the model explicitly describes ALM actions such as investment and financing activities. Funding changes and the changes to the central bank's financial situation because of price fluctuations in the asset market express the impact of ALM actions on operational collapse.

2.2 Agents

In the M-A model, each financial institution

$$a\_i(\mathrm{i} = 1, \ldots, N),\ A = \{a\_i | \mathrm{i} = 1, \ldots, N\}$$

has a simplified balance sheet as follows (Table 1): Each balance sheet consists of (1) cash CA_i; (2) marketable assets $MA_i^{bookvalue}$; (3) non-marketable assets $nonMA_i$; (4) debt D_i; (5) equity E_i; (6) short-term investment SI_i; and (7) short-term financing SF_i.

The financing gap and fund surplus/shortage institutions are then defined as follows:

$$Gap_i = D_i + E_i - CA_i - nonMA_i - MA_i^{bookvalue}$$
$$C^{surplus} = \{a\_i | Gap_i > = 0\}$$
$$C^{shortage} = \{a\_i | Gap_i < 0\}.$$

Note that the status of each financial institution remains unchanged.

We define unrealized profit or loss (UP_i), the capital adequacy ratio (CAR_i), income profit (IP_i) and ROE (ROE_i) in the following:

$$UP_i = MA_i^{t\,marketvalue} - MA_i^{bookvalue}$$
$$CAR_i = (E_i + UP_i + \Sigma CP_i)/\left(nonMA_i + MA_i^{bookvalue}\right)$$
$$IP_i = \beta * MA_i^{bookvalue} + \gamma * nonMA_i$$
$$ROE_i = IP_i/E_i,$$

where β and γ denote the rates of return for marketable and non-marketable assets, respectively. Furthermore, $MA_i^{t\,marketvalue} = \left(MA_i^{bookvalue}/P_0\right) * P_t$, where P_t is the market price of marketable assets in step t.

Table 1. Balance sheet items of commercial financial institutions

Debit	Credit
Cash *CA*	Debt *D*
Non-marketable asset *nonMA*	Equity *E*
Marketable asset $MA^{bookvalue}$	Short term financing *SF*
Short term investment *SI*	

Additionally, financial institutions face demands that they maintain a minimum capital adequacy ratio ($CAR\text{-}demand_i$). Where institutions facing a shortage of capital have ordered that funds be supplied to institutions with a capital surplus, *CAR-demand* is the lowest capital adequacy ratio surplus requested of institutions facing a capital shortage.

2.3 Network

Financial institution *a_i* engages in short-term investment and funding with other financial institutions

$$W_i^{Interbank} = \{a_{-j} | m_{-ij} = 1\}$$

and tries to eliminate the funding gap between itself and connected institutions. Then, $M = (m_{-ij})$ is the adjacency matrix of the I/B network.

2.4 Financing Behavior

Financing of Shortage Institutions by Surplus Institutions

(Step1) Cash-hungry institution agents make requests for financing from cash-wealthy institutions that lend to the interbank network and evenly split their own financing gap. The minimum order size is within a specified range.

Cash-hungry institution a_{-i} makes requests for financing to cash-wealthy institution $a_{-j}\left(amount_i^j > 0\right)$

$$\text{Order}\left(\text{i, j}, amount_i^j\right)$$

Additionally, a_{-i} denotes all cash-wealthy institutions that lend to the interbank network and obey the instruction to evenly split the value of their own financing gap. Here,

$$\left(a_{-i}, a_{-j}\right) \in C^{surplus} \times C^{shortage}$$

δ: minimum order size $\in$ Z, $amount_i^j = \max(\text{ceil}(Gap_i^{\prime}/\#\left(C_i^{surplus} Interbank\right)),$ $\delta) \in$ Z.

(Step2) a_{-j} checks the financial condition and amount of self-funding available for a_{-i} to perform the contract judgment.

- *Contractual conditions:*

$$CAR_i >= CAR{-}demand_j \text{ and}$$
$$amount_i^j = < Gap_j' - \Sigma amount^{j\, other\ implemented-orders}$$

- *non-Contractual conditions:*
  Other than those above.

Then, transactions to which a_j has already committed $amount^{j\ other\ implemented\text{-}orders}$.

(Step1′) This step is an alternate to Step1. If non-contract, a_i will (1) change the order destination to another cash-wealthy institution a_k, and/or (2) reduce the order amount.

$$\text{Order}\left(\text{i, j}, amount_i^j\right) \rightarrow \text{Order}'\left(\text{i, k}, amount_i^k\right)$$
$$\text{Order}\left(\text{i, j}, amount_i^j\right) \rightarrow \text{Order}'\left(\text{i, j}, amount_i^{\prime j}\right)$$
$$\text{Order}\left(\text{i, j}, amount_i^j\right) \rightarrow \text{Order}'\left(\text{i, k}, amount_i^{\prime k}\right)$$

Then, $k \in C^{surplus} \cap W_i^{Interbank}, amount_i' = \text{floor}\ (amount_i/2) > = \delta.$

(Step3) If a_i cannot meet $Gap_i' > = 0$, the result is collapse.

Other Short-Term Investments and Funding

Assuming cross-trades (both built transactions) are carried out [11], we consider the following short-term funding or investment transactions between financial institutions. A certain percentage of the balance sheet amount then becomes the upper limit.

Between cash-hungry– cash-hungry and cash-wealthy– cash-wealthy institutions: a_i is the upper limit of l times between a_j of the same status, and generates the following order:

$$\text{Order}\left(\text{i, j}, amount_i^j\right) \text{ or Order}\left(\text{j, i}, amount_j^i\right)$$

Between cash-wealthy – cash-hungry institutions:
a_i is the upper limit of l times between a_j of the same status, and generates the following order:

$$\text{Order}\left(\text{i, j}, amount_i^j\right)$$

Then,

$$amount_i = \text{ceil}\left(\left(CA_i + MA_i^{bookvalue} + nonMA_i\right) * \varepsilon\right),$$
$$l = \text{floor}\left(\left(CA_i + MA_i^{bookvalue} + nonMA_i\right) * \zeta / \text{ amount}_i\right).$$

2.5 Investment Behavior

Outline

Financial institutions make investment decisions regarding marketable assets at each step under the following management constraints: (a) capital adequacy ratio; (b) VaR; (c) ROE; and (d) market outlook.
Here, the range of possible values of marketable assets is as follows:

$$0 < MA_i^{\prime\, bookvalue} < = MA_i^{bookvalue} + CA_i.$$

Additionally, the balance of non-marketable assets is constant, untreated lending behavior.

Formulation of Management Contracts and Market Outlook

Various constraints and market outlook are defined as shown in Table 2.

$$VaR_i = MA_i\sqrt{n}(r_{ave} - \eta\sigma_m)$$

Furthermore, r_{ave} and σ_m are calculated from the daily return of P_t (sample period: m days).

$$r_{ave} = \frac{1}{m}\sum P_t/P_{t-1}$$

$$\sigma_m = \sqrt{\frac{1}{m}\sum (P_t/P_{t-1} - r_{ave})^2}$$

Investment Decision-Making

Under the above constraints, the financial institution decides to buy, sell, or hold the market assets in accordance with the algorithm shown in Fig. 2.

Decision Regarding Buying and Selling Amount

If buying or selling has been selected in the previous section, the amount of buying and selling is described as follows:

Buying amount: Compare current holdings of marketable assets calculated from the upper limit of the VaR constraint to buy $\theta\%$ of the difference.
Selling amount: A sale resulting from conflict with the capital adequacy ratio constraints and VaR constraints should comprise only those marketable assets sufficient to meet each of the constraints. Additionally, a sale resulting from the conflict with the ROE budget constraint should sell sufficient marketable assets to exceed the budget.

Table 2. Management contract and market outlook

Contracts	Formulation
(a) CAR	CAR_i > CAR-demand$_i$
(b) VaR	VaR < $E_i\zeta$
(c) ROE Budget	IP_i > Budget target y_i
(d) Market outlook	$f = r_{exp} - \lambda_i\, \sigma_m$

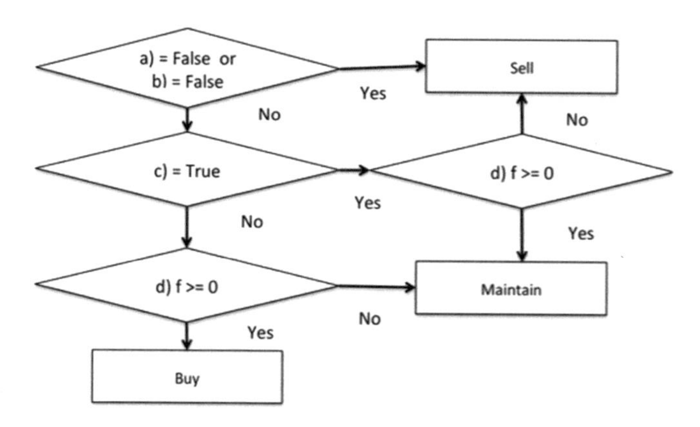

Fig. 2. Flowchart of investment decisions

2.6 The Effects of Bankruptcy

Individual Bankruptcy

If the capital adequacy ratio is equal to or less than the threshold value, or if the funding gap is not filled, for example owing to a failure of cash flow, financial institution a_i will experience bankruptcy:

$$CAR_i < \alpha \text{ or } Gap'_i < 0.$$

Chain Reaction Collapse

If financial institution a_i experiences bankruptcy, financial institutions a_j that are involved in short-term operations with the first financial institution are regarded as non-exposed with regard to these investments, and it is assumed that the capital cancels out:

$$E'_j = E_j - SI^i_j.$$

If the following conditions are satisfied, financial institution a_j also suffers bankruptcy and a chain reaction collapse occurs:

$$E'_j < 0 \text{ or } Gap'_j < 0 \text{ or } CAR_j < \alpha.$$

2.7 Evaluation

This paper considers situations with and without BS restrictions to evaluate the loads of individual financial institutions and the financial system. Specifically, the number of surviving financial institutions (number of failed financial institutions) is used as a macro indicator.

Furthermore, as a macro index related to the liquidity of marketable assets [3], the trading volume of marketable assets of each financial institution are adopted and defined as the total trade volume as follows:

$$Gross\_amount_t = \sum_i buy\_amount_i^t + \sum_i |sell\_amount_i^t|$$

3 Analysis of Model Behavior

This numerical experiment employs the following steps to implement individual attempts to undertake investment behavior in accordance with the operating limitations described in Sect. 2.5 (Table 2). Furthermore, we compare the cases between without limiting the parameter occurrence range of marketable assets (Case 1: without a balance sheet [BS] restriction) a with a BS restriction (Case 2: with a BS restriction). We will indicate the overall trend of the stability of the system in each case and show the results of the individual and overall trials on liquidity.

3.1 Price Time Series of Marketable Assets

The price of risky assets is assumed to follow in the discretized stochastic differential equation below [12]:

$$P_{t,j} = P_{t-1,j} + r_f P_{t-1,j}\Delta t + \sigma P_{t-1,j}\tilde{\varepsilon}\sqrt{\Delta t}$$

where t is time step, j is the trail number, $P_{t,j}$ is the price of a marketable asset (j times, step t) (=100), r_f is the risk free rate, σ is volatility, and $\tilde{\varepsilon} \sim N(0, 1)$. In this simulation, we set 1 step = 1 day = 1/250 year and $\Delta t = 1/250$, T = 125 (assuming 6 months is the budget-closing period for a bank account). Additionally, taking into account long-term government bond yield levels and stock markets in each country, $r_f = 2\%$ and $\sigma = 25\%$. In total, 100,000 sample paths were generated and marketable asset prices in the final step are used to adopt the lowest price time series (Fig. 3).

3.2 Common Settings

Table 3 provides the parameters used in our experiment. We prepared two patterns of parameter ranges for marketable assets, corresponding to the two cases (with or without BS constraints). In this simulation, we focus on financial institutions that constitute an important and core network in the system. Thus, we refer to the [13] and set the number of financial institutions to 20 companies (#1–#20, including 10 each of surplus and

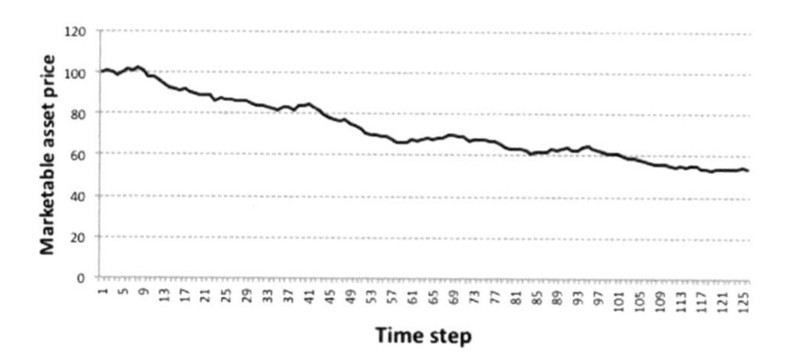

Fig. 3. Price time series of marketable assets employed in this simulation. A total of 100,000 trials and the adopted price time series became the lowest price.

Table 3. Parameter set used in this simulation

Parameters	Value
Number of institutions N	20
Interbank network $W^{Interbank}$	Complete graph
Cash CA	U(10, 25)
Non-marketable asset $nonMA$	100
Marketable asset MA	U(20, 50)
Financing gap Gap	0.05 BS amount <= Gap <= 0.10 BS amount
Capital adequacy ratio CAR	U(0.12, 0.22)
CAR-demand $CAR\text{-}demand$	U(0, 0.03)
Risk aversion λ	U(0.0, 1.0)
Budget target y	U(1.5, 2.0)

shortage institutions), and the I/B network is a complete graph. Regarding the financing gap, since surplus institutions outnumber shortage institutions, surplus institutions were subjected to the following adjustment:

$$(\text{Financing gap adjustment}) = \Sigma(\text{nonMAi} + \text{MAibookvalue})^{*}\ \iota/\#\ \text{Csurplus}.$$

Additionally, for balance sheet items outside the table, we set the capital based on a given capital adequacy ratio. Liabilities taking into account the financing gap were determined by back calculation such that credit and debit match. We then set short-term investment and funding in the following: α: 0%; β: 1.2%; γ: 1.0%; δ: 1; ε: 5%; ζ: 50%; η: 2.33; θ: 10%; ι: 10%; ndays: 10; mdays: 16.

3.3 Stability of Financial System

The numerical experiments were pursued 100 times for each of for the two cases with and without BS restrictions based on the parameters shown in Table 3. Figure 4 shows a box-and-whisker plot for every five steps for the number of remaining financial institutions representing the stability of the system (top panel: Case 1, bottom panel:

Case 2). The case without a BS restriction shows that the variation of the remaining number of institutions is large and that there is a greater distribution in the range where the remaining number is small. The average number of remaining institutions at the final step for Case 1 is 12.8 while that is 19.7 for Case 2. The case with the BS restriction has a lower number of failed financial institutions. Thus, it can be said that the stability of the financial system can be improved by imposing a BS restriction.

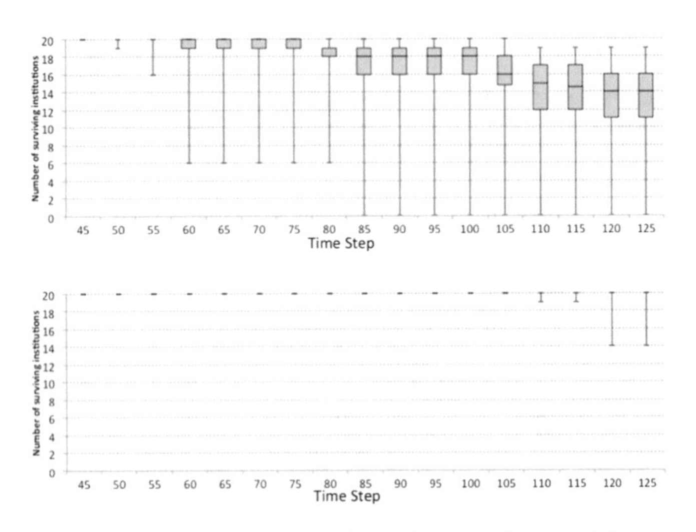

Fig. 4. Box-and-whisker plot of the number of financial institutions surviving at each five-step interval (top panel: Case 1 without a BS restriction; bottom panel: Case 2 with a BS restriction). It can be seen in the case without a BS restriction that there is a greater variation in the residual number and a greater distribution throughout the range where the number of residuals is small. Note that time steps below 45 are omitted because no bankruptcies occur at this stage.

3.4 Liquidity

Individual Trials

In this section, we confirm the impact on the liquidity of marketable assets for each case (with/without BS restrictions). As in the previous section, we used the same initial parameter set (Table 3) and compared the results for the corresponding one-sample trial. Figure 5 shows the trends for the gross trading volume "Gross amount" of marketable assets (which is a macro index of liquidity) for Case 1: without a BS restriction and Case 2: with a BS restriction. Regarding to the common tendency for each case, for each financial institution, (1) the initial decision to increase the marketable assets because of the ROE/budget constraint resulted in an overall increase in purchase volume from the start and (2) as a result of the increase in VaR (because of declines in market price and increased volatility) the amount of sales increased due to the temporary VaR constraints. In contrast, the total value of the total trading volume at the final step is 1,326 in the case without a BS restriction and 480 with a BS restriction.

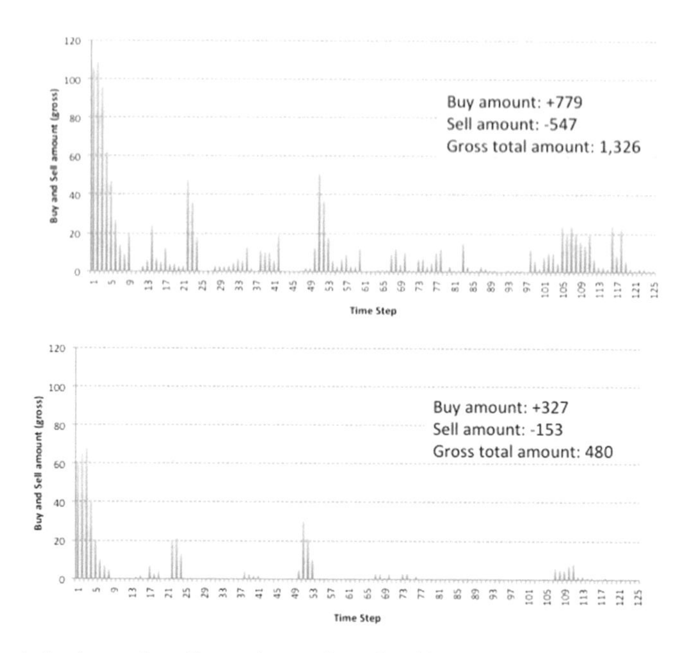

Fig. 5. Trends in the total trading volume of marketable assets (above: Case 1 without a BS restriction, below: Case 2 with a BS restriction). Because of ROE · budget constraints, the decision to increase marketable assets resulted in an increase in purchase volume. Thereafter, due to declines in market price and an increase in VaR because of an increase in volatility, there were also circumstances where the amount sold increased because of a violation of temporary VaR constraints. In contrast, the total value of the total trading volume up to the final step was smaller in the case with a BS restriction.

Overall Trials Based on the individual trials in the previous section, we carried out the simulation 100 times for each of the two cases (Case 1: without a BS restriction and Case 2: with a BS restriction) by generating the parameters as shown in Table 3.

Table 4 shows the results. As with the sample trials in the previous section, the total value of the trading volume at the final step for the case with a BS restriction was limited to just 40% of the value in the case without a BS restriction. In this regard, imposing the BS restriction suggests that the trading behavior of each financial institution is suppressed and the liquidity of marketable assets may decline.

Table 4. Average value and standard deviation of the total trading volume of marketable assets. As with the sample trials in the previous section, the total trade volume at the final step in the case with a BS restriction is smaller than that of the case without BS restriction, representing a 40% difference.

Statistics	Case	
	Without BS restriction	With BS restriction
Average	1217.5	442.1
S.D.	202.2	49.3

4 Concluding Remark

In this paper, we expanded a simulation model of systemic risk that expresses the financial regulation and financing/investment behavior of financial institutions proposed by the authors [8–10]. We also proposed a model that expresses the stability of the financial system and the liquidity of marketable assets.

In addition, under various management constraints and market trends, we conducted a scenario analysis for two cases where a BS restriction is not assumed (Case 1: without a BS restriction) and then assumed (Case 2: with a BS restriction), depending on the parameter occurrence range of the marketable assets.

The overall trend regarding the stability of the financial system shows that there were fewer failed financial institutions under imposed BS restrictions, and that the stability of the financial system can be enhanced under such restrictions.

On the other hand, in the individual and overall trials on the liquidity of the marketable assets, the BS restriction case was limited with respect to total trade volume in comparison to cases without BS restrictions. Thus, the liquidity of marketable assets could be reduced by imposing a BS restriction.

These results are consistent with the earlier study [3] that was also concerned with declining market liquidity. However, our evaluation also showed that the regulation strengthening process has improved the stability of financial institutions.

References

1. Sahara, Y.: Trend of International Financial Regulatory Reform, 9th edn. Mizuho Research Institute (2015). (in Japanese)
2. Miyauchi, A.: Economics of financial crisis and basel regulation, pp. 1–2. Keiso-shobo (2010). (in Japanese)
3. Tomiyasu, H.: Regulatory cost increase and liquidity decline. Secur. Anal. J. **54**(2), 35–46 (2016). (in Japanese)
4. Eisenberg, L., Noe, H.: Systemic risk in financial systems. Manage. Sci. **47**(2), 236–249 (2001)
5. Nier, E., Yang, J., Yorulmazer, T., Alentorn, A.: Network models and financial stability. J. Econ. Dyn. Control **31**(6), 2033–2060 (2007)
6. May, R., Arinaminpathy, N.: Systemic risk: the dynamics of model banking system. J. R. Soc. Interface **7**(46), 823–838 (2010)
7. Georg, C.-P.: The effect of interbank network structure on contagion and common shocks. J. Bank. Financ. **37**(7), 2216–2228 (2013)
8. Kikuchi, T., Takahashi, H., Terano, T.: The propagation of bankruptcies of financial institutions — an agent model of financing behavior and asset price fluctuations. In: The 9th International Workshop on Agent-based Approach in Economic and Social Complex Systems (2015)
9. Kikuchi, T., Kunigami, M., Yamada, T., Takahashi, H., Terano, T.: Analysis of the influences of central bank financing on operative collapses of financial institutions using agent-based simulation. In: The 40th Annual International Computers, Software & Applications Conference, the 3rd International workshop on Social Services through Human and Artificial Agent Models. IEEE (2016)

10. Kikuchi, T., Kunigami, M., Yamada, T., Takahashi, H., Terano, T.: Does negative interest rate policy stabilize the interbank network? In: Social Simulation Conference 2016 (2016)
11. Kuroda, H., Kato, I.: Tokyo Money Market, Totan Research ed., 7th edn. Yuhikaku (2009). (in Japanese)
12. Luenberger, D.G.: Investment Science. Oxford University Press, New York (1997)
13. FSB Homepage. http://www.fsb.org/wp-content/uploads/2015-update-of-list-of-global-systemically-important-banks-G-SIBs.pdf. Accessed 1 Dec 2017

Author Index

A
Abdullah, Noryusliza, 23
Alanis, Arnulfo, 215, 275
Aoki, Ryoma, 321

B
Benlloch-Dualde, José-V., 220
Bermudez, Eugenia, 253
Borova, Monika, 231
Bucki, Robert, 187

C
Caro, Maricela Sevilla, 261, 269
Chen-Burger, Yun-Heh, 134
Cisneros, Ricardo Rosales, 261
Cortina Rodríguez, Guillermo, 231
Cristani, Matteo, 123, 144, 164
Čunko, Krešimir, 76

D
Darman, Rozanawati, 23
del Carmen Osuna Millán, Nora, 269

F
Flores-Sanchez, Carlos, 275

G
Garza, Arnulfo Alanís, 253
Ghedira, Khaled, 97
Governatori, Guido, 123

H
Hafit, Hanayanti, 23
Halaška, Michal, 177
Härting, Ralf-Christian, 287
Hassan, Mohd Khairul Azmi, 134
Henskens, Frans, 44
Hurtado, Carlos, 209, 215

I
Inuzuka, Nobuhiro, 13
Ishino, Yoko, 331
Ivanović, Mirjana, 110

J
Janošová, Karolína, 231
Jarvis, Dennis, 3
Jarvis, Jacqueline, 3
Jevtić, Dragan, 76
Jezic, Gordan, 34
Juarez-Ramirez, Reyes, 253

K
Khaleefah, Shihab Hamad, 23
Kikuchi, Takamasa, 341
Kotulski, Leszek, 156
Krivic, Petar, 57
Ktata, Farah Barika, 97
Kunigami, Masaaki, 341
Kurahashi, Setsuya, 311
Kusek, Mario, 57

L
Lemus-Zúñiga, Lenin-G., 220, 231, 243
Lorenzo-Sáez, Edgar, 220, 243
Luzuriaga, Jorge E., 231, 243

M
Magdaleno, Sergio, 253
Maldonado-Mahauad, Jorge, 220

G. Jezic et al. (Eds.): KES-AMSTA-18 2018, SIST 96, pp. 355–356, 2019.
https://doi.org/10.1007/978-3-319-92031-3

Manrique, Esperanza, 215
Massimkanova, Zhazira A., 199
Mateo Pla, Miguel Ángel, 231, 243
Matsui, Tohgoroh, 13
Meza-Fregoso, Juan, 275
Miura, Masashi, 67
Miyawaki, Masaya, 13
Moreno, Hilda Beatriz Ramírez, 209, 261, 269
Moriyama, Koichi, 13
Mostafa, Salama A., 23
Mustapha, Aida, 23
Mutoh, Atsuko, 13

N
Neumaier, Pascal, 287
Noorunnisa, Salma, 3
Núñez, Sergio Octavio Vázquez, 209

O
Oliver-Villanueva, José-Vicente, 243
Olivieri, Francesco, 123, 144, 164
Osuna-Millan, Nora, 275

P
Paul, David, 44
Pla, Miguel A. Mateo, 220

Q
Quezada, Ángeles, 253

R
Ramirez, Beatriz, 215
Ramírez, Margarita Ramírez, 209, 215, 261, 269
Reichstein, Christopher, 287
Rojas, Esperanza Manrique, 209, 261, 269
Rosales, Ricardo, 275

S
Salgado-Soto, Consuelo, 275
Samet, Donies, 97
Samigulina, Galina A., 199
Scannapieco, Simone, 164
Shibata, Masashi, 299
Shiroishi, Hidetoshi, 67
Skocir, Pavle, 34, 57
Soic, Renato, 34
Soto, Consuelo Salgado, 261
Šperka, Roman, 177
Stantić, Dejan, 110
Suchánek, Petr, 187

T
Takahashi, Hiroshi, 341
Takahashi, Masakazu, 299
Terano, Takao, 321, 341
Tomazzoli, Claudio, 123, 144, 164

U
Ud Din, Fareed, 44
Urchueguía, Javier F., 243

V
Valle, Valeria Alexandra Haro, 220
Vazquez, Sergio Octavio, 215
Vidaković, Jovana, 110
Vidaković, Milan, 110
Vuković, Marin, 76

W
Wallis, Mark, 44
Watson, Marcus, 3
Wojnicki, Igor, 156

Y
Yail, Bogart, 253
Yamada, Takashi, 341

Z
Zargayouna, Mahdi, 87
Zeddini, Besma, 87
Zorzi, Margherita, 144

Printed in the United States
By Bookmasters